GLOBAL WARMING
IS THE SOLUTION

Third Edition

Dr. Auke Schade

nemonik-thinking.org

Copyright

Third Edition
Published 1 July 2016
@ nemonik-thinking.org
ISBN **978-0-473-36423-6**

Abstract

This study presents a bilateral synthesis of artificial global warming and natural global cooling. Mainstream climatology lacks scientific integrity and statistical methodology. Peer review is changed into peer pressure and objectors are labelled *"Deniers"*. Proper statistical analyses are replaced by graphs and non-causal correlation analyses that are based on the last 166 years, while 420,000 years of Antarctic data are mainly discarded. Furthermore, climatology ignores that 400 ppm of CO_2 predicts a global temperature of 11.5 °C, rather than the current 1.3 °C. It focuses on artificial global warming and overlooks the threat of natural global cooling. It also ignores the solar expert Professor Zharkova, who predicts a mini ice-age by 2030, which is likely to turn global warming into global cooling. The current study compared the Antarctic temperatures during the last 10,000 years (baseline 0.00 °C) with the global temperature of 1.3 °C. This common definition of global warming failed to reach statistical significance. However, the Antarctic temperatures during the last 420.000 years support the notion that we live in a glacial period of -8.9 °C, rather than in an interglacial period of 0.00 °C. In that case, the artificial global warming would be 10.2 °C, rather than 1.3 °C. This alternative definition of global warming is statistically significant. Furthermore, it is supported by the current CO_2 level of 400 ppm and the significant duration and stability of the current interglacial. Consequently, decreasing the CO_2 level could cause a global disaster threatening the survival of humanity. The increased thermal range and the precarious balance between artificial global warming and natural global cooling could also explain the current climatological instability.

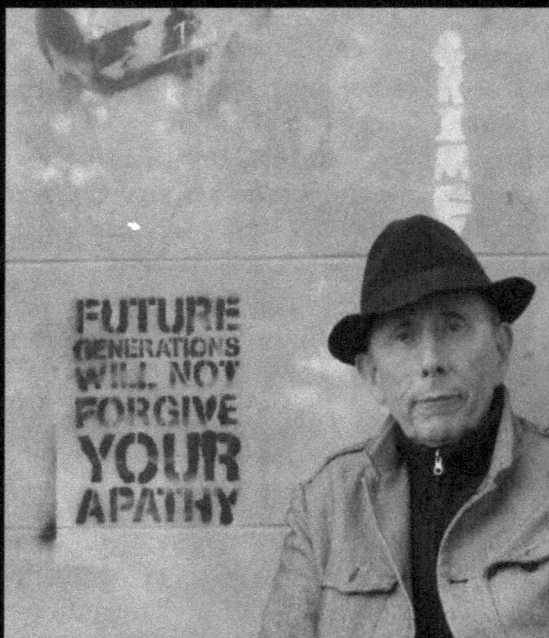

NEMONIK THINKING

FUTURE GENERATIONS WILL NOT FORGIVE YOUR APATHY

I agree with the unknown graffiti artist

CLIMATE CHANGE

Dr Auke Schade

My life started during the devastation of World War II. As a teenager, I worked as a carpenter and studied building engineering at night school. During the seventies, I became a financial manager for a multinational corporation, ran my own business, and studied economics in my spare time. My interest in the psychology of management extended to the interaction between the mind, body, and reality. In 1980, I immigrated to New Zealand where I obtained a doctorate in psychology from the University of Auckland. My mission is to make people the smartest thinkers they can be, which has led me to the development of nemonik thinking.[i]

Reality shows that humanity's way of thinking is failing dramatically. As a result, the next generation is facing overpopulation, dwindling resources, nuclear warfare, industrial pollution, climate change, etc. Therefore, they have to become the best thinkers they can be.

Download free eBooks and videos
@ nemonik-thinking.org

---

i Appendix: Nemonik Thinking.

Notes

CONTENTS

CONTENTS...7

CLIMATE CHANGE 11

STATISTICAL ANALYSES 34

Leaders beware 34

Vostok °C (423 Kyrs) 37

Vostok °C depth-scale.....................................37

Vostok °C millennial-scale..............................40

Vostok °C base (Kyr1 = 0.00 °C)...................43

Vostok °C trend 45

Vostok °C detrended millennial-scale............47

Duration glacials..51

Duration interglacials.......................................52

Duration current interglacial...........................55

Glacial decline...57

Glacial decline window....................................58

Interglacial thermal stability 61

Vostok °C (10 Kyrs) 66

Vostok °C depth-scale.....................................66

Vostok °C semi-annual-scale...........................68

Vostok °C window-scale...................................72

Global °C (1880-2014) 76

Global °C annual-scale.....................................76

Global and Vostok °C annual-scale 79

Global °C backward extension..........................81

Global and Vostok °C combined 85

Global and Vostok °C calibrated 87

Global °C window-scale 91

Global versus Vostok °C window-scale 94

Vostok CO2 / Vostok °C (423 Kyrs) 105

Vostok CO2 105

Vostok synchronizing CO2 and °C 107

Vostok CO2 versus °C 111

Hawaii CO2 / HAWAII °C (1959-2014) 115

Hawaii °C calibrated 115

Hawaii CO2 121

Hawaii CO2 versus °C 124

Hawaii CO2 / Global °C (1959-2014) 128

Correlation CO2 and °C 128

CO2 predicts temperature 132

Bilateral Climate-Change 139

Vostok CO2 predicts Vostok temperature 139

Hawaiian CO2 predicts global temperature 143

Thermal gap 147

Onset thermal gap 153

Thermal effect CO2 reduction 162

APPENDICES 174

Bibliography 174

Glossary 176

Lists 186

List of Figures 186

List of Tables 200

List of Datasets..204

List of Tests 212

List of Equations..217

Statistics 220

Datasets 223

Dataset 1..223

Dataset 2..225

Dataset 5..282

Dataset 8..291

Dataset 15..459

Dataset 18..463

Dataset 22..469

Dataset 27..471

Abstracts other books..481

Think Smarter..481

Glossary...482

Dictionary..483

Lao Zi's Dao De Jing 484

Sun Zi'S The Art of War...............................485

Declaration of Independence 486

Endnotes 487

Notes

_____

_____

_____

_____

_____

_____

_____

_____

_____

_____

_____

_____

_____

_____

_____

CLIMATE CHANGE

The majority of the population seem to believe that a catastrophic global warming is in progress. Collective groupthink and individual cognitive dissonance are likely to maintain that position by ridiculing, ostracizing, and labelling opponents. I am a seventy-four years old psychologist who does not know the first thing about climatology. I could not predict tomorrow's weather if my life would depend on it. However, the problem of climate change is not about the climate, but about humanity's failing way of thinking, which I happily claim as my field of expertise. Climate change is just one of the many symptoms.

Despite previous warnings by Edward de Bono, the educational system still propagates an incomplete version of a 2,500 years old way of thinking that has been never updated. Even worse, this ancient way of thinking has been corrupted by incompetent thinkers turning Aristotle's rational thinking upside down into rationalizations. They draw their conclusion first and then select fitting evidence. They use Socrates' critical thinking to defeat their opponents in debates, rather than to find the truth by criticizing their own thinking. Winning becomes more important than success. In addition, Confucius developed collective thinking in order to provide justice for individuals. However, incompetent thinkers abuse it to subdue and control individuals. This control suppresses the creativity that is required to adjust the collective to the forever-changing reality. Furthermore, intuitive-emotional thinking is still ridiculed, because it does not fit the conventional 'rational' paradigm.

Albert Einstein pointed out that no one can solve problems with the same way of thinking that creates them. Therefore, I have developed nemonik thinking.[1] Nemonik thinking is the operating manual for your brain that you should have received at birth. It is a smarter way of thinking that aims to maximize your success by evaluating 17 nemoniks, which are memorized keywords describing all the perceived aspects of your mind, reality, and their interaction. In accord to Laozi, success is obtaining what you seek and escaping what you suffer. To maximize your success, nemonik thinking mobilizes your hidden genius, accelerates your thinking, improves your memory, and reveals strengths, weaknesses, opportunities, and threats. Furthermore, it creates questions and ideas, and reduces your stress levels. It is like playing a musical keyboard with 17 keys producing an infinite repertoire of smart strategies. Nemonik thinking is unique because it is the first comprehensive and transferable way of thinking.

The results of applying nemonik thinking to climate-change suggest that both the proponents and opponents of global warming are right. Their arguments continue because humanity's way of conventional thinking is failing. This study supports the notion that carbon dioxide (CO_2) is a precious substance that is critical for the survival of humanity. Nevertheless, it is also a hazardous substance threatening the environment in other ways.

The conventional unilateral climate-change hypothesis holds that industrial greenhouse gasses, such as carbon dioxide (CO_2), increase the global temperature by trapping solar heat in the atmosphere.[2] Allegedly, the recent increase in global temperature melts the global ice deposits, which poses a huge threat to humanity. Consequently, the obvious solution to climate change is the reduction of greenhouse gasses. However, the unilateral hypothesis is about artificial global warming, rather than climate change. The unilateral hypothesis cannot explain why the increase in global temperature has failed to reach statistical significance. Neither can it account adequately for the current instability of the climate.

The public focus on the potentially devastating effects of global warming is understandable. After all, the ice is melting and seawater levels are rising. Therefore, most resources are directed to ameliorating the symptoms of global warming, rather than testing the hypothesis of global warming. This situation is maintained by groupthink and cognitive dissonance. Nemonik thinking suggests that these cognitive defences are likely to obstruct the development of climatology.

The nemonik accelerator is a dynamic tool to eliminate mental stagnation—it is like Hegel's dialectic of knowledge on steroids.[3] If there is only a thesis available, then nemonik thinkers accelerate their thinking by aiming for the antithesis. If there is both a thesis and an antithesis, then they aim for the synthesis of thesis and antithesis. The synthesis turns into a new thesis and the acceleration is repeated. Previously, the nemonik accelerator was unknowingly used by Albert Einstein to synthesize Sir Isaac Newton's thesis that light is a particle with Thomas Young's antithesis that light is a wave. Hence, the development of knowledge is an infinite journey without resting places.[4]

Despite the mass acceptance of the unilateral climate-change thesis of global warming, the Vostok data show evidence for an antithesis of global cooling. As explained, if a reliable thesis is opposed by a reliable antithesis, then the nemonik accelerator predicts that the solution is a synthesis.[5] Hence, to develop climatology, the apparently conflicting concepts of global warming and global cooling have to be synthesized.

The bilateral climate-change hypothesis is proposed in order to prevent stagnation in climatology. That synthesis holds that the thermal effect of artificial global warming has compensated for the thermal effect of natural global cooling. Low natural glacial temperatures are increased by heat that is trapped by artificial domestic and industrial atmospheric greenhouse gasses such as CO_2. As a result, those huge hidden thermal opposites destabilize the weather, while the observable thermal average remains relatively stable. Hence, this bilateral climate-change hypothesis explains why the average global temperature has not increased significantly, while the climate has become unstable.

The bilateral hypothesis predicts that the increasing thermal extremes pose the direct and immediate threat to humanity, rather than the average temperature. Short-term high temperatures could do long-lasting damage to the global ice deposits. For example, melting one metre of ice in Vostok during a warm period would take 17 years of precipitation to be replaced, even if it gets extremely cold after the meltdown. This duration will increase rapidly with the depth of the ice. It will take 52 years to replace two metres of molten ice. Three metres of molten ice would not be replaced within our lifetime.[6] Hence, even with a stable thermal average the ice would gradually disappear.

Based on the bilateral hypothesis, the solution to climate change is a careful management of greenhouse gasses in order to maintain the sensitive thermal balance. Hence, nemonik thinking has changed the problem by providing the hypothesis that artificial global warming is the solution for natural global cooling. It is emphasized that a hypothesis is an untested, but testable description of reality.

NASA reported that the global temperature in 2014 was 0.67 degrees Celsius (°C) above their arbitrary short-term baseline.[7] In contrast, the results of this study indicate that the global temperature in 2014 was 1.32 °C above the long-term baseline.[8] Despite that apparent increase, the popular belief that the current global temperature has reached an all-time high is unscientific. The analyses of publically available data show that the uninterrupted slope of the increase, duration, and magnitude of the global temperature during the last 135 years has been within the natural limits.[9] Therefore, the results reject the unilateral hypothesis in favour of the bilateral one.[10]

In 2015, the carbon dioxide (CO_2) concentration has reached about 400 parts per million (ppm).[12] It could be argued that this concentration is very small in comparison to the other substances comprising the atmosphere. That ratio does not sound alarming. After all, 0.0004 is indeed a very small number.[13]   However, let me put that into my laymen's perspective.

The atmospheric pressure at ground-level is about 1,000 millibars.[14] That pressure equals a water column of about ten metres (10,000 millimetres). Now imagine that the atmosphere cooled down and condensed into a ten metres deep liquid covering the entire planet. In comparison to the depth of that liquid, the thickness of the CO_2 layer would be ((400 / 1,000,000 ppm) x 10,000 mm) = (400 / 100) = 4 millimetres.[15] Hence, our ten metres of liquid atmosphere would be covered with 4 millimetres of CO_2. Does that sound alarming?

If not, then imagine that all our oceans were covered with 4 mm of oil. As the oceans are much deeper than the liquefied atmosphere, that oil slick would not be even close to 400 ppm. Nevertheless, all marine life would vanish and humanity would follow suit. Alternatively, imagine a 4 mm CO_2-foil wrapped around our entire planet. Wait—you don't have to imagine that, because that is reality.

In 2014, the atmospheric CO_2 concentration in Hawaii reached a maximum of 398.6 parts per million (ppm).[16]   In contrast, the average CO_2 concentration in Vostok was only 234.0 ppm during the last 420,000 years, and 266.2 ppm during the last 10,000 years concerning the current interglacial.[17] Hence, the current CO_2 concentration is 164.6 ppm (70%) higher than the long-term average, and 132.4 ppm (50%) higher than the average of the current interglacial.[18]

Despite the apparently high Hawaiian CO_2 concentration, to claim that it is at an all-time high would be unscientific. The Vostok dataset contains only seven CO_2 values for the last 10,000 years.[19] Consequently, the mean data interval is about 1,300 years. This long duration hides crucial information about annual CO_2 concentrations.[20] In addition, regression towards the mean is likely to have reduced the extracted CO_2 values. Consequently, the historical maximum of the annual CO_2 concentrations is unknown. Hence, the data cannot support the notion that the current annual CO_2 concentration is at an unnaturally all-time high. Although crucial for the validity and reliability of the statistical analyses, my resources are insufficient to reconstruct additional CO_2 concentrations from the Vostok ice-core.

Advocates of either the unilateral or the bilateral hypothesis, assume that atmospheric CO_2 drives the temperature. In accord, the results show significant positive correlations between the historical Vostok CO_2 concentrations and temperatures, as well as, between the recent Hawaiian annual CO_2 concentrations and the global mean temperatures.[21]

However, such simple correlation analyses cannot test the cause-effect relationship between CO_2 and temperature. A significant correlation means only that CO_2 and temperature change in synchrony. Hence, a significant correlation could be caused by CO_2 driving temperature; temperature driving CO_2; or confounding variables driving both CO_2 and temperature.

To evaluate whether CO_2 drives temperature, this study introduces a method that uses the CO_2 concentration of a particular year to predict the temperature of the following year. The method predicted correctly the direction of 45 of the 55 annual changes in global temperature (82%).[22] The significance of that result provides support for the notion that the CO_2 concentration indeed drives the temperature.

The validation of the proposed method requires additional data-pairs obtained under different climatological conditions. Unfortunately, an adequate extension of the Hawaiian CO_2 and global temperature datasets will take considerable time— just to double the size will take 56 years! The following sections will explain why such a delay is luxury humanity cannot afford.

Thermal gap

The thermal gap is the difference between the estimated and actual global annual atmospheric temperature deviations.

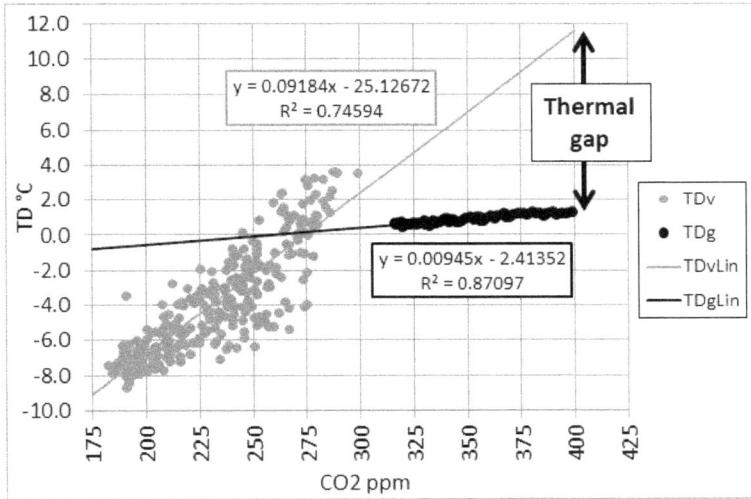

Figure 45 (Dataset 27)—thermal gap.[23] Figure 45 shows the Vostok temperatures from 412,096 BCE to 1989 (TDv), and the global annual temperatures from 1959 to 2014 (TDg), as a function of CO_2 concentrations.[24] Furthermore, it shows the linear functions of TDv (TDvLin) and TDg (TDgLin).

Figure 45 shows that the slope of the Vostok long-term natural temperature deviations (TDvLin) is about ten times steeper than the slope of the global short-term artificial temperature deviations (TDgLin). The significance of this difference suggests that it is systematic, rather than random. As a result, the global linear function is a spurious predictor of natural temperatures.

The linear function of the Vostok temperatures (TDvLin) predicts a global temperature of 11.48 °C for the recent Hawaiian CO_2 concentration of 398.6 ppm.[25] In contrast, based on the same CO_2 concentration, the linear function of the global temperatures (TDgLin) predicts a global temperature of 1.35 °C.[26] Hence, a comparison of the two predictions shows a thermal gap in the predicted global temperature of 10.13 °C (11.48 – 1.35 °C).[27]

Many people consider the observed global temperature of 1.32 °C in 2014 already as a serious threat to the environment and the survival of humanity. However, in comparison, the magnitude of the thermal gap is gigantic. Imagine the disaster that would be caused by an increase of 10.13 °C.

As discussed in the section Statistical Analyses, the data suggest that the thermal gap cannot be attributed to a difference between artificial and natural CO_2; natural or artificial pollutants; changes in the thermal effect of CO_2; unreliability of the Vostok predictor (TDvLin); thermal ceiling effect of CO_2; delayed thermal response; or melting ice working as a heat-sink.

The last known long-term Vostok data-pair of CO_2 (284.7 ppm) and temperature (-0.64 °C) contributing to the Vostok linear function is dated 353 BCE.[28] Therefore, it is proposed that since 353 BCE, one or more confounding variables have significantly altered the slope of the linear function between the CO_2 concentration and temperature.[29]

If the thermal effect of CO_2 would decrease with increasing magnitude, then the threat of global warming would be limited. Hence, the unilateral hypothesis implies that an increase in CO_2 will cause a proportional increase in temperature. On the other hand, the bilateral hypothesis predicts that an increase in glacial conditions will mask the thermal effect of CO_2.[30] The bilateral prediction is supported, because the global temperature failed to reach statistical significance. The resulting thermal gap rejects the unilateral hypothesis and supports the bilateral one. Nevertheless, under bilateral conditions, a zero global temperature will not reduce the threat of climate change. A synchronized increase of CO_2 and glacial conditions will increase the thermal gap. Therefore, the likelihood will increase that the balance will be disturbed and that either extreme will cause a disaster.

The hypothesis is proposed that our ancestors controlled unknowingly the climate with domestic greenhouse gasses long before the industrial revolution. This hypothesis is supported by the significant interglacial thermal stability; significant interglacial duration; and the early onset of the thermal gap at about 1,500 BCE.

The thermal stability during the last 10,000 years of the current interglacial is significantly larger than the thermal stability of any other period of 10,000 years, during the preceding period of about 400,000 years.[31] This supports the notion that a new confounding variable has stabilized the interglacial atmospheric temperature.

The increasing temperature at the beginning of the current interglacial fostered the migration of humanity from the equatorial to the Polar Regions. The resulting population increase could be the new stabilizing factor of the climate. For example, it is likely that our ancestors would have lit fires in response to low temperatures. In turn, the released CO_2 could increase the atmospheric temperature, which then stopped the need for fires. Hence, the release of artificial greenhouse gasses was inversely related to the natural temperature. In this way, our ancestors might have stabilized unintentionally the atmospheric temperature.

It could be argued that the amount of domestic CO_2 would be insufficient to affect the climate. However, the huge thermal gap that is presented in this study supports the notion that the recently observed thermal effect of CO_2 is confounded. The long-term Vostok data suggest that the natural thermal effect of CO_2 is ten times stronger. Hence, the interglacial thermal stability could support the hypothesis that our ancestors stabilized unknowingly the climate with domestic greenhouse gasses.

The four previous interglacials show that their average duration is about 3,500 years. Furthermore, the critical interglacial duration is about 8,500 years.[32] Hence, the duration of the current interglacial is 6,500 to 1,500 years longer than expected.[33] Therefore, the significant duration of the current interglacial supports the notion that the interglacial conditions are extended by the release of artificial greenhouse gasses. At the onset of the thermal gap about 3,500 years ago, the origin of such gasses could not have been industrial. Hence, the duration of the current interglacial supports the hypothesis that our ancestors extended the interglacial conditions with domestic greenhouse gasses.

The results provide support for the notion that the thermal gap started about somewhere between 5,300 and 1500 BCE.[34] This suggests that artificial CO_2 was released into the atmosphere that cannot be attributed to industrialization. Nevertheless, huge amounts of industrial greenhouse gasses were recently released into the atmosphere. These gasses are likely to disturb the domestic thermal balance, because their release is unrelated to the actual temperature.

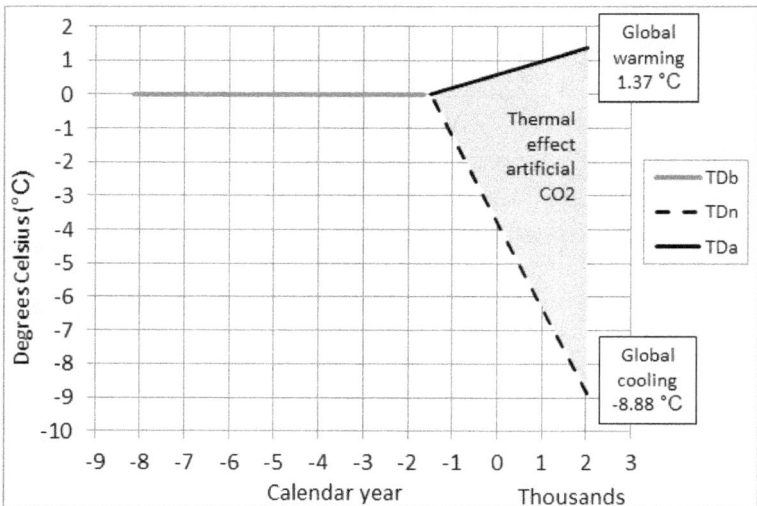

Figure 49 shows the estimated thermal effect of artificial CO_2 during the current interglacial.[35] (TDb)—Vostok base

Kyr1 = 0.00 degrees Celsius (°C); (TDn)—estimated natural glacial decline of atmospheric temperature deviations; (TDa)—observed artificial atmospheric temperature deviations.

Figure 49 presents the components of the bilateral climate-change hypothesis. This hypothesis is supported by the significant thermal stability and duration of the current interglacial, magnitude of the thermal gap, and the onset of that thermal gap thousands of years before the industrial revolution.

The bilateral hypothesis is positive for humanity, because it implies that we did not cause global warming. Unknowingly, we compensated for the cold of the glacial cold. We have not destroyed our planet. No guilt trip necessary. We just try to survive in a harsh environment and learn from our mistakes.

Furthermore, the bilateral hypothesis suggests that artificial greenhouse gasses provide the opportunity to manage our climate with CO_2. Such management is crucial for the survival of humanity, because the data predict that a glacial period of about 97,000 years has already started. Hence, artificial global warming might be the only solution for natural global cooling. However, the bilateral hypothesis raises also the question whether we have enough CO_2 to last that long.

Despite the advantages, the bilateral hypothesis requires rigorous testing based on datasets containing more Vostok CO_2 concentrations during the last 10,000 years than currently available.

Reducing CO2

The short history of dealing with climate change shows that global action is lethargic at the best of times. Global leaders have been discussing the dangers of the rising industrial CO_2 concentration for decades and still the level keeps rising steadily. Economic, military, ideological, religious, and political forces divide humanity and delay global action. Right now, humanity has run out of time. Whatever direction is selected, once the global train is set in motion it will be hard to stop it and turn it around in time.

In August 2015, President Obama announced his plan to decrease the CO_2 concentrations with about 30% by 2025 in order to reduce the global temperature of 1.32 °C.[36] However, this study suggests that the greatest threat to human survival is the hidden thermal gap of 10.13 °C, rather than the observable global temperature of 1.32 °C.[37] If his plan fails and the unilateral hypothesis is true, then the thermal gap could create an overheated planet. The current CO_2 concentration of 398.6 ppm could increase the global temperature with up to 10.13 °C.

Such a high temperature would raise the sea levels significantly flooding our harbours, cities, fertile land, and industries. The heat would turn the remaining fertile areas into deserts and destroy living-space, fresh water supplies, global food production, and infrastructures. The need for domestic cooling would increase the demand for energy. In addition, the equatorial heat would force a devastating mass migration towards the cooler Polar Regions. The recent migration from Syria towards Europe would be insignificant in comparison to the billions of refugees forced by climate change to move.

On the other hand, if Obama's plan succeeds and the bilateral hypothesis is true, then the thermal gap suggests that his actions will backfire. In that case, the decrease in temperature will be determined by a mixture of the Vostok long-term and the global short-term linear functions describing the relationship between CO_2 and temperature. The slope of the Vostok long-term linear function is ten times steeper than the global one that is flattened by glacial conditions.[38] Therefore, the bilateral hypothesis predicts that Obama's 30% reduction in CO_2 concentration would decrease the global temperature with 5.51 °C. Compared to the alleged global warming of 1.32 °C, this is a huge decrease.

In addition, in July 2015, Professor Valentina Zharkova, a solar researcher at the University of Northumbria, predicted a drastic drop in temperatures around 2030 due to a change in solar activity.[39] Hence, the combined effect of reduced CO2 concentrations and solar energy could plunge humanity into a devastating glacial period.[40]

The data show, that a glacial period normally last about 97,000 years.[41] Hence, such conditions are likely to create long-term problems. Falling global temperatures would increase the demand for greenhouse gasses and, therefore, the value of oil and coal. The need for domestic warming would drive the demand for energy even higher. Ironically, CO2 would change from a negative to a positive substance fostering the survival of humanity. Polluting villains would become environmental heroes.

Snow and ice will cover solar panels, and immobilize windmills, hydro generators, and transport. The largest threat is that glacial temperatures will freeze the fresh water reservoirs. The resulting lack of water would force millions of people almost directly out of the cities with no place to go. Furthermore, low glacial temperatures combined with low CO2 concentrations would destroy the global food production. Consequently, the glacial cold would force a devastating mass migration from both Polar Regions towards the warmer equatorial region. The fragile global infrastructure would be destroyed forcing all-out wars for living-space. Within a few years, humanity could be forced to the brink of extinction.

Ironically, humanity is threatened by the human mind. Climate change could become the worst example of groupthink in human history. Nevertheless, it is only one of the many lethal threats resulting from humanity's failing way of thinking. It has created a chaotic world of diminishing resources, economic stagnation, famine, global urbanisation, growing deserts, industrial pollution, information overload, mental escapism, monoculturism, nuclear threats, overpopulation, Pandora's genetic box, poverty, protracted warfare, religious conflicts, resistant microbes, substance abuse, terrorism, etc.

Thinking

Despite the outside symptoms, the real problem is inside the human mind. Our subconscious keeps telling our conscious that it is not part of the problem. It makes us conscious believe that our personal way of thinking is okay— if just the eight billion other people would improve their thinking. The educational system is reinforcing that belief by promoting, institutionalizing, and propagating that outdated way of thinking. Ironically, at the first sign of failure, the victims demand even more education. They ignore that the educational system has created the conditions that cause their failure, because it has created the world we live in. If that world is wrong, then the educational system is wrong. More of the same is not better, because we cannot solve problems with the same way of thinking that has created them.

Many conventional thinkers are born to win. They are great experts, critical thinkers, problem solvers, debaters, rule makers, educators, and unbiased intellectuals. The global educational thinking propagates and maintains their way of thinking by rewards and punishments. In contrast, nemonik thinking holds that the value of each ability depends exclusively on the situation. Even such worthy abilities as propagated by the educational system could turn against us.

Although conventional thinkers focus on winning, they might not succeed. Two-and-halve thousand years ago, Lao Tzu defined success as obtaining what you seek and escaping what you suffer. Our mass educational system is biased towards winning. To deal with the large number of students, it has to examine them in a standardized way and scale their results towards a Gaussian distribution. In reality, each student receives a z-score that is dependent on the results of other students.

Students with high-standardized marks will pass their exam, while students with low ones will fail. The average student will pass in a class with low marks. On the other hand, that same student is likely to fail in a class with high marks. Consequently, without having much choice, the educational system is based on competition and the student has to win that competition in order to earn approval for further education or to obtain a job. The educational system teaches student to win, rather than to succeed. Ironically, Lao Tzu's success is corrupted to obtaining a job and escaping the educational system.

Winning is easy. You only have to find or create an opponent and beat him. With that philosophy, the reason for the competition is unimportant. It becomes competition for competition's sake. Like in exams, it is winning that counts. The student is measured in comparison to his peers. Hence, any exam is an external validation. In contrast, Lao Tzu's success is an internal validation. You have to know what success means for you, because the reason is vital. If you want to become a doctor, but you become the richest person on earth, then you have still failed internally. That internal failure will erode the satisfaction provided by your wealth. Therefore, you have to know what you want to obtain and what you want to escape. You have to know yourself. You have to know your needs and skills. Therefore, nemonik thinking will improve your life by reminding you to strive for success, rather than winning.

Reality shows that humanity's conventional way of thinking is failing. Climate change is just one of the many symptoms. The educational system is failing, because it propagates conventional thinking. Due to its huge momentum, the educational system delays the development of thinking. Most important, you are likely to fail if you apply conventional thinking, because the mind of conventional thinkers is scarred.

SCARRED is a nemonik acronym that stands for—**S**tatic, rather than dynamic thinking; **C**riticizing, rather than critical thinking; **A**nswering, rather than questioning; **R**ationalizing, rather than rational thinking; **R**ighteous, rather than collective; **E**ducated, rather than wise; **D**etached, rather than compassionate. Those are the seven pitfalls of conventional thinking.

Static versus dynamic thinking

Rewards reinforce the repetition of any winning strategy, while pain causes the avoidance of any failed strategy. Therefore, successful strategies are repeated, while unsuccessful ones are avoided. Hence, it is natural to become a static thinker. Static thinkers become experts in a static environment by a bias towards a particular cognitive strategy.

The static approach is rampant in law, politics, and accountancy. The self-inflicted rules become a net that constricts conventional thinkers and delays their adjustments to reality. On the other hand, the universe is dynamic. There are no good or bad strategies. The value of each strategy depends exclusively on the actual situation. What worked yesterday might fail today, while what failed yesterday might bring success today. The only constant in the universe is change and, therefore, static thinkers will ultimately fail.

Static thinking is a weakness you can ill afford, because it decreases the likelihood of your success. Becoming a dynamic thinker is a conscious process and the exhaustive options provided by nemonik thinking are the tools. Nemonik thinkers are dynamic and set their sails to the forever-changing wind.

Criticizing versus critical thinking

Socrates developed a questioning way of critical thinking in order to find the eternal truth. However, the development of conventional thinking has corrupted critical thinking. Assertive criticism is applied to discredit and berate others, rather than to find the truth. This corrupted approach creates conflict and aggression.

Criticizing could make each side of a debate a loser. The eloquent winner of the debate might lose, because the loser of the debate might hold the truth. Nevertheless, the educational system rewards that crippling way of thinking in debating contests and moot arguments. The truth is lost in an idiotic search for external popularity and righteousness.

For nemonik thinkers, success is paramount. They might lose many battles in order to obtain what they seek and escape what they suffer. Nemonik thinkers turn inside and apply the nemonik accelerator to their own ideas. They search for the synthesis if both sides of the argument are valid. If any side is invalid, then no amount of debate could make a valid and reliable argument. No matter how eloquent, debating such spurious arguments is a weakness wasting precious time and resources that you should use to succeed.

Answering versus questioning

Conventional thinkers tend to focus on finding the right answers and solutions. The educational system is an enthusiastic promoter of this approach. Right from the start, questions are phrased so that the answer of the student is either right or wrong. The application of multiple choices limits even the depth of their answers. That is like asking— what is your favourite colour? Red or blue? Unfortunately, I like green! Hence, the freedom needed for creativity is inherently inhibited by the system. Despite its obvious weaknesses, the system is maintained, because narrowing the possible answers allows the effective mass grading of the multitude of students.

Students are likely to succeed in the educational system even if they find the right answers on the wrong questions asked by their teachers. Hence, the educational system trains students to find answer and solutions, rather than questions or problems. That might be why geniuses have problems to fit in that system. It is in their very nature to question established knowledge.

Albert Einstein is one of the many geniuses who were rejected by the educational system. Not understanding Einstein's questioning way of thinking, his tutor called him even a 'lazy dog'. As a result, a superb creative thinker like Einstein was initially forced to accept a job as a clerk in the bureaucracy of a Swiss patent office.

The 17 nemoniks of nemonik thinking will open your mind in order to discover the questions that are vital for your personal success.

Rationalizing versus rational thinking

Aristotle's introduced a rational way of thinking based on logic and reason. True statements or premises would lead inevitably to a valid and reliable conclusion. Rational thinking initiated amazing phenomena such as the renaissance, industrialization, and computerization.

Conventional thinking has corrupted rational thinking into the rationalization of predetermined opinions and beliefs. Rationalization turns rational thinking upside down. It determines first the conclusion and then searches for statements to prove that conclusion. Ultimately, the defence of personal opinions and beliefs replaces rational thinking.

Similar to criticism, rationalization nurtures debating classes and moot arguments. This approach wastes precious resources on arguments for argument's sake without any focus on the truth. It promotes popularity over wisdom. However, science is not a democracy.

Nemonik thinkers realize themselves that each thesis, antithesis, or synthesis is only one of the many steps towards the elusive truth. Libraries are filled with rejected information that was once considered the eternal truth. Nemonik thinkers focus on their success and, therefore, reject ownership of any thesis, antithesis, or synthesis. Owning such information equals mental stagnation. To rationalize your opinions for the sake of being 'right', is a weakness you cannot afford, because it decreases the likelihood of success.

Righteous versus collective thinking

Initially, rules were created to regulate and strengthen the cooperation between the members of a collective. In accord, Confucius suggested—do not do to others what you do not want others to do to you.

The educational system has taught an entire generation of conventional thinkers that they are entitled. Allegedly, they have rights to education, security, healthcare, good income, house, job, etc. They are conditioned to win, which drives them to use the rules to their advantage. Consequently, conventional thinkers have learned to claim privileges and reject obligations.

Being seen to be right has become more important than being right. Accountants help the rich to evade taxes, while lawyers help criminals to avoid the punishments for their crimes. However, tax evasion and criminality weaken the society. Hence, they might win, but they will not succeed. They only climb in the mast of a sinking Titanic that they steered themselves into an iceberg.

One might think that I overstate my case, but look at the millions of refugees on this planet. These refugees include the previously rich and their financial advisors who had the power to prevent the impending disaster. The refugees include also the criminals and their lawyers who accelerated the decline. Their lives are destroyed and their loved ones killed, because they allowed their societies to break down. As the graffiti on the wall says—"Future generations will not forgive your apathy."

Nemonik thinkers take care of their collective. Not because they are do-gooders, but because it increases their likelihood of long-term success. They act selfish and, therefore, their actions will last.

Educated versus wise

The main focus of the educational system is on memorizing information, rather than processing information. Although crucial for individual and collective success, phenomena such as thinking, intuition, creativity, and wisdom are notoriously difficult to teach and quantify. As a result, they play no important role in the educational system. For evaluative reasons, educational achievement is firmly based on the filling and evaluation of memory. In computer terminology, education is focussed on the hard-drive, rather than the CPU.

Nemonik thinkers cultivate equally their memory and thinking. It is hard to think without having information stored in the memory. On the other hand, it is ineffective to store information without processing it. Furthermore, nemonik thinkers use both the rational and affectorial part of their mind. They rely on rational facts, logic, reason, and rules, but also on affectorial intuitions, emotions, creativity, and wisdom. Therefore, their mind is dynamic in a dynamic universe. They will succeed where conventional thinkers fail, because they adjust to reality.

Detached versus compassionate

Conventional thinkers have corrupted Aristotle's way of thinking and use its rational detachment to avoid a confrontation with the consequences of their own actions.

Rich people detach themselves from the poor by blaming the victims for their poverty. They maintain—It is just business. Accountants detach themselves from the taxpayers by pointing out that they are just doing their job and act within the law. Criminals detach themselves from their victims by saying—it is nothing personal. It is a world in which dog eats dog. Their lawyers detach themselves from the victims by arguing that a criminal has the right of a fair trial.

Nemonik thinkers are no do-gooders. They strive to develop genuine compassion for their own sake. As their compassion is for their own sake, it will last. Like karma, compassion returns compassion. Compassion is the basic force that creates and maintains a collective. Its strength cannot be underestimated. History has shown repeatedly that people will sacrifice their lives out of compassion for their collective.

What next?

It is beyond the scope of this book to explain the details of nemonik thinking.[42] Nevertheless, the nemoniks could be used to talk about climate change.

The three temporal nemoniks are—act, wait, and prepare. 'Act' prompts the mind to change or move matter in space and time. 'Wait' prompts the mind to delay an action until it is the right time for that action. 'Prepare' prompts the mind to get ready for action.

If nemonik thinkers are prepared for action, then they wait for the right time to execute that action. Action for the sake of action is counterproductive. However, waiting without being fully prepared is also counterproductive. Furthermore, waiting or preparing for the sake of avoiding the risk of action is counterproductive. Nemonik thinkers let time work in their advantage in order to obtain additional information, improve their position, and discover alternative solutions. They act when their position is likely to weaken and wait when it will strengthen.

From a nemonik point of view, humanity applies a counterproductive strategy to the threat of climate change. It has been waiting for decades, while it is still unprepared. The conventional knowledge about climate change is inadequate, because this study 'reveals' a huge thermal gap. That gap 'accepts' the proposed bilateral climate-change hypothesis and 'rejects' the conventional unilateral climate-change hypothesis. The thermal gap is the climate game-changer.

The thermal gap suggests that 'action' to reduce artificial CO_2 could create a frozen planet, while 'waiting' could create a boiling planet when the glacial decline turns around. The pressure is on, because either event could happen within a few years.

Nemonik thinking suggests that humanity cannot 'act' or 'wait'. The ongoing debate between global warming and cooling shows that scientists failed to solve the problem of climate change with their 'objective mindmodes'. As a result, the leaders do not know whether to 'accept' or 'reject' the thesis or antithesis. Therefore, they cannot use their 'collective mindmodes' to make rules that correct the situation. As a result, everyone is forced to rely on unreliable opinions generated by their 'reactive mindmodes'. This results in stagnation and a widespread confusion.

The three material nemoniks are—accumulating, preserving, and disposing. The unilateral hypothesis supports the nemonik to 'dispose' CO_2, while the bilateral hypothesis supports the nemonik to 'preserve' CO_2. We might even have to 'accumulate' or produce CO_2 in order to compensate for the inevitable long-term glacial cold. Hence, immediate 'preparation' in the form of research is vital. We just do not know enough to consider any action.

Humanity has run out of time and needs all the help it can get. Conventional thinking has corrupted an already incomplete way of thinking that is two-and-halve thousand years old. Humanity's thinking has to be upgraded to nemonik thinking as soon as possible. I agree with Edward de Bono that the art of thinking should become part of the educational curriculum.

If you believe that I overstate my case, then ask yourself whether humanity will succeed. Ask yourself whether you will succeed. Will you really find what you seek and escape what you suffer? Alternatively, will you try to win the competition and climb the mast of the sinking Titanic? Do you let an outdated system and a corrupted way of thinking ruin your life and dreams? Do not! Become a thinker, rather than a victim. Upgrade your mind, because you will need it to survive. Join the ranks of nemonik thinkers. Act now.[ii]

---

ii Appendix: Nemonik Thinking.

Notes

STATISTICAL ANALYSES

LEADERS BEWARE

Statistics is a sophisticated tool for making decisions. Nevertheless, statisticians are not decision makers. It is simply not their job. If statisticians fail to disclose fully the weaknesses and limitations of their methodology, then they cease to be statisticians and become concealed decision makers. If decision makers accept blindly statistical advice, then they have ceased to be part of the decision making process and abandoned their responsibility as leaders.

The idea of a computable 'statistical significance' is the root of many problems. Statisticians compute objectively the probability of making an incorrect decision (p). However, knowingly or unknowingly, decision makers should determine subjectively the risk criterion, which is the accepted probability of making an incorrect decision (α).

Without setting the risk criterion, one cannot make a statistical decision. Therefore, statisticians accepted traditionally a probability of making five incorrect decisions in every one hundred decisions, which sets the risk criterion at $\alpha = 0.05$.

The problem with statistical significance can be illustrated with Russian roulette. This 'game' is played with a six-shooter that has one bullet in the cylinder. The cylinder is spun and the barrel is placed against the head. Do not try this at home! Pulling the trigger provides the probability of making an incorrect and fatal decision of $p = 1/6 = 0.167$. Consequently, the probability of making an incorrect decision ($p = 0.167$) is larger than the accepted probability of making an incorrect decision ($\alpha = 0.050$). Therefore, statisticians would advise you not to pull the trigger. The risk to get shot is above the generally accepted level of risk.

In case of a gun with twenty blanks and one life bullet randomly placed in the magazine, the probability of making an incorrect decision would be p = 1 / 21 = 0.048. Hence, the probability of making an error is smaller than the accepted probability of making an error (p = 0.048 < α = 0.050). The statistical criterion is met, but only a fool would pull the trigger.

Statistical significance is not determined by objective computations alone. The risk criterion is determined by the subjective decision to accept or reject the computed risk. Statisticians can only inform you about your chances of making an incorrect decision, but you remain the decision maker! You are the one who will die from pulling the trigger. Therefore, you have to set the risk criterion, which should balance your potential risk and pay-out.

Making decisions about global warming is more complex than playing Russian roulette. Even if you decide not to make a decision, then you might still make the wrong decision. If it is decided that the hypothesis of industrial global warming is true, then humanity is ill prepared for that catastrophe. In contrast, if it is decided that the hypothesis of industrial global warming is false, then humanity has already wasted many precious resources on the prevention of an illusion. Any incorrect decision about this threat is likely to cause severe suffering for humanity. It is the task of statisticians to provide reliable statistics and full disclosure of weaknesses so that leaders have a scientific basis for their decisions. Leaders beware. Never let a statistician pull the trigger.

Advocates of global warming might try to avoid the mistake of rejecting the true hypothesis that industrial global warming is a threat to humanity and set their α-level at a lax 0.05. On the other hand, their opponents might try to avoid the mistake of accepting the false hypothesis that industrial global warming is a threat to humanity and set their α-level at a strict 0.001. Although both groups apply objective data and statistics, their arguments are fuelled by their reactive, rather than objective mindmodes. They argue about the amount of acceptable risk, which relies on the subjective art of persuasion, rather than on the objective science of statistics.

Although the use of statistics in this study is a necessity, complexity is avoided as far as possible. As shown in the associated files, the statistics in this study can be computed with Microsoft's Excel. The descriptions and examples in the appendix 'Statistics' might be helpful for readers unfamiliar with statistics.

VOSTOK °C (423 KYRS)

VOSTOK °C DEPTH-SCALE

Method

The aim of this section is to evaluate the atmospheric temperatures, during the last 423 Kyrs, which were reconstructed from an ice-core drilled in Vostok Antarctica. A general aim of this study is to compare those past Vostok temperatures with the present global temperatures. Modification of Dataset 1 provided Dataset 2, which is the basis for the analyses in this section.

Dataset 1 (VostokDeutTempDepth423Kyrs).[43] This dataset contains the following Vostok variables: (Depth corrected)—ice depth from 3,310 metres to 0 metre in steps of 1 metre; (Ice Age (GT4))—ice age in years from 422,766 years before present (BP) to the present calendar year 1989; (deut)—depth-scale deuterium concentrations in ‰ Standard Mean Ocean Sea Water (SMOW); and (DeltaTS)—depth-scale atmospheric temperature deviations in degrees Celsius (°C) relative to the Vostok base 1850-1989 = 0.00 °C.

Citation for Dataset 1: (Petit, et al., 1999); (Petit, Vostok Ice Core Data for 420,000 Years, 2001); and (NOAA/NGDC Paleoclimatology Program). Details about this dataset are included in appendix: Dataset 1. Dataset 2 was created by removing the variable 'deut' and changing the labels of the remaining variables of Dataset 1.

Dataset 2 (VostokTempDepth423Kyrs).[44] This dataset contains the following Vostok variables: (Depth)—ice depth from 3,310 metres to 0 metre in steps of 1 metre; (Age)—ice age in years from 422,766 years before present (BP) to the present calendar year 1989; and (TDv)—depth-scale atmospheric temperature deviations in degrees Celsius (°C) relative to the Vostok base 1850-1989 = 0.00 °C. This dataset is included in appendix: Dataset 2.

Temperature deviations are expressed in degrees Celsius (°C) relative to an arbitrarily selected base temperature, rather than in absolute degrees Celsius relative to the objective base of zero degrees Celsius.[45] The advantage of temperature deviations is that changes in temperature can be compared directly. For example, a change in the Antarctic temperature from -29.00 to -30.00 in absolute degrees Celsius is difficult to compare with a change in the African temperature from 16.00 to 17.00 in absolute degrees Celsius. In contrast, those changes would provide a comparable temperature deviation of 1.00 °C in both Antarctica and Africa. However, the bases of both temperature deviations should be the same. The atmospheric temperature deviations comprising Dataset 2 are relative to the arbitrary base from calendar year 1850 to 1989 with a mean of 0.00 °C.

Results

Dataset 2 provides the statistics for the Vostok depth-scale temperature deviations: mean = -4.52; median = -5.10; standard deviation = 2.90; maximum = 3.23; minimum = -9.39; and range = 12.62 °C; n = 3,311.[46]

The statistics for the data intervals are: mean = 128; median = 85; standard deviation = 113; maximum = 663; minimum = 21; and range = 642 years; n = 3,303.[47]

The duration of the most recent data interval equals 21 years ranging from 160 to 139 years BP.[48] In contrast, the duration of the oldest data interval equals 630 years ranging from 422,451 to 421,821 years BP.[49]

The statistics for the temperature-peaks (TP) are: mean = 2.56; median = 2.20; standard deviation = 0.59; maximum = 3.23; and minimum = 2.06; and range = 1.17 °C; n = 5.[50] Furthermore, the statistics for the peak-to-peak durations (PTP) are: mean = 100,587; median = 98,732; standard deviation = 17,156; maximum = 120,222; minimum = 84,663; and range = 35,559 years; n = 4.[51]

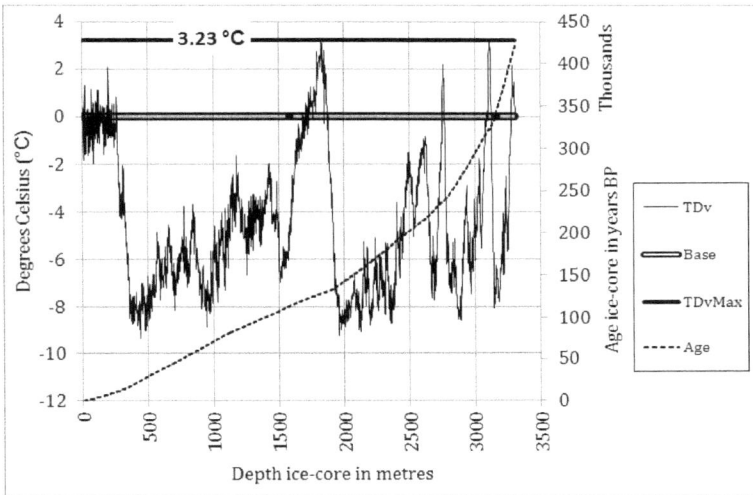

Figure 1 (Dataset 2)—Vostok depth-scale temperatures.[52] This graph shows: (TDv)—depth-scale atmospheric temperature deviations from 422,766 years before present (BP) to the present calendar year 1989, in degrees Celsius (°C) relative to the Vostok base 1850-1989 = 0.00 °C; (TDvMax)—maximum TDv = 3.23 °C; (Age)—ice age from 422,766 years BP to the present calendar year 1989.[53]

Comments

The pressure of the ice-sheet and glacial flow have gradually compressed the time-scale of Dataset 2 with depth. As a result, each metre of ice provides a mean temperature deviation that is based on a different data interval ranging from 21 to 663 years. Figure 1 illustrates that compression accelerates the increase in ice age with depth. As a result, compression distorts the temporal thermal pattern.

Regression towards the population mean is likely to decrease the range of the sample means with an increase of the data interval. Due to the different data intervals, this regression is likely to confound comparisons between the mean temperature deviations within Dataset 2.

The effect of different data intervals on the overall statistics is illustrated by comparing the mean of the raw temperature deviations (-4.52 °C) with the mean of the temperature deviations weighted with data intervals (-4.79 °C).

The statistics of the temperature peaks support the notion that the duration of the thermal cycles in Vostok is about 100,000 years with a fluctuation of about thirteen degrees Celsius. However, due to regression towards the mean and the relatively long and different data intervals, the annual thermal fluctuation is likely to be larger.

To enhance the evaluation of the temporal thermal pattern, the depth-scale of Dataset 2 will be converted into a millennial-scale.

VOSTOK °C MILLENNIAL-SCALE

Method

The aim of this section is to evaluate the characteristics of the Vostok mean temperature deviations with equal data intervals. Modification of Dataset 2 provided Dataset 3, which is the basis for the analyses in this section.

Dataset 2 (VostokTempDepth423Kyrs). This dataset contains the following Vostok variables: (Depth)—ice depth from 3,310 metres to 0 metre in steps of 1 metre; (Age)—ice age in years from 422,766 years before present (BP) to the present calendar year 1989; and (TDv)—depth-scale atmospheric temperature deviations in degrees Celsius (°C) relative to the Vostok base 1850-1989 = 0.00 °C.

The depth-scale temperature deviations in Dataset 2 were weighted with the duration of their data-intervals.

Equation 1 (Dataset 2)—computes the millennial thermal product: $\sum Tp = (TD * DI)$.[54] In which, $(\sum Tp)$—sum thermal products for an observation period of 1,000 years; (TD)—depth-scale temperature deviation in degrees Celsius (°C) relative to the Vostok base Kyr1 = 0.00 °C; and (DI)—depth-scale data interval in years.

Equation 2 (Dataset 2)—computes the millennial temperature deviations: TDw = TP / 1000.[55] In which (TDw)—millennial weighted mean temperature deviation in degrees Celsius (°C) relative to the Vostok base Kyr1 = 0.00 °C; (Tp)—thermal product Equation 1; and (1,000)—duration of the observation period in years. If a data interval overlapped the boundary between two millennial periods, then TP was allocated proportionally to each millennium.

Dataset 3 (VostokTempMill423Kyrs).[56] This dataset contains the following Vostok variables: (Age)—ice age in millennia or Kyrs from 423 Kyrs before present (BP) to the present calendar year 1989; (TDv)—millennial-scale atmospheric temperature deviations in degrees Celsius (°C) relative to the Vostok base 1850-1989 = 0.00 °C.

Results

Dataset 3 provides the statistics for the Vostok millennial temperature deviations: mean = -4.78; median = -5.47; standard deviation = 2.77; maximum = 3.03; minimum = -8.76; and range = 11.78 °C; n = 423.[57]

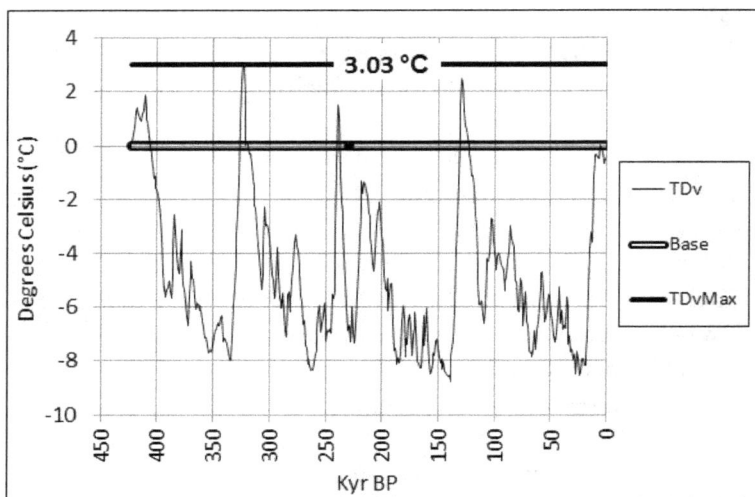

Figure 2 (Dataset 3)—Vostok millennial-scale temperatures.[58] This graph shows: (TDv)—millennial-scale atmospheric temperature deviations from Kyr423 to Kyr1, in degrees Celsius (°C) relative to the Vostok base 1850-1989 = 0.00 °C; and (TDvMax)—maximum TDv = 3.03 °C.[59]

Comments

Figure 2 facilitates the visual inspection of the thermal cycles in Vostok during the last 423 Kyrs.

The statistical results show that the obtained mean of the millennial-scale temperature deviations (-4.78 °C) is almost the same as the weighted mean of the depth-scale temperature deviations (-4.79 °C).[60] This support the notion that the conversion from depth- to millennial-scale is reliable.

Unfortunately, the confounding effects of the long and unequal data intervals on the reliability of statistical analyses cannot be undone. Furthermore, the millennial conversion to equal data intervals increases the reliability of the mean temperature deviations at the cost of the variance. In accord, the conversion decreases the standard deviation of the temperature deviations from 2.90 to 2.77 °C; the range from 12.62 to 11.78 °C; and the maximum from 3.23 °C (Dataset 2) to 3.03 °C (Dataset 3). This shows that the regression towards the overall mean of the millennial data will confound comparisons with the standard deviations, maxima, minima, and ranges of annual data.

Dataset 3 is more reliable than Dataset 2, because it comprises mean millennial temperatures. Nevertheless, the base of Dataset 3 (1850-1989 = 0.00 °C) is arbitrary and too short for reliable and valid long-term interpretations.

VOSTOK °C BASE (KYR1 = 0.00 °C)

Method

As mentioned previously, the temperature deviations comprising Dataset 2 are relative to the arbitrary base from calendar year 1850 to 1989 with a magnitude of 0.00 °C. To reflect adequately the statistical centre (mean) and spread (standard deviation) of such temperature deviations, the selected base should represent a reliable standard or 'normal temperature'. However, the base was determined by the bad conditions of the Vostok ice-core during the first few metres, rather than by a careful selection of the base period. Consequently, the arbitrarily selected base does not necessarily reflect a 'normal temperature', while a base of 139 years is too short for long-term evaluations concerning 423 Kyrs. Therefore, it is the aim of this section to enhance the validity and reliability of the thermal base. Modification of Dataset 3 provided Dataset 4, which is the basis for the analyses in this section.

Dataset 3 (VostokTempMill423Kyrs). This dataset contains the following Vostok variables: (Age)—ice age in millennia or Kyrs from 423 Kyrs before present (BP) to the present calendar year 1989; (TDv)—millennial-scale atmospheric temperature deviations in degrees Celsius (°C) relative to the Vostok base 1850-1989 = 0.00 °C.

The base of Dataset 3 was converted from (1850-1989 = 0.00 °C) to the mean temperature of the most recent millennia (Vostok Kyr1 = 0.00 °C).

Dataset 4 (VostokTempMillBase423Kyrs).[61] This dataset contains the following Vostok variables: (Age)—ice age in millennia or Kyrs from 423 Kyrs before present (BP) to the present calendar year 1989; (TDv)—millennial-scale atmospheric temperature deviations in degrees Celsius (°C) relative to the Vostok base Kyr1 = 0.00 °C.

Results

Dataset 4 provides the statistics for the Vostok millennial temperature deviations: mean = -4.32; median = -5.00; standard deviation = 2.77; maximum = 3.49; minimum = -8.29; and range = 11.78 °C; n = 423.[62]

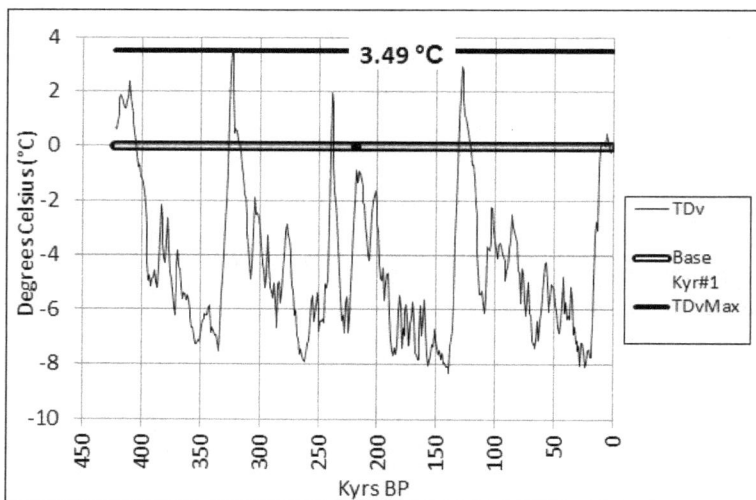

Figure 3 (Dataset 4)—Vostok temperatures base Kyr1.[63] This graph shows: (TDv)—millennial-scale atmospheric temperature deviations from Kyr423 to Kyr1, in degrees Celsius (°C) relative to the Vostok base Kyr1 = 0.00 °C; (TDvMax)—maximum TDv = 3.49 °C.[64]

Comments

In this section, the base for the Vostok temperature deviations was changed from (1850-1989 = 0.00 °C) to (Kyr1 = 0.00 °C). This new base reflects an average temperature that has been considered 'normal' for the last 1,000 years. As a result, the temperature deviations comprising Dataset 4 are comparable millennial temperatures relative to a reliable and valid base period. Furthermore, the new base increased the maximum millennial temperature deviation with 0.46 °C, from 3.03 °C (Figure 2) to 3.49 °C (Figure 3). Despite the improvements, a thermal trend could still confound the statistics of the temperature deviations.

VOSTOK °C TREND

Method

The aim of this section is to evaluate the long-term trend in the Vostok millennial temperature deviations comprisingDataset 4.

Dataset 4 (VostokTempMillBase423Kyrs). This dataset contains the following Vostok variables: (Age)—ice age in millennia or Kyrs from 423 Kyrs before present (BP) to the present calendar year 1989; (TDv)—millennial-scale atmospheric temperature deviations in degrees Celsius (°C) relative to the Vostok base Kyr1 = 0.00 °C.

Results

Dataset 4 provides the statistics for the Vostok millennial temperature deviations: mean = -4.32; median = -5.00; standard deviation = 2.77; maximum = 3.49; minimum = -8.29; and range = 11.78 °C; n = 423.[65]

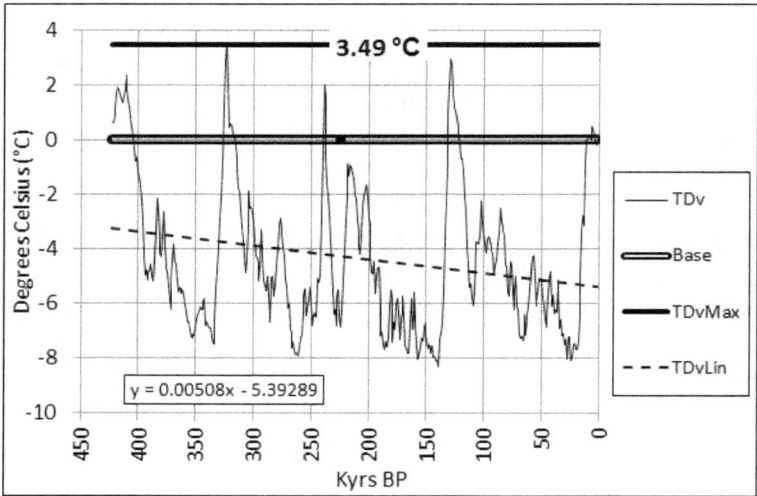

Figure 4 (Dataset 4)—Vostok thermal trend.[66] This graph shows: (TDv)—millennial-scale atmospheric temperature deviations from Kyr423 to Kyr1, in degrees Celsius (°C) relative to the Vostok base Kyr1 = 0.00 °C; (TDvMax)— maximum TDv = 3.49 °C; and (TDvLin)—linear trend of TDv = 0.00508 °C/Kyr.[67]

Comments

Figure 4 shows a linear downward trend of -0.00508 °C/Kyr in the Vostok temperature deviations. This trend is likely to confound the visual inspection of the thermal cycles shown in that graph. In addition, the downwards trend in Figure 4 decreases the likelihood that previous maxima of the temperature deviations will reoccur. Hence, the trend in Dataset 4 has to be eliminated in order to improve the comparison of cycles and the predictive strength of the data. The detrending will make past temperatures relevant to today's temperatures.

VOSTOK °C DETRENDED MILLENNIAL-SCALE

Method

It is the aim of this section to detrend the Vostok millennial temperature deviations of the last 423 Kyrs in order to make them relevant to today's temperatures. Modification of Dataset 4 provided Dataset 5, which is the basis for the analyses in this section.

Dataset 4 (VostokTempMillBase423Kyrs). This dataset contains the following Vostok variables: (Age)—ice age in millennia or Kyrs from 423 Kyrs before present (BP) to the present calendar year 1989; (TDv)—millennial-scale atmospheric temperature deviations in degrees Celsius (°C) relative to the Vostok base Kyr1 = 0.00 °C. The Vostok temperature deviations in Dataset 4 were detrended with 0.00508 °C/Kyr.[68]

Dataset 5 (VostokTempMillBaseDetrend423kyrs).[69] This dataset contains the following Vostok variables: (Age)—ice age in millennia or Kyrs from 423 Kyrs before present (BP) to the present calendar year 1989; (TDv)—detrended millennial-scale atmospheric temperature deviations in degrees Celsius (°C) relative to the Vostok base Kyr1 = 0.00 °C. This dataset is included in appendix: Dataset 5.

Results

Dataset 5 provides the statistics for the Vostok detrended millennial temperature deviations: mean = -5.39; median = -6.07; standard deviation = 2.70; maximum = 2.30; minimum = -9.22; and range = 11.52 °C/Kyr; n = 423.[70] The maximum detrend correction was 2.15 °C at Kyr423.[71]

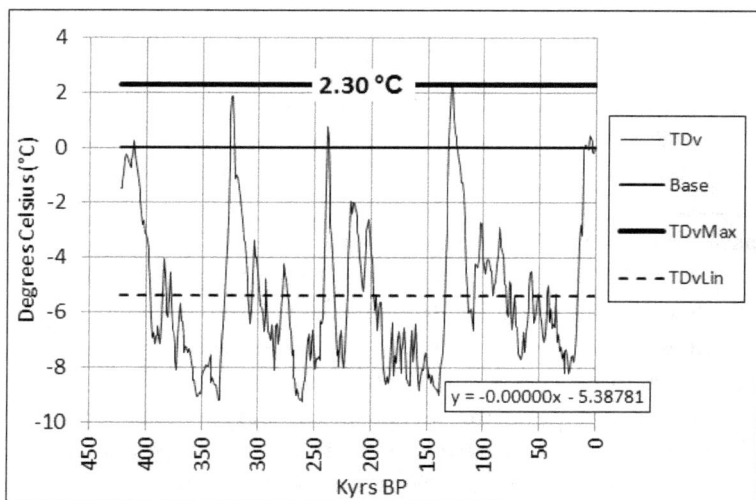

Figure 5 (Dataset 5)—Vostok detrended temperatures.[72] This graph shows: (TDv)—detrended millennial-scale atmospheric temperature deviations from Kyr423 to Kyr1, in degrees Celsius (°C) relative to the Vostok base Kyr1 = 0.00 °C; (TDvMax)—maximum TDv = 2.30 °C; and (TDvLin)— residual linear trend of TDv = 0.00000 °C/Kyr.[73]

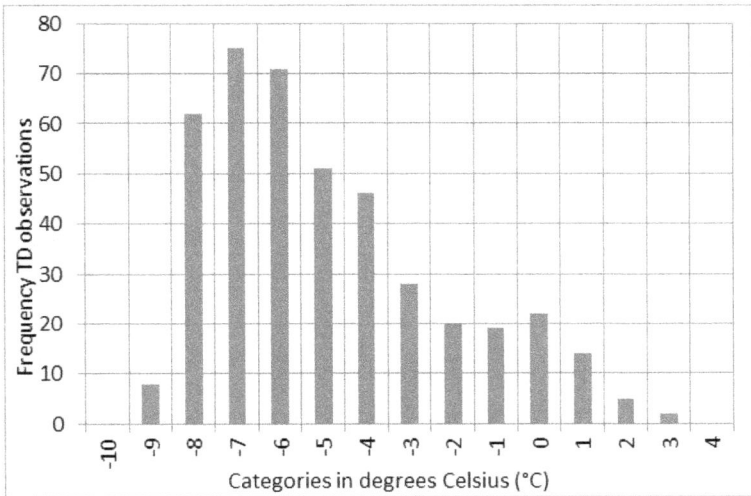

Figure 6 (Dataset 5)—histogram Vostok detrended temperatures.[74] This graph shows the frequency distribution of the Vostok detrended millennial temperature deviations from Kyr423 to Kyr1, in degrees Celsius (°C) relative to the Vostok base Kyr1 = 0.00 °C.

The histogram shows that the frequency distribution of the temperature deviations is positively skewed towards the higher temperatures. In addition, the distribution is bimodal because it peaks around -7.00 and 0.00 °C.

Table 1				
	Degrees Celsius (°C) during 423 Kyrs			
	TDvd	TDvm	TDvmb	TDv
mean	-4.52	-4.78	-4.32	-5.39
median	-5.10	-5.47	-5.00	-6.07
STD	2.90	2.77	2.77	2.70
max	3.23	3.03	3.49	2.30
min	-9.39	-8.76	-8.29	-9.22
range	12.62	11.78	11.78	11.52
n	3311	423	423	423

Table 1—modification Vostok temperature deviations during 423 Kyrs.[75] This table shows the statistics for the following variables: (TDvd (Dataset 2))—depth-scale atmospheric temperature deviations, during the last 423 Kyrs, in degrees Celsius (°C) relative to the Vostok base 1850-1989 = 0.00 °C; (TDvm (Dataset 3))—millennial-scale atmospheric temperature deviations, during the last 423 Kyrs, in degrees Celsius (°C) relative to the Vostok base 1850-1989 = 0.00 °C; (TDvmb (Dataset 4))—millennial-scale atmospheric temperature deviations, during the last 423 Kyrs, in degrees Celsius (°C) relative to the Vostok base Kyr1 = 0.00 °C; (TDv (Dataset 5))—detrended millennial-scale atmospheric temperature deviations, during the last 423 Kyrs, in degrees Celsius (°C) relative to the Vostok base Kyr1 = 0.00 °C.

Comments

The graph shows that the residual linear trend of the temperature deviations equals 0.00 °C / Kyr, which suggests that the detrending procedure was successful. Table 1 shows that detrending the temperature deviation decreased the maximum from 3.49 °C (Figure 4) to 2.30 °C (Figure 5). The skewed and bimodal frequency distribution of the Vostok temperature deviations suggests that parametric statistical analyses are inappropriate for the evaluation of Dataset 5.

The detrended millennial temperature deviations provide a reliable basis to compare the temperatures during the previous interglacials with the temperatures of the current interglacial. Nevertheless, the transformation from depth-scale to millennial-scale has caused a regression towards the mean that reduced the maximum temperature deviations in this dataset. This effect could confound the comparison between Vostok and global maximum temperatures.

DURATION GLACIALS

Method

The aim of this section is to evaluate the duration of glacial periods that occurred during the last 423 Kyrs. A glacial period is a long-term climatological period with mean millennial temperature deviations below the Vostok base $Kyr1 = 0.00$ °C. The start or finish of a glacial period is the year where a straight line, connecting two millennial temperature deviations at adjacent millennial midpoints, crosses the Vostok base $Kyr1 = 0.00$ °C. The analyses in this section are based on Dataset 5.

Dataset 5 (VostokTempMillBaseDetrend423kyrs). This dataset contains the following Vostok variables: (Age)—ice age in millennia or Kyrs from 423 Kyrs before present (BP) to the present calendar year 1989; (TDv)—detrended millennial-scale atmospheric temperature deviations in degrees Celsius (°C) relative to the Vostok base $Kyr1 = 0.00$ °C. This dataset is included in appendix: Dataset 5.

Results

Table 2		
Start YBP	Finish YBP	Duration years
122,992	9,927	113,065
237,362	130,676	106,687
321,192	238,853	82,339
409,963	325,290	84,673
mean		96,691
median		95,680
STD[76]		15,475
maximum		113,065
minimum		82,339
range		30,726
n[77]		4

Table 2 (Dataset 5)—durations of glacials during 423 Kyrs.[78] This table shows the start and finish of glacial periods in years before the present calendar year 1989 (YBP); their duration in years; and the statistics of those glacials.

Comment

Four glacial periods have occurred during the previous 423 Kyrs with a mean duration of 96.7 Kyrs.

DURATION INTERGLACIALS

Method

The aim of this section is to evaluate the duration of the interglacial periods during the last 423 Kyrs. An interglacial period is a long-term climatological period with mean millennial temperature deviations equal or above the Vostok base Kyr1 = 0.00 °C. The start or finish of an interglacial period is the year where a straight line, connecting two millennial temperature deviations at adjacent millennial midpoints, crosses the Vostok base Kyr1 = 0.00 °C. The analyses in this section are based on Dataset 5.

Dataset 5 (VostokTempMillBaseDetrend423kyrs). This dataset contains the following Vostok variables: (Age)—ice age in millennia or Kyrs from 423 Kyrs before present (BP) to the present calendar year 1989; (TDv)—detrended millennial-scale atmospheric temperature deviations in degrees Celsius (°C) relative to the Vostok base Kyr1 = 0.00 °C. This dataset is included in appendix: Dataset 5.

Equation 3 (Dataset 5)—computes the critical duration of the interglacials: CD = DIx + (STD * 1.645).[79] In which, (CD)—critical interglacial duration; (DIx)—mean duration of the previous interglacials in years; (STD) = standard deviation of previous interglacial durations in years; and (1.645)— critical z-score at $\alpha = 0.05$.

If the z-score of the current interglacial duration is larger than 1.645, then that duration is significant longer than the durations of previous interglacials ($p < 0.05$). This means that the duration is due to a systematic effect, rather than random variation.

Results

Table 3		
Start YBP	Finish YBP	Duration years
130,675	122,993	7,682
238,852	237,363	1,489
325,289	321,193	4,096
410,996	409,964	1,032
mean		3,575
median		2,792
STD[80]		3,053
critical duration		8,597
maximum		7,682
minimum		1,032
range		6,650
n		4

Table 3 (Dataset 5)—durations of interglacials during 423 Kyrs.[81] This table shows the start and finish of interglacial periods in years before the present calendar year 1989 (YBP); their duration in years; and the statistics of those interglacials.

Table 3 (Dataset 5) shows the start and finish of the previous interglacial periods in years before the present calendar year 1989 (YBP); the duration of interglacial periods in years; and the statistics of interglacial periods during the previous 423 Kyrs. Furthermore, the results show that the critical interglacial duration equals 8,597 years ($\alpha = 0.05$; $z = 1.645$).

Comments

Four interglacial periods have occurred during the past 423 Kyrs with a mean duration of 3,575 years. Furthermore, the maximum duration of the previous interglacials equals 7,682 years, which is shorter than the critical duration of 8,597 years.

DURATION CURRENT INTERGLACIAL

Method

It is the aim of this section to test the hypothesis that the duration of the current interglacial is significantly longer than the previous interglacials during the last 423 Kyrs. The analyses in this section are based on Dataset 5.

Dataset 5 (VostokTempMillBaseDetrend423kyrs). This dataset contains the following Vostok variables: (Age)—ice age in millennia or Kyrs from 423 Kyrs before present (BP) to the present calendar year 1989; (TDv)—detrended millennial-scale atmospheric temperature deviations in degrees Celsius (°C) relative to the Vostok base Kyr1 = 0.00 °C. This dataset is included in appendix: Dataset 5.

Results

Table 4		
Start YBP	Finish YBP	Duration years
9,926	0	9,926
maximum previous interglacials		7,682
critical duration		8,597
z-score (9,926 years)		2.08
p-value (z-score)		0.0188
n		4

Table 4 (Dataset 5)—duration of the current interglacial.[82] This table shows the start and finish of the current interglacial period in years before the present calendar year 1989 (YBP); and its duration in years. Furthermore, it shows the maximum duration of the previous interglacials; critical interglacial duration; z-score of the current interglacial duration; p-value of the current interglacial duration; and the number of previous interglacials.

None of the previous interglacial durations exceeds the critical duration of 8,597 years. Hence, based on non-parametric statistics, the probability of a Type I error (p)equals (n / N). In which, n = number of previous interglacials longer than the current interglacial; N = number of previous interglacials. Hence, (p = 0/4 = 0.0000 < α 0.05; n = 4). In accord, based on parametric statistics, the probability of a Type I error (p) equals (z = 2.08; p = 0.0188 < α = 0.05; n = 4).

Comments

If the duration of glacial and interglacial periods is determined by a regular effect such as the orbit of the planet, then the irregularities in the durations might reflect inaccuracies in the reconstructed data. On the other hand, the data might be accurate and the irregular durations might be due to irregularities in the cause such as anomalies in solar activity.

The current interglacial is 2,244 years longer than any previous interglacial and 1,329 years longer than the critical duration.[83] Hence, the current interglacial duration is significantly longer than the mean duration of the previous interglacials. This result supports the alternative hypothesis (Ha) that the long duration of the current interglacial is a systematic effect.

The current interglacial duration cannot be explained by random variation around the mean. Therefore, it is likely that a new variable has appeared within the last 10,000 years that extended the duration of the current interglacial with a period ranging from 1,329 to 8,894 years.[84] The significant long duration of the current interglacial might contribute to the present melting of polar ice and glaziers. The reason for this long duration will be evaluated in the following sections.

GLACIAL DECLINE

Method

The aim of this section is to evaluate the expected slope of the glacial decline at the end of the current interglacial. A glacial decline is the onset of a glacial period, which is measured by a decrease in millennial temperature deviations (TD) ranging from the interglacial thermal peak to an arbitrary limit of -4.00 degrees Celsius relative to the Vostok base Kyr1 = 0.00 °C. The analyses in this section are based on Dataset 5.

Dataset 5 (VostokTempMillBaseDetrend423kyrs). This dataset contains the following Vostok variables: (Age)—ice age in millennia or Kyrs from 423 Kyrs before present (BP) to the present calendar year 1989; (TDv)—detrended millennial-scale atmospheric temperature deviations in degrees Celsius (°C) relative to the Vostok base Kyr1 = 0.00 °C. This dataset is included in appendix: Dataset 5.

Results

Dataset 5 provides the statistics for the Vostok millennial temperature deviations: mean = -5.39; median = -6.07; standard deviation = 2.70; maximum = 2.30; minimum = -9.22; and range = 11.52 °C/Kyr; n = 423.[85]

The statistics for the glacial declines presented in Figure 7 are: mean = -0.52; median = -0.40; standard deviation = 0.32; maximum = -0.30; minimum = -1.00; and range = 0.70 °C/Kyr; n = 4.[86]

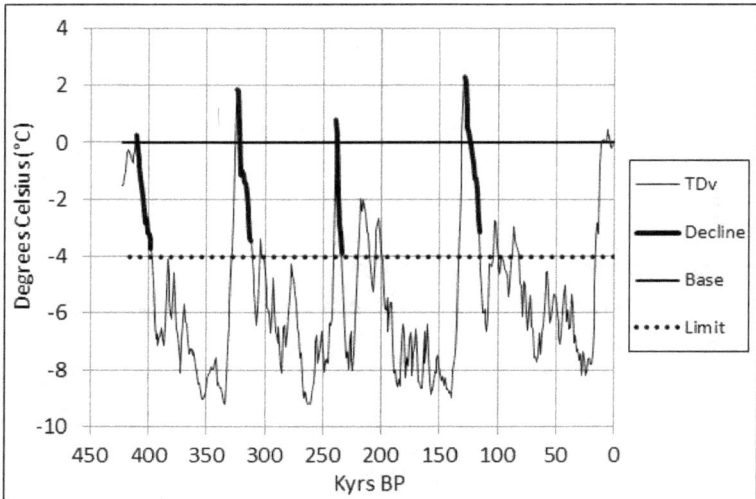

Figure 7 (Dataset 5)—Vostok glacial declines.[87] This graph shows: (TDv)—detrended millennial-scale atmospheric temperature deviations from Kyr423 to Kyr1, in degrees Celsius (°C) relative to the Vostok base Kyr1 = 0.00 °C; (Decline)—thermal glacial decline; and (Limit)—arbitrary end of thermal decline = -4.00 °C.

Comments

The results predict that after reaching its peak, the current interglacial is likely to decline with an average of 0.52 degrees Celsius per Kyr (°C/Kyr).

GLACIAL DECLINE WINDOW

Method

It is the aim of this section to merge the previous findings into a glacial decline window. The glacial decline window shows the estimated temporal and thermal limits of the expected decrease in temperature at the end of the current interglacial. The analyses in this section are based on Dataset 5.

Dataset 5 (VostokTempMillBaseDetrend423kyrs). This dataset contains the following Vostok variables: (Age)—ice age in millennia or Kyrs from 423 Kyrs before present (BP) to the present calendar year 1989; (TDv)—detrended millennial-scale atmospheric temperature deviations in degrees Celsius (°C) relative to the Vostok base Kyr1 = 0.00 °C. This dataset is included in appendix: Dataset 5.

Results

Table 5			
	TD-CIG (°C)	D-PIG (years)	GD-PIG (°C/Kyr)
mean	0.08	3,575	-0.52
median	0.04	2,792	-0.40
STD	0.21	3,053	0.32
maximum	0.46	7,682	-0.30
minimum	-0.21	1,032	-1.00
range	0.67	6,650	0.70
n	10	4	4

Table 5 (Dataset 5)—statistics of the glacial decline window.[88] This table shows the statistics for the following variables: (TD-CIG)—mean TDv last 10 Kyrs; (TDv)—detrended millennial-scale atmospheric temperature deviations in degrees Celsius (°C) relative to the Vostok base Kyr1 = 0.00 °C; (D-PIG)—durations of the previous interglacials in years; (GD-PIG) glacial declines ending the previous interglacials in °C/Kyr.

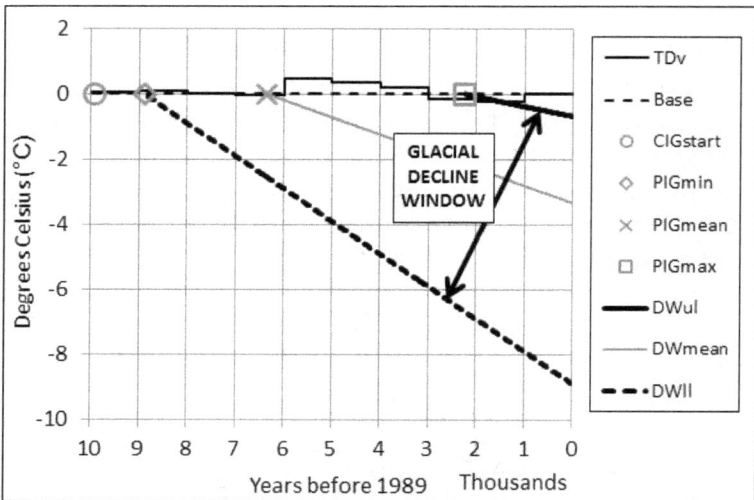

Figure 8 (Dataset 5)—Vostok glacial decline window.[89] This graph shows: (TDv)—detrended millennial-scale atmospheric temperature deviations from year 10,000 before present (BP) to the present calendar year 1989, in degrees Celsius (°C) relative to the Vostok base Kyr1 = 0.00 °C,; (CIGstart)—start current interglacial period at 9,926 years BP;[90] (PIGmin)—minimum duration previous interglacials = 1,032 years;[91] (PIGmean)—mean duration previous interglacials = 3,575 years;[92] (PIGmax)—maximum duration previous interglacials = 7,682 years;[93] (DWul)—upper-limit of the decline window = -0.30 °C/Kyr;[94] (DWmean)—mean decline = -0.52 °C/Kyr;[95] and (DWll)—lower-limit of the decline window = -1.00 °C/Kyr (DWll).[96]

Based on 423 Kyrs of natural temperature deviations, the glacial decline window predicts that the temperature deviations during calendar year 1989 would range between -0.68 and -8.88 °C.[97]

Comments

A decline-window is developed to determine the boundaries of the decrease in temperature that could be expected at the onset of a glacial period. That window predicts a lower Vostok temperature for the last millennium than observed. Therefore, the following sections will evaluate that potential thermal gap in more detail.

INTERGLACIAL THERMAL STABILITY

Method

It is the aim of this section to test the hypothesis that the thermal stability during the current interglacial of 10 Kyrs is significantly larger than the thermal stability during the preceding 404 Kyrs. The analyses in this section are based on Dataset 5.

Dataset 5 (VostokTempMillBaseDetrend423kyrs). This dataset contains the following Vostok variables: (Age)—ice age in millennia or Kyrs from 423 Kyrs before present (BP) to the present calendar year 1989; (TDv)—detrended millennial-scale atmospheric temperature deviations in degrees Celsius (°C) relative to the Vostok base Kyr1 = 0.00 °C. This dataset is included in appendix: Dataset 5.

The Vostok millennial temperature deviations of Dataset 5 were subjected to a sliding window with a duration of 10 Kyrs, shifting from Kyr423 to Kyr1, in steps of 1 Kyr. For each step, the mean temperature deviation was computed (TDx). In addition, the thermal stability index (TSI) of the corresponding temperature deviations was computed.

Equation 4 (Dataset 5)—computes the thermal stability index: TSI = 1 / STD.[98] In which, (TSI)—Thermal Stability Index; (STD)—standard deviation of TDv; and (TDv)—detrended millennial-scale atmospheric temperature deviations in degrees Celsius (°C) relative to the Vostok base Kyr1 = 0.00 °C.

STD increases with an increase in the variation of the temperature deviations around their mean. In contrast, TSI increases with a decrease in the variation of the temperature deviations around their mean. Therefore, TSI is a direct measure of thermal stability.[99]

Dataset 5 (VostokTempMillBaseDetrend423kyrs). This dataset contains the following Vostok variables: (Age)—ice age in millennia or Kyrs from 423 Kyrs before present (BP) to the present calendar year 1989; (TDv)—detrended millennial-scale atmospheric temperature deviations in degrees Celsius (°C) relative to the Vostok base Kyr1 = 0.00 °C. This dataset is included in appendix: Dataset 5.

Results

Table 6		
	TD-CIG (10 Kyrs)	TSI-SW (404 Kyrs)
mean	0.08	1.33
median	0.04	1.18
STD	0.21	0.72
maximum	0.46	**4.02**
minimum	-0.21	0.26
range	0.67	3.77
n	10	404
TSI-CIG	**4.75**	NA

Table 6 (Dataset 5)—sliding window of thermal stability.[100] This table shows the statistics for the following variables: (TD-CIG)—detrended millennial-scale atmospheric temperature deviations, during the current interglacial of 10 Kyrs, in degrees Celsius (°C) relative to the Vostok base Kyr1 = 0.00 °C; (TSI-SW)—thermal stability indexes as obtained with a sliding window of 10 Kyrs. This window shifted from Kyr414 to Kyr11 in steps of 1 Kyr;[101] (TSI-CIG)—thermal stability index of the current interglacial during the last 10 Kyrs.

The TSI-CIG is significantly larger than the mean of the TSI-SW values (4.75 versus 1.33; z-score 4.72; p < 0.0002).[102] Alternatively, TSI-CIG is larger than the maximum TSI-SW value, which excludes the possibility of making Type I errors. Hence, the TSI-CIG of the current interglacial is significantly larger than the TSI-SW values during the last 404 Kyrs (p = 0.0000; n = 404).[103]

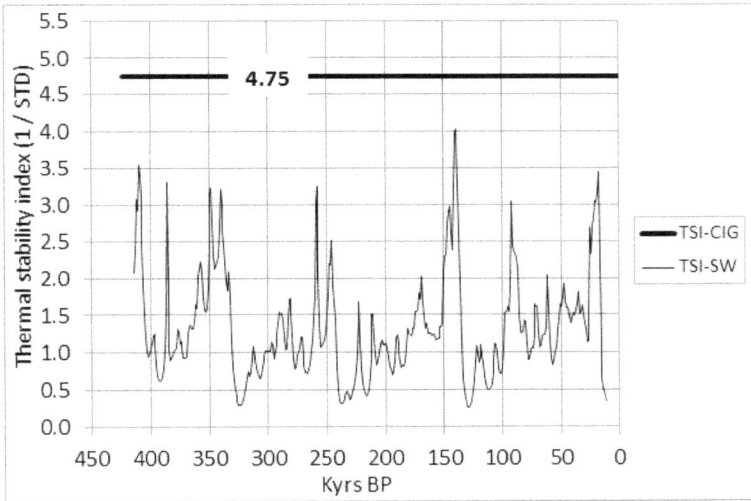

Figure 9 (Dataset 5)—Vostok thermal stability index (TSI).[104] This graph shows: (TSI-CIG)—thermal stability index during the current interglacial of 10 Kyrs; and (TSI-SW)—thermal stability indexes during the observation period from Kyr414 to Kyr11, as obtained with a sliding window of 10 Kyrs.

The data show that the TSI-SW decreased abruptly at the beginning of the current interglacial. During the period from Kyr20 to Kyr11, the thermal stability index equals 0.35.[105] However, during the adjacent current interglacial period from Kyr10 to Kyr1, the thermal stability index increased to 4.75.[106]

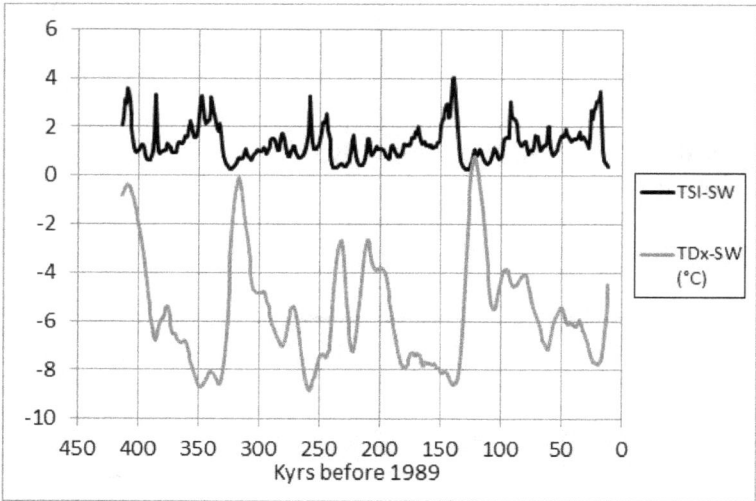

Figure 10 (Dataset 5)—Vostok thermal stability index versus temperature.[107] This graph shows: (TSI-SW)—Vostok thermal stability indexes from Kyr414 to Kyr11, as obtained with a sliding window of 10 Kyrs; and (TDx-SW)—Vostok mean temperature deviations as obtained with a sliding window of 10 Kyrs from Kyr414 to Kyr11.

Dataset 5 provides the statistics for the TDx-SW values: mean = -5.60; median = -6.03; standard deviation = 2.27; maximum = 0.77; minimum = -8.87; and range = 9.64 °C; n = 404.[108] Furthermore, the statistics for the TSI-SW values are: mean = 1.33; median = 1.18; standard deviation = 0.72; maximum = 4.02; minimum = 0.26; and range = 3.77 °C; n = 404.[109]

Test 1 (Dataset 5)—randomized correlation TDxSWr and TSISW (RCT).[110] In which: (TDxSW)—Vostok ranked mean millennial temperature deviation obtained with a sliding window of 10 Kyrs; and (TSISW)—Vostok ranked thermal stability index (1/STD) of the millennial temperature deviation obtained with a sliding window of 10 Kyrs; observation period from Kyr423 to Kyr11. At each iteration, the correlation coefficient was computed after TDxSW was randomized relative to TSISW. Using ranked data avoids the statistical problems associated with asymmetric frequency distributions, unequal variances, and outliers. The chosen observation period reduces the confounding effects of human activity on the variables.

Test 1 provided the following statistics for the randomized correlation coefficients: mean = 0.00; median = 0.00; standard deviation = 0.05; maximum = 0.20; minimum = -0.18; and range = 0.38 random r; n = 404; original r = -0.50; R = 0.25; p = 0.0000; i = 10,000.[111]

Comments

The thermal stability during the 10 Kyrs of the current interglacial is significantly larger than the thermal stability during the preceding 413 Kyrs. This supports the notion that the thermal stability during the current interglacial is a systematic, rather than random effect. This suggests that a new climatological variable appeared at the beginning of the current interglacial, which stabilized the millennial temperature within a narrow range of 0.67 °C.[112]

The significant negative correlation between the mean temperature deviations and the thermal stability indexes suggest that the thermal stability during the high temperatures of the current interglacial should be low, rather than high. This contradiction is further support for the notion that a new variable appeared during the current interglacial.

VOSTOK °C (10 KYRS)

VOSTOK °C DEPTH-SCALE

Method

It is the aim of this section to change the depth-scale of Dataset 2 to a semi-annual-scale in order to create a valid and reliable basis for the evaluation of the annual global temperature.[113] Modification of Dataset 2 provided Dataset 6, which is the basis for the analyses in this section.

Dataset 2 (VostokTempDepth423Kyrs). This dataset contains the following Vostok variables: (Depth)—ice depth from 3,310 metres to 0 metre in steps of 1 metre; (Age)—ice age in years from 422,766 years before present (BP) to the present calendar year 1989; and (TDv)—depth-scale atmospheric temperature deviations in degrees Celsius (°C) relative to the Vostok base 1850-1989 = 0.00 °C.

The mean temperature deviation of a particular data interval should remain the same despite the conversion from a depth-scale to a semi-annual-scale. Consequently, the intermediate semi-annual temperature deviations cannot be computed by linear interpolation of the depth-scale values. Therefore, the mean temperature deviation of each data interval was copied to each year of that interval. In that way, the mean temperature deviation remains the same. In addition, each depth-scale temperature deviation contributes to the overall statistics of the semi-annual temperature deviations in proportion to the duration of its data interval.

Dataset 6 (VostokTempAnnualRaw10Kyrs).[114] This dataset contains the following Vostok variables: (Age)—in years from 10,000 years before present (BP) to the present calendar year 1989; and (TDv)—Vostok semi-annual atmospheric temperature deviations in degrees Celsius (°C) relative to the Vostok base 1850-1989 = 0.00 °C.

Results

Dataset 6 provides the statistics for the Vostok semi-annual temperature deviations: mean = -0.36; median = -0.39; standard deviation = 0.60; maximum = 2.06; minimum = -1.97; and range = 4.03 °C; n = 10,000.[115]

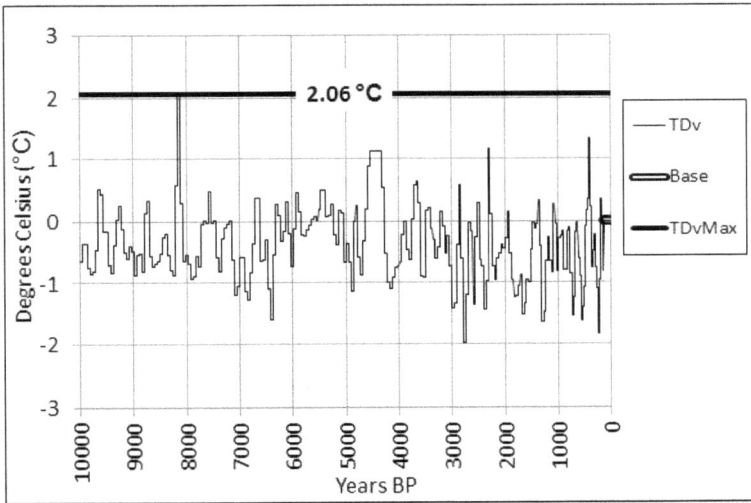

Figure 11 (Dataset 6)—Vostok annual temperature deviations 10 Kyrs.[116] This graph shows: (TDv)—Vostok semi-annual atmospheric temperature deviations from 10,000 years before present (BP) to the present calendar year 1989, in degrees Celsius (°C) relative to the Vostok base 1850-1989 = 0.00 °C,; (TDvMax)—maximum TDv = 2.06 °C.[117]

Comments

Each depth-scale temperature deviation of Dataset 2 contributes in proportion to the duration of its data interval to the overall statistics of the semi-annual temperature deviations comprising Dataset 6. Nevertheless, the semi-annual data are not comparable with the millennial temperature deviations because their basis and trends are still different.

VOSTOK °C SEMI-ANNUAL-SCALE

Method

It is the aim of this section to improve the compatibility between the semi-annual temperature deviations of Dataset 6 and the millennial temperature deviations of Dataset 5. Modification of Dataset 6 provided Dataset 7, which is the basis for the analyses in this section.

Dataset 6 (VostokTempAnnualRaw10Kyrs). This dataset contains the following Vostok variables: (Age)—in years from 10,000 years before present (BP) to the present calendar year 1989; and (TDv)—Vostok semi-annual atmospheric temperature deviations in degrees Celsius (°C) relative to the Vostok base 1850-1989 = 0.00 °C.

The base of the semi-annual temperature deviations comprising Dataset 6 was changed from (Vostok 1850-1989 = 0.00 °C) to (Vostok Kyr1 = 0.00 °C). For that reason, each temperature deviation was increased with the same correction as used previously for improving the base of the millennial temperature deviations (0.46 °C).[118] Furthermore, the semi-annual temperature deviations were detrended with the same slope that was used previously for detrending the millennial temperature deviations (0.00508 °C/Kyr).[119]

Dataset 7 (VostokTempAnnualDetrend10Kyrs).[120] This dataset contains the following Vostok variables: (Age)—in years from 10,000 years before present (BP) to the present calendar year 1989; (TDv)—Vostok semi-annual atmospheric temperature deviations in degrees Celsius relative to the Vostok base 1850-1989 = 0.00 °C; and (TDvd)—detrended semi-annual atmospheric temperature deviations in degrees Celsius (°C) relative to the Vostok base Kyr1 = 0.00 °C.

Results

Table 7				
	Degrees Celsius (°C) during 10 Kyrs			
	TDvd	TDva	TDvab	TDv
mean	-0.36	-0.36	0.10	0.07
median	-0.38	-0.39	0.07	0.04
STD	0.61	0.60	0.60	0.60
max	2.06	2.06	2.52	2.48
min	-1.97	-1.97	-1.51	-1.52
range	4.03	4.03	4.03	4.00
n	238	10,000	10,000	10,000

Table 7—modification Vostok temperature deviations during 10 Kyrs.[121] This table shows the statistics for the following variables: (TDvd (Dataset 2))—Vostok depth-scale atmospheric temperature deviations, during the last 10 Kyrs, in degrees Celsius (°C) relative to the Vostok base 1850-1989 = 0.00 °C; (TDva (Dataset 6))—Vostok semi-annual atmospheric temperature deviations, during the last 10 Kyrs, in degrees Celsius (°C) relative to the Vostok base 1850-1989 = 0.00 °C; (TDvab)—Vostok semi-annual temperature deviations, during the last 10 Kyrs, in degrees Celsius (°C) relative to the Vostok base Kyr1 = 0.00 °C; (TDv)— detrended semi-annual atmospheric temperature deviations, during the last 10 Kyrs, in degrees Celsius (°C) relative to the Vostok base Kyr1 = 0.00 °C.

The detrended maximum temperature deviation in Table 7 (TDv) of 2.48 °C is associated with the temperature deviation (TDvd) of 2.06 °C and a data interval of 44 years in Dataset 2.[122]

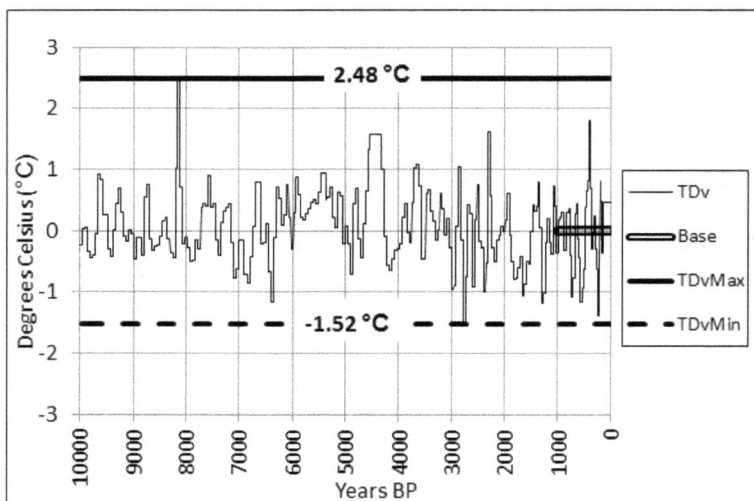

Figure 12 (Dataset 7)—Vostok detrended annual temperatures 10 Kyrs.[123] This graph shows: (TDv)—Vostok detrended semi-annual atmospheric temperature deviations from 10,000 years before present (BP) to the present calendar year 1989, in degrees Celsius (°C) relative to the Vostok base Kyr1 = 0.00 °C; (TDvMax)—maximum TDv = 2.48 °C;[124] and (TDvMin)—minimum TDv = -1.52 °C.[125]

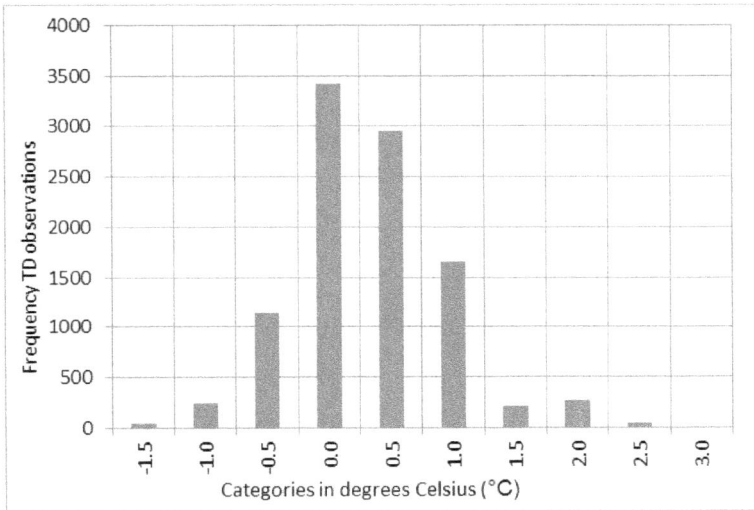

Figure 13(Dataset 7)—histogram Vostok detrended annual temperatures 10 Kyrs.[126] This graph shows the frequency distribution of the detrended Vostok semi-annual atmospheric temperature deviations from 10,000 years before present (BP) to the present calendar year 1989, in degrees Celsius (°C) relative to the Vostok base Kyr1 = 0.00 °C.

The histogram shows that the frequency distribution of the temperature deviations is positively skewed towards the higher temperatures. This is in accord with the millennial temperature deviations during the last 423 Kyrs as presented in Figure 6 (Dataset 5)—histogram Vostok detrended temperatures. Furthermore, the frequency peak at 0.00 °C in Figure 13 is similar to the frequency peak at 0.00 °C in Figure 6. This high frequency is caused by the relatively long duration of the current interglacial with a mean of 0.07 degrees Celsius.[127]

Comments

Figure 12 shows that the modified maximum temperature deviation of 2.48 °C was reached more than 8,000 years ago and the minimum of -1.52 °C almost three-thousand years ago. NASA reported in Dataset 9 a maximum global annual temperature of 0.67 °C during the calendar year 2014. These statistics cannot support the hypothesis that recent industrial activity induced an artificial climate change in either direction.[128]

The maximum detrend correction of 0.050 °C is unlikely to have affected significantly the results.[129] The skewed and bimodal frequency distribution of the Vostok temperature deviations, suggest that parametric statistical analyses are inappropriate for the evaluation of Dataset 7.

In contrast to the depth-scale temperature deviations, each semi-annual temperature deviation contributes proportionally to the overall statistics of the dataset. Nevertheless, these semi-annual temperature deviations are not true annual values. For example, the detrended maximum annual temperature deviation of 2.48 °C is associated with a 44 years depth-scale data interval.[130] Hence, regression has reduced the true annual values towards the mean of that data interval. This innate inaccuracy of the semi-annual data is likely to confound comparisons within and outside the dataset.

VOSTOK °C WINDOW-SCALE

Method

The aim of this section is to improve 'within comparisons' between the Vostok semi-annual temperature deviations. Dataset 2 provides the statistics for the original data intervals of the Vostok depth-scale temperature deviations during the last 10 Kyrs: mean = 43; median = 46; standard deviation = 8; maximum = 51; minimum = 21; and range = 31 years; n = 231.[131] The differences in the data intervals suggest that regression towards the mean could confound 'within comparisons' of the Vostok data. Modification of Dataset 7 provided Dataset 8, which is the basis for the analyses in this section.

Dataset 7 (VostokTempAnnualDetrend10Kyrs). This dataset contains the following Vostok variables: (Age)—in years from 10,000 years before present (BP) to the present calendar year 1989; (TDv)—Vostok semi-annual atmospheric temperature deviations in degrees Celsius relative to the Vostok base 1850-1989 = 0.00 °C; and (TDvd)—detrended semi-annual atmospheric temperature deviations in degrees Celsius (°C) relative to the Vostok base Kyr1 = 0.00 °C.

The Vostok semi-annual temperature deviations of Dataset 7 (TDv) were subjected to a sliding window (SW), with a duration of 50 years, shifting from 10,000 years before present (BP) to the present calendar year 1989, sliding in steps of 1 year (window-scale). For each step, the mean temperature deviation was computed (TDvSW).

Dataset 8 (VostokWindow10Kyrs).[132] This dataset contains the following Vostok variables: (Age)—in years from 10,000 years before the present (BP) to the present calendar year 1989; (TDv)—semi-annual Vostok atmospheric temperature deviations; and (TDvSW)—Vostok window-scale atmospheric temperature deviations, which are the mean temperature deviations obtained with a sliding window of 50 years. Temperature deviations are in degrees Celsius (°C) relative to Vostok base Kyr1 = 0.00 °C. This dataset is included in appendix: Dataset 8.

Results

Dataset 7 provides the statistics for the Vostok semi-annual temperature deviations: mean = 0.07; median = 0.04; standard deviation = 0.60; maximum = 2.48; minimum = -1.52; and range = 4.00 °C; n = 10,000.[133] Dataset 8 provides the statistics for the Vostok mean temperature deviations obtained with the 50 years sliding window: mean = 0.07; median = 0.04; standard deviation = 0.55; maximum = 2.30;

minimum = -1.46; range = 3.76; and variance = 0.30 °C; n = 9,951.[134]

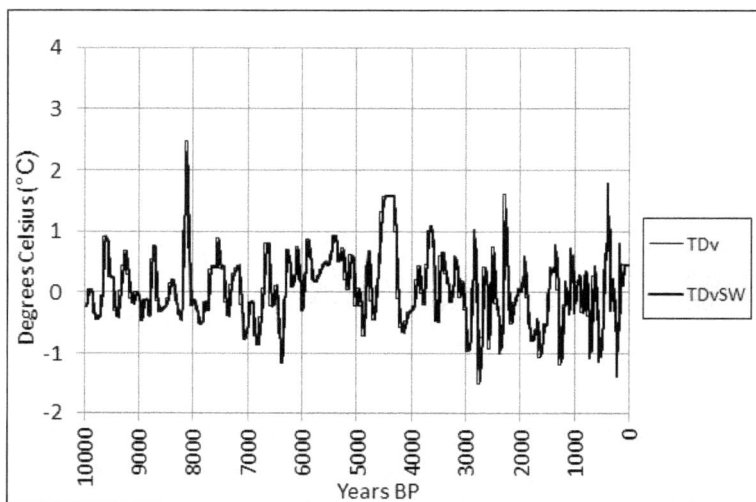

Figure 14 (Dataset 8)—Vostok window-scale temperature deviations 10 Kyrs. This graph shows: (TDv)—semi-annual Vostok atmospheric temperature deviations; and (TDvSW)— Vostok window-scale atmospheric temperature deviations, which are the mean temperature deviations obtained with a sliding window of 50 years. The observation period ranges from 10,000 years before the present (BP) to the present calendar year 1989, while the temperature deviations are expressed in degrees Celsius (°C) relative to Vostok base Kyr1 = 0.00 °C.[135]

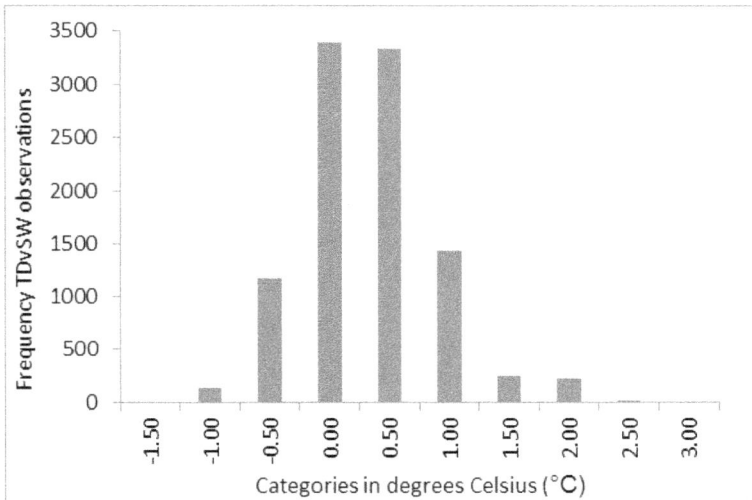

Figure 15 (Dataset 8)—histogram Vostok window-scale temperature deviations 10 Kyrs.[136] This graph shows the frequency distribution of: (TDvSW)—Vostok window-scale atmospheric temperature deviations, which are the mean temperature deviations obtained with a sliding window of 50 years. The observation period ranges from 10,000 years before the present (BP) to the present calendar year 1989, while the temperature deviations are expressed in degrees Celsius (°C) relative to Vostok base Kyr1 = 0.00 °C.

The histogram shows that the frequency distribution of the temperature deviations is positively skewed towards the higher temperatures.

Comments

The statistics and graph of this section suggest that the transformation from semi-annual temperature deviations to window-scale temperature deviations did not evoke considerable differences. Nevertheless, the window-scale temperature deviations are comparable within Dataset 8 and could be used for comparisons outside that dataset.

GLOBAL °C (1880-2014)

GLOBAL °C ANNUAL-SCALE

Method

The aim of this section is to evaluate the global annual temperature during the last 135 years. Modification of Dataset 9 provided Dataset 10, which is the basis for the analyses in this section.

Dataset 9 (GlobalTempMonthly).[137] This dataset contains the monthly and annual global land-ocean temperature index in 0.01 degrees Celsius (°C) relative to the base period: 1951-1980, during the observation period from calendar year 1880 to 2014.

Citation Dataset 9: GISTEMP Team, 2015: GISS Surface Temperature Analysis (GISTEMP). NASA Goddard Institute for Space Studies. Dataset accessed 2015-05-03 at http://data.giss.nasa.gov/gistemp/. Hansen, J., R. Ruedy, M. Sato, and K. Lo, 2010: Global surface temperature change, Rev. Geophys., 48, RG4004, doi: 10.1029/2010RG000345.

Dataset 10 (GlobalTempAnnual).[138] This dataset contains the following global variables: (Year)—calendar years from 1880 to 2014; (TDg)—global annual atmospheric temperature deviations in degrees Celsius (°C) relative to the global base temperature 1951-1980 = 0.00 °C.

Results

Dataset 10 provides the statistics for the global annual temperature deviations: mean = -0.01; median = -0.07; standard deviation = 0.30; maximum = 0.67; minimum = -048; and range = 1.15 °C; n = 135.[139]

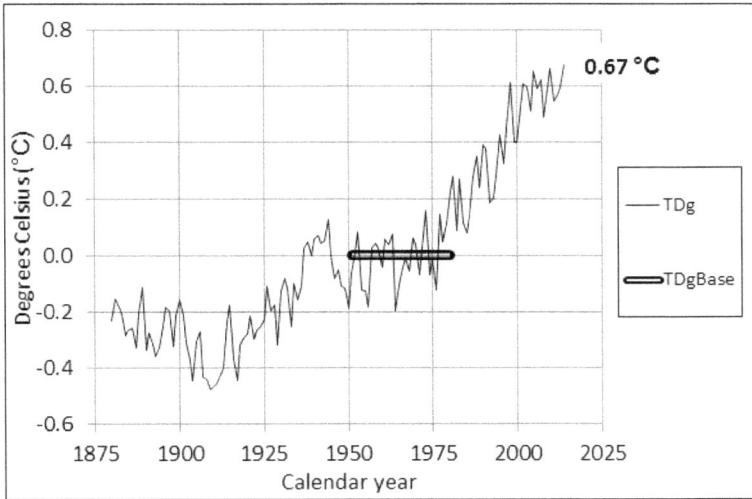

Figure 16 (Dataset 10)—global temperatures base 1951-1980.[140] This graph shows: (TDg)—global annual atmospheric temperature deviations from calendar year 1880 to 2014, in degrees Celsius (°C) relative to the global base 1951-1980 = 0.00 °C; and (TDgBase)—mean TDg from 1951-1980 = 0.00 °C.

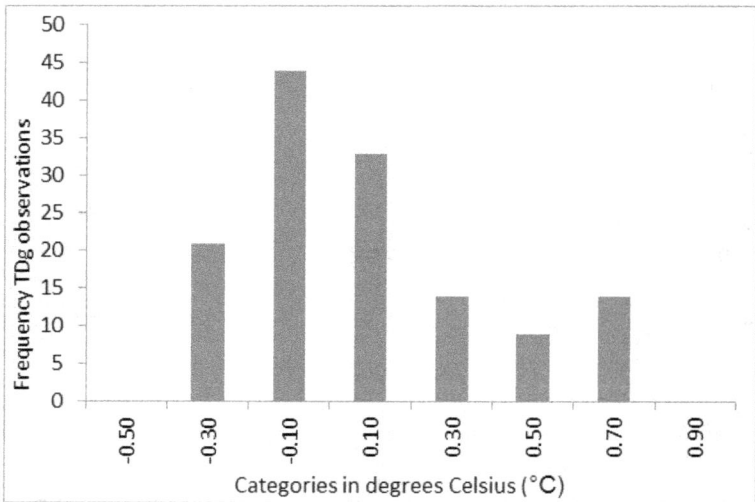

Figure 17 (Dataset 10)—histogram global temperatures base 1951-1980.[141] This graph shows the frequency distribution of the global annual atmospheric temperature deviations (TDg) from calendar year 1880 to 2014, in degrees Celsius (°C) relative to the global base 1951-1980 = 0.00 °C.

The histogram shows that the frequency distribution of the temperature deviations is positively skewed towards the higher temperatures.

Comments

NASA reported that the global annual temperature deviations reached a maximum of 0.67 °C during calendar year 2014. However, the arbitrary and relative short base (global 1951-1980 = 0.00 °C) makes any interpretation of that maximum meaningless. In order to evaluate the global temperature deviations, the base of the global dataset has to be adjusted to the Vostok base Kyr1 = 0.00 °C. In that way, the global temperature deviations become directly comparable to the Vostok temperature deviations.

Figure 17 shows that the frequency distribution of the TDg observations is strongly and positively skewed towards the higher temperatures. This suggests that parametric statistical analyses are inappropriate for the evaluation of Dataset 10.

GLOBAL AND VOSTOK °C ANNUAL-SCALE

Method

It is the aim of this section to combine the global annual temperature deviations and the Vostok semi-annual temperature deviations. The analyses in this section are based on Dataset 10 and Dataset 11. Truncation of Dataset 7 to 1,000 years provided Dataset 11.

Dataset 7 (VostokTempAnnualDetrend10Kyrs). This dataset contains the following Vostok variables: (Age)—in years from 10,000 years before present (BP) to the present calendar year 1989; (TDv)—Vostok semi-annual atmospheric temperature deviations in degrees Celsius relative to the Vostok base 1850-1989 = 0.00 °C; and (TDvd)—detrended semi-annual atmospheric temperature deviations in degrees Celsius (°C) relative to the Vostok base Kyr1 = 0.00 °C.

Dataset 10 (GlobalTempAnnual). This dataset contains the following global variables: (Year)—calendar years from 1880 to 2014; (TDg)—global annual atmospheric temperature deviations in degrees Celsius (°C) relative to the global base temperature 1951-1980 = 0.00 °C.

Dataset 11 (VostokTempAnnualKyr1).[142] This dataset contains the following Vostok variables: (Age)—in years from 1,000 years before present (BP) to the present calendar year 1989; and (TDv)—Vostok detrended semi-annual atmospheric temperature deviations in degrees Celsius (°C) relative to the Vostok base Kyr1 = 0.00 °C.

Results

Dataset 10 provides the statistics for the global annual temperature deviations (TDg): mean = -0.01; median = -0.07; standard deviation = 0.30; maximum = 0.67; minimum = -048; and range = 1.15 °C; n = 135.[143] Furthermore, Dataset 11 provides the statistics for the Vostok semi-annual temperature deviations (TDv): mean = -0.01; median = 0.21; standard deviation = 0.63; maximum = 1.79; minimum = -1.38; and range = 3.17 °C; n = 1,000.[144]

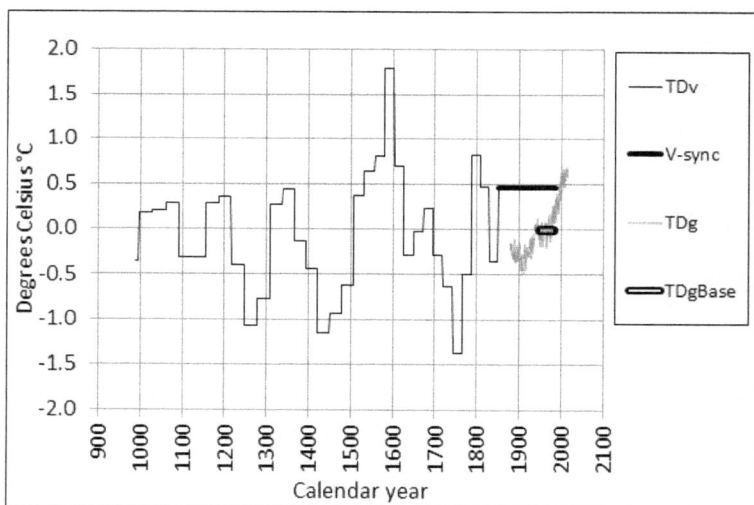

Figure 18—Vostok temperatures.[145] This graph shows: (TDv (Dataset 11))—Vostok detrended semi-annual atmospheric temperature deviations from 1,000 years before present (BP) to the present calendar year 1989, in degrees Celsius (°C) relative to the Vostok base Kyr1 = 0.00 °C; (V-sync)—Vostok mean temperature deviation from calendar year 1850 to 1989 = 0.46 °C;[146] (TDg (Dataset 10))—global annual atmospheric temperature deviations from calendar year 1880 to 2014, in degrees Celsius (°C) relative to TDgBase; and (TDgBase)—global mean temperature

deviation from calendar year 1951 to 1980, in degrees Celsius (°C) = 0.00 °C.[147]

Comments

The Vostok base (Kyr1 = 0.00 °C) and the global base (1951-1980 = 0.00 °C) are different and, therefore, the Vostok and global temperature deviations presented in Figure 18 are not comparable.

The basis of Dataset 10 and Dataset 11 has to be synchronized to foster the comparison of Vostok and global temperature deviations. The Vostok and global datasets have in common the observation period from 1880 to 1989. This period could be used as a synchronization period. However, the original Vostok base (1850-1989) cannot be divided, because Dataset 2 provides only one temperature deviation for the entire period (0.00 °C).[148] To synchronize the basis of dataset Dataset 14 and Dataset 10, the global temperature deviations have to be estimated backwards from 1880 to 1850.

GLOBAL °C BACKWARD EXTENSION

Method

It is the aim of this section to extend the global temperature deviations backwards from the onset of 1880 to 1850. Modification of Dataset 10 provided Dataset 12, which is the basis for the analyses in this section.

Dataset 10 (GlobalTempAnnual). This dataset contains the following global variables: (Year)—calendar years from 1880 to 2014; (TDg)—global annual atmospheric temperature deviations in degrees Celsius (°C) relative to the global base temperature 1951-1980 = 0.00 °C.

The data in Dataset 10 were extended backwards from 1880 to 1850 with the means of a second order polynomial function and a linear function of the temperatures.

Dataset 12 (GlobalTempAnnualExtended).[149] This dataset contains the following global variables: (Year)—Calendar year 1850 to 2014; (TDg)—global annual atmospheric temperature deviations from 1880 to 2014, in degrees Celsius (°C) relative to the global base 1951-1980 = 0.00 °C; (ExtPol)—second order polynomial function of TDg from 1850 to 1879; (ExtLin)—linear function of TDg from 1850 to 1879; and (TDgExt)—backwards extension of TDg comprising the means of ExtPol and ExtLin from 1850 to 1879.

Results

Dataset 12 provides the statistics for the global annual temperature deviations from 1880 to 2014 (TDg): mean = -0.01; median = -0.07; standard deviation = 0.30; maximum = 0.67; minimum = -048; and range = 1.15 °C; n = 135.[150] Furthermore, the statistics for the extended global annual temperature deviations from 1850 to 2014 (TDgExt) are: mean = -0.08; median = -0.15; standard deviation = 0.31; maximum = 0.67; minimum = -048; and range = 1.15 °C; n = 165.[151]

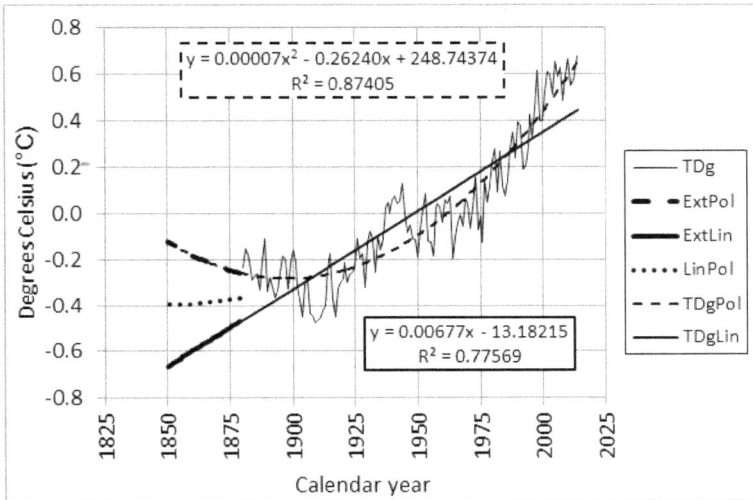

Figure 19 (Dataset 12)—extended global temperatures.[152]
This graph shows: (TDg)—global annual atmospheric
temperature deviations from calendar year 1880 to 2014, in
degrees Celsius (°C) relative to the global base 1951-1980 =
0.00 °C; (ExtPol)—TDgPol extended to 1850; (ExtLin)—
TDgLin extended to 1850; (LinPol)—backwards extension of
TDg from 1850 to 1879, comprising the means of ExtPol
and ExtLin; (TDgPol)—second order polynomial function of
TDg; and (TDgLin)—linear function of TDg.

Table 8				
	1880-2014	1850-2014		
	TDg	ExtPol	ExtLin	ExtMean
mean	-0.01	-0.04	-0.11	-0.08
median	-0.07	-0.13	-0.15	-0.15
STD	0.30	0.28	0.35	0.31
max	0.67	0.67	0.67	0.67
min	-0.48	-0.48	-0.67	-0.48
range	1.15	1.15	1.34	1.15
n	135	165	165	165

Table 8 (Dataset 12)—extended global temperature deviations °C.[153] This table shows the statistics for the following variables: (TDg)—global annual atmospheric temperature deviations from 1880 to 2014, in degrees Celsius (°C) relative to the global base 1951-1980 = 0.00 °C; (ExtPol)—second order polynomial function of TDg from 1850 to 1879; (ExtLin)—linear function of TDg from 1850 to 1879; and (TDgExt)—backwards extension of TDg comprising the means of ExtPol and ExtLin from 1850 to 1879.

Comments

Figure 19 shows that the polynomial function (ExtPol) and the linear function (ExtLin) both explain adequately the variance in the global temperature deviations (R = 0.87 and 0.78).[154] Those functions could be seen as the upper and lower limits of the mean global temperature deviations during the observation period from 1850 to 1880. Therefore, the mean of those functions (ExtMean) was used to extend the global temperatures backwards from 1880 to 1850.

Table 8 shows that the mean of the extended temperature deviations from 1880 to 2014 is only 0.07 °C lower than the mean the temperature deviations from 1850 to 2014.[155] Hence, it is unlikely that the backward extension affects significantly the comparison between global and Vostok temperatures. Nevertheless, further research should evaluate the accuracy of the extension.

After the extension of the global dataset, the Vostok and global datasets share the same observation period from 1850 to 1989. This period is used to synchronize those datasets.

GLOBAL AND VOSTOK °C COMBINED

Method

It is the aim of this section to combine the Vostok and global temperature deviations in one dataset. The analyses in this section are based on Dataset 13. Dataset 13 was created by combining Dataset 11 and Dataset 12.

Dataset 11 (VostokTempAnnualKyr1). This dataset contains the following Vostok variables: (Age)—in years from 1,000 years before present (BP) to the present calendar year 1989; and (TDv)—Vostok detrended semi-annual atmospheric temperature deviations in degrees Celsius (°C) relative to the Vostok base Kyr1 = 0.00 °C.

Dataset 12 (GlobalTempAnnualExtended). This dataset contains the following global variables: (Year)—Calendar year 1850 to 2014; (TDg)—global annual atmospheric temperature deviations from 1880 to 2014, in degrees Celsius (°C) relative to the global base 1951-1980 = 0.00 °C; (ExtPol)—second order polynomial function of TDg from 1850 to 1879; (ExtLin)—linear function of TDg from 1850 to 1879; and (TDgExt)—backwards extension of TDg comprising the means of ExtPol and ExtLin from 1850 to 1879.

Dataset 13 (VostokGlobalTempCombined).[156] This dataset contains the following variables: (Year)—calendar years from 989 to 2014; (TDv)—Vostok detrended semi-annual atmospheric temperature deviations from 989 to 1989, in degrees Celsius (°C) relative to the Vostok base Kyr1 = 0.00 °C; (TDvSyn)—TDv from 1850 to 1989; (TDgExt)—

extended global annual atmospheric temperature deviations from 1850 to 2014, in degrees Celsius (°C) relative to the global base 1951-1980 = 0.00 °C; and (TDgSyn)—TDgExt from 1850 to 1989.

Results

Dataset 13 provides the statistics for the Vostok detrended semi-annual atmospheric temperature deviations (TDv): mean = -0.01; median = 0.21; standard deviation = 0.63; maximum = 1.79; minimum = -1.38; and range = 3.17 °C; n = 1,000.[157] Furthermore, it provides the statistics for the global extended annual temperature deviations (TDg): mean = -0.08; median = -0.15; standard deviation = 0.31; maximum = 0.67; minimum = -0.48; and range = 1.15 °C; n = 165.[158]

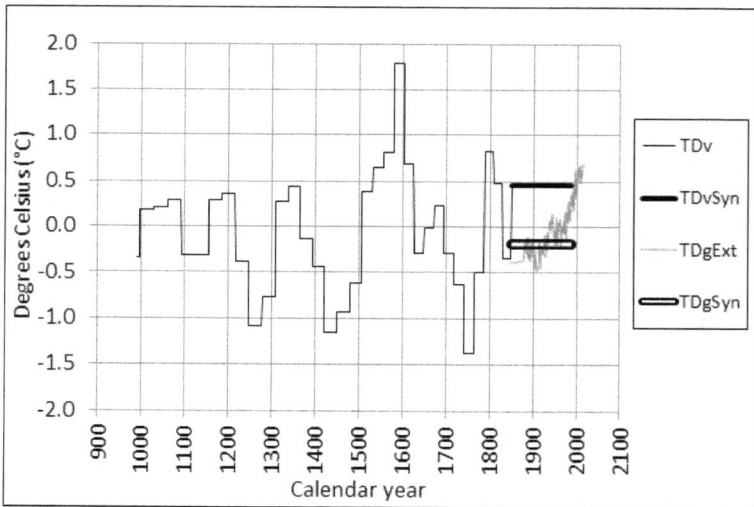

Figure 20 (Dataset 13)—Vostok and global synchronization periods.[159] This graph shows: (TDv)—Vostok detrended semi-annual atmospheric temperature deviations from 989 to 1989, in degrees Celsius (°C) relative to the Vostok base Kyr1 = 0.00 °C; (TDvSyn)—mean TDv

from 1850 to 1989 = 0.46 °C;[160] (TDgExt)—extended global annual atmospheric temperature deviations from 1850 to 2014, in degrees Celsius (°C) relative to the global base 1951-1980 = 0.00 °C; and (TDgSyn)—mean TDgExt from 1850 to 1989 = -0.18 °C.[161]

The Vostok synchronization temperature (TDvSyn) is 0.64 °C (0.46 – (-0.18) °C) higher than the global synchronization temperature (TDgSyn).[162]

Comments

The magnitudes of the Vostok and global temperature deviations during the synchronization period from 1850 to 1989 have to be equalized in order to make the Vostok temperatures compatible with the global temperatures.

GLOBAL AND VOSTOK °C CALIBRATED

Method

It is the aim of this section to calibrate the extended global annual temperature deviations to the Vostok semi-annual temperature deviations by synchronizing the common base period. Modification of Dataset 13 provided Dataset 14, which is the basis for the analyses in this section.

Dataset 13 (VostokGlobalTempCombined). This dataset contains the following variables: (Year)—calendar years from 989 to 2014; (TDv)—Vostok detrended semi-annual atmospheric temperature deviations from 989 to 1989, in degrees Celsius (°C) relative to the Vostok base Kyr1 = 0.00 °C; (TDvSyn)—TDv from 1850 to 1989; (TDgExt)—extended global annual atmospheric temperature deviations from 1850 to 2014, in degrees Celsius (°C) relative to the global base 1951-1980 = 0.00 °C; and (TDgSyn)—TDgExt from 1850 to 1989.

The statistics of Dataset 13 show that the mean of the Vostok semi-annual temperature deviations during the synchronization period from 1850 to 1989 (TDvSyn) equals

0.46 °C.[163] On the other hand, the mean of the global annual temperature deviations during the synchronization period from 1850 to 1989 (TDgSyn) equals -0.18 °C.[164] To adjust the global temperature to the Vostok temperature, Dataset 14 was created by adding 0.64 °C ((0.46) - (-0.18)) to the extended global annual temperature deviations (TDgExt) of Dataset 13.[165]

Dataset 14 (VostokGlobalTempSynchronized).[166] This dataset contains the following variables: (Year)—calendar years from 989 to 2014; (TDv)—Vostok semi-annual atmospheric temperature deviations from 989 to 1989; (TDg)—extended global annual atmospheric temperature deviations from calendar year 1850 to 2014; (Syn)—means of TDv and TDg during synchronization period from calendar year 1850 to 1989. TDg and TDv are expressed in degrees Celsius (°C) relative to the Vostok base Kyr1 = 0.00 °C.

Results

Dataset 14 provides the statistics for the Vostok semi-annual temperature deviations (TDv): mean = -0.01; median = 0.21; standard deviation = 0.63; maximum = 1.79; minimum = -1.38; and range = 3.17 °C; n = 1,000.[167] In addition, it provides the statistics for the extended global annual temperature deviations (TDg): mean = 0.56; median = 0.49; standard deviation = 0.31; maximum = 1.32; minimum = 0.16; and range = 1.15 °C; n = 165.[168]

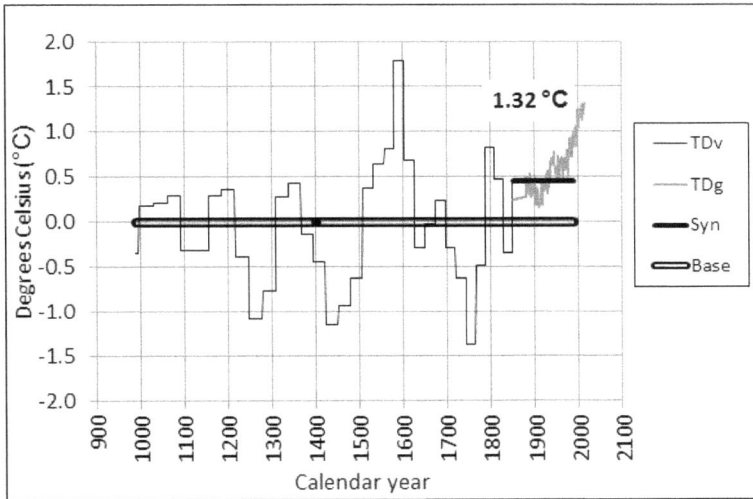

Figure 21 (Dataset 14)—Vostok and global temperatures synchronized.[169] This graph shows: (TDv)—Vostok semi-annual atmospheric temperature deviations from 989 to 1989; (TDg)—extended global annual atmospheric temperature deviations from calendar year 1850 to 2014; (Syn)—means of TDv and TDg during synchronization period from calendar year 1850 to 1989 = 0.46 °C;[170] maximum global temperature deviation in 2014 = 1.32 °C.[171] TDg and TDv are expressed in degrees Celsius (°C) relative to the Vostok base Kyr1 = 0.00 °C.

Dataset 2 provides the statistics for the data intervals of the Vostok temperature deviations during the last 10 Kyrs: mean = 43; median = 46; standard deviation = 8; maximum = 51; minimum = 21; and range = 31 years; n = 231.[172]

Comments

In the previous sections, the standard or 'normal' atmospheric temperature deviation was redefined as the mean temperature deviation observed in Vostok during the last 1,000 years before calendar year 1989 (Vostok base Kyr1 =

0.00 °C). This long-term Vostok base has been used to modify the global temperatures. It is emphasized that this calibration is an approximation, because the observation period of the global temperature is 165 years, rather than the 1,000 years of the Vostok base. Nevertheless, it is the best calibration available.

Despite having the same base, the Vostok and global temperature deviations are still not comparable. Dataset 14 comprises Vostok semi-annual temperature deviations and global annual temperature deviations. The semi-annual temperature deviations are based on the means of multi-annual data intervals ranging from 21 to 51 years.[173] As mentioned previously, unequal regression towards the means is likely to confound 'within comparisons' of the Vostok data. Therefore, the temperature deviations of Dataset 7 were converted to the thermal means of a sliding window with a duration of 50 years comprising Dataset 8.

In addition, regression towards the mean is likely to confound comparisons between the Vostok semi-annual and global annual data. To make the global annual temperature deviations comparable to the Vostok window-scale temperature deviations in Dataset 8, the global annual temperature deviations have also to be converted into the means of a sliding window with a duration of 50 years.

The maximum global temperature deviation of 1.32 °C is considerably higher than the maximum reported by NASA of 0.67 °C.[174] This difference is predominantly due to different base periods. More importantly, the hypotheses should be tested that the recent global window-scale temperature deviations are significantly higher than the Vostok window-scale temperature deviations during the current interglacial period.

GLOBAL °C WINDOW-SCALE

Method

The aim of this section is to improve the comparison between the Vostok window-scale and global annual-scale temperature deviations. Modification of Dataset 14 with a sliding-window of 50 years provided Dataset 15. Dataset 15 is the basis for the analyses in this section.

Dataset 14 (VostokGlobalTempSynchronized). This dataset contains the following variables: (Year)—calendar years from 989 to 2014; (TDv)—Vostok semi-annual atmospheric temperature deviations from 989 to 1989; (TDg)—extended global annual atmospheric temperature deviations from calendar year 1850 to 2014; (Syn)—means of TDv and TDg during synchronization period from calendar year 1850 to 1989.

The global annual temperature deviations (TDg) comprising Dataset 14 were subjected to a sliding window with a duration of 50 years, shifting from calendar year 1880 to 2014 in steps of 1 year. For each step, the mean temperature deviation was computed (TDgSW).

Dataset 15 (GlobalWindow°C).[175] This dataset contains the following global variables: (Year)—calendar years from 1880 to 2014; (TDg)—global annual-scale atmospheric temperature deviations from calendar year 1880 to 2014; (TDgSW)—global window-scale atmospheric temperature deviations from calendar year 1929 to 2014, which are the mean temperature deviations obtained with a window of 50 years sliding in steps of 1 year. TDg and TDgSW are in degrees Celsius (°C) relative to the Vostok base Kyr1 = 0.00 °C. This dataset is included in the appendix: Dataset 15.

Results

Dataset 15 provides the statistics for the global annual-scale temperature deviations from 1880 to 2014 (TDg): mean = 0.63; median = 0.57; standard deviation = 0.30; maximum

= 1.32; minimum = 0.16; and range = 1.15 °C; n = 135.[176] In addition, it provides the statistics for the global window–scale temperature deviations from 1929 to 2014 (TDgSW): mean = 0.59; median = 0.58; standard deviation = 0.16; maximum = 0.93; minimum = 0.35; range = 0.58; and variance = 0.03 °C; n = 86.[177]

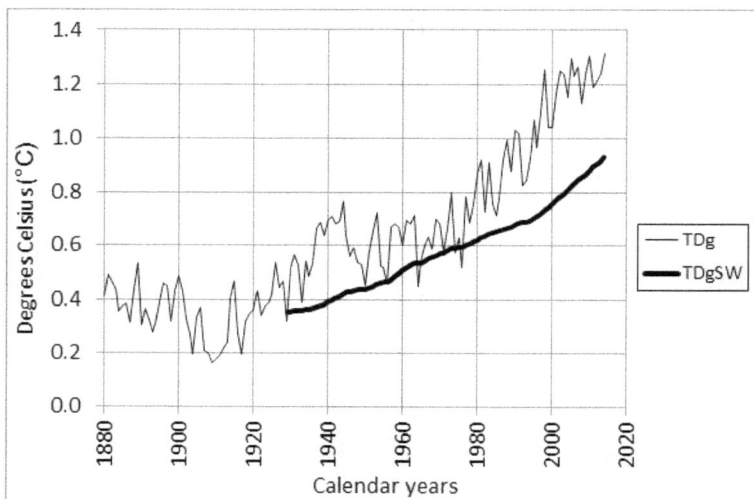

Figure 22 (Dataset 15)—global annual- and window-scale temperatures.[178] This graph shows: (TDg)—global annual-scale atmospheric temperature deviations from calendar year 1880 to 2014; and (TDgSW)—global window-scale atmospheric temperature deviations from calendar year 1929 to 2014, which are the mean temperature deviations obtained with a window of 50 years sliding in steps of 1 year. TDg and TDgSW are in degrees Celsius (°C) relative to the Vostok base Kyr1 = 0.00 °C.

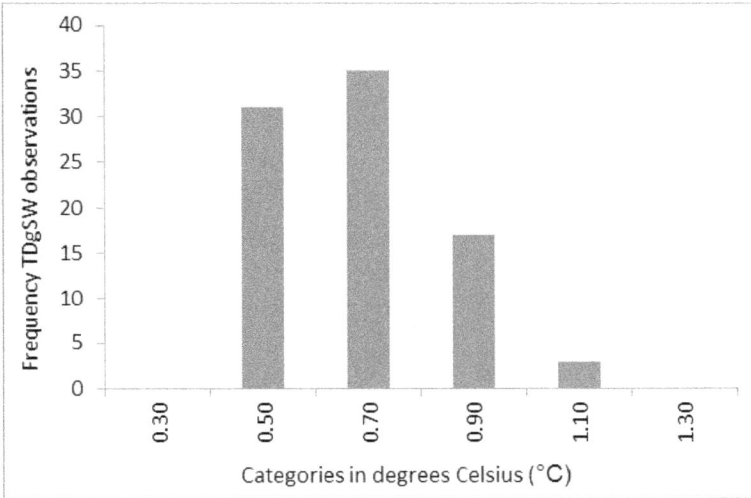

Figure 23 (Dataset 15)—histogram global window-scale temperatures.[179] This graph shows: (TDgSW)—frequency distribution of the global window-scale temperature deviations from calendar year 1929 to 2014, in degrees Celsius (°C) relative to the Vostok base Kyr1 = 0.00 °C.

The histogram shows that the frequency distribution of the temperature deviations is positively skewed towards the higher temperatures.

Comments

The global window-scale temperature deviations are comparable with the Vostok window-scale temperature deviations, because they have the same base (Vostok Kyr1 = 0.00 °C) and the same sliding window-scale of 50 years.

The statistics and the graph in this section show that the mean global temperature deviation has decreased from the annual-scale of 1.32 to the window-scale of 0.93 °C. Furthermore, the results show that the global window-scale temperature deviations have steadily increased during the last 86 years. However, this might be random variation around the mean. In the next section the significance of that increase will be evaluated.

GLOBAL VERSUS VOSTOK °C WINDOW-SCALE

Method

The first aim of this section is to evaluate the hypothesis that the magnitude of the global temperature deviations during the last 86 years is significantly higher than the magnitude of the Vostok temperature deviations during the last 10,000 years.

The second aim of this section is to evaluate the hypothesis that the linear increase in global temperature deviations during the last 86 years is significantly steeper than the linear increases in Vostok temperature deviations during the last 10,000 years.

The third aim of this section is to evaluate the hypothesis that the duration of the increase in global temperature deviations during the last 86 years is significantly longer than the durations of the increases in Vostok temperature deviations during the last 10,000 years.

The analyses in this section are based on Dataset 16. This dataset was created by combining the Vostok window-scale temperature deviations (TDvSW) of Dataset 8 with the global window-scale temperature deviations (TDgSW) of Dataset 15.

Dataset 8 (VostokWindow10Kyrs). This dataset contains the following Vostok variables: (Age)—in years from 10,000 years before the present (BP) to the present calendar year 1989; (TDv)—semi-annual Vostok atmospheric temperature deviations; and (TDvSW)—Vostok window-scale atmospheric temperature deviations, which are the mean temperature deviations obtained with a sliding window of 50 years.

Dataset 15 (GlobalWindow°C). This dataset contains the following global variables: (Year)—calendar years from 1880 to 2014; (TDg)—global annual-scale atmospheric temperature deviations from calendar year 1880 to 2014; (TDgSW)—global window-scale atmospheric temperature deviations from calendar year 1929 to 2014, which are the mean temperature deviations obtained with a window of 50 years sliding in steps of 1 year. TDg and TDgSW are in degrees Celsius (°C) relative to the Vostok base Kyr1 = 0.00 °C.

Dataset 16 (VostokGlobalWindow°C).[180] This dataset contains the following variables: (Year)—calendar years from 8011 BCE to 2014; (TDvSW)—Vostok window-scale atmospheric temperature deviations from calendar year 7962 BCE to 1989; (TDgSW)—global window-scale atmospheric temperature deviations from calendar year 1929 to 2014. Window-scale temperature deviations are mean temperature deviations obtained with a window of 50 years sliding in steps of 1 year, which are expressed in degrees Celsius (°C) relative to the Vostok base Kyr1 = 0.00 °C.

Non-parametric test

Figure 15 and Figure 23 show that the frequency distributions of the Vostok and global window-scale temperature deviations are positively skewed towards the higher temperatures. Furthermore, the differences between Dataset 8 and Dataset 15 are large both in regard to sample size (9,951 versus 86) and thermal variance (0.30 versus 0.03 °C).[181] Hence, the data suggest that the assumptions underlying parametric statistical analyses are breached, which is likely to compromise the results. Therefore, non-parametric analyses are more appropriate for the analyses of Vostok Dataset 8 and global Dataset 15.

If decision makers would accept the hypothesis that the current global temperature is so extreme that it is the result of industrial pollution, then statisticians could compute the probability of that decision being incorrect with a simple non-parametric probability test. For that test, the mean, slope, and duration of the current global window-scale temperature deviations are used as criteria.

Each Vostok temperature deviation, from 7962 BCE to 1989, that has been higher, increased faster, or lasted longer than the corresponding global criterion has to be counted as a Type I error. The non-parametric probability test includes exhaustive comparisons between the thermal means, slopes, and durations of the Vostok 50 years samples (n = 9,951) and the corresponding global criteria. Therefore, the non-parametric probability is the result of thousands of tests, rather than a single one. The exhaustive sampling method is preferred over the total randomization method, because the time-series and, therefore, the variance within the samples remains intact.

The non-parametric probability test is simple, because it makes no assumptions. It is robust and has a high face-value, because it counts the number of extremes and compares that to the total number of observations. The statistics of the Vostok and global window-scale temperature deviations are comparable, because they were both obtained with sliding windows of 50 years and are both relative to the Vostok base Kyr1 = 0.00 °C. The equal duration of the Vostok and global windows, improves the ratio between the Vostok and global thermal variance within each sub-comparison. Using short sub-comparisons reduces the confounding effects of skewed frequency distributions and differences between the size and variance of Vostok Dataset 8 and global Dataset 15.

Results

(1) Thermal magnitudes

Dataset 16 provides the statistics for the Vostok window-scale temperature deviations (TDvSW): mean = 0.07; median = 0.04; standard deviation = 0.55; maximum = 2.30; minimum = -1.46; and range = 3.76 °C; n = 9,951.[182] In addition, it provides the statistics for the global window–scale temperature deviations (TDgSW): mean = 0.59; median =

0.58; standard deviation = 0.16; maximum = 0.93; minimum = 0.35; and range = 0.58 °C; n = 86.[183]

The maximum global window-scale temperature deviation of 0.93 °C (TDgSW) is lower than the maximum Vostok window-scale temperature deviation of 2.30 °C (TDvSW). To denote the global maximum as artificial, it should exceed the natural Vostok maximum. Hence, the lower global maximum cannot support the hypothesis for artificial global warming. However, the maximum TDgSW of 0.93 °C is higher than both the mean and median TDvSW of respectively 0.07 and 0.04 °C. This difference supports the hypothesis for natural global warming. Nevertheless, to support the hypothesis of industrial global warming, the probability that this difference can be explained by natural variation around the mean should be insignificant (p < α).

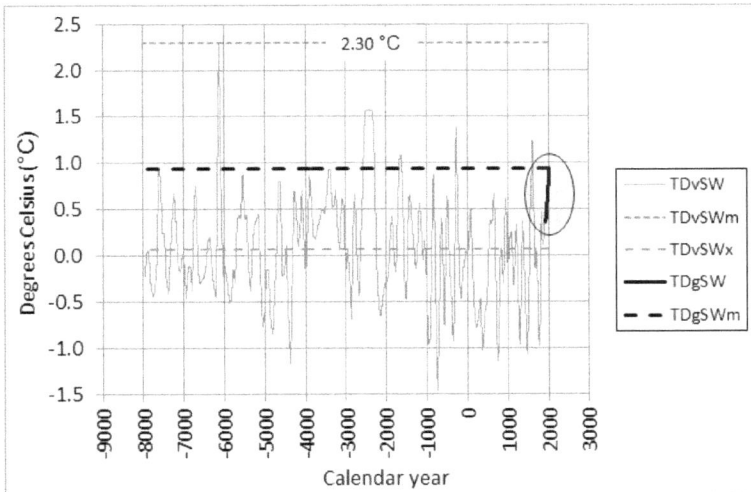

Figure 24 (Dataset 16)—Vostok and global window-scale temperatures.[184] This graph shows: (TDvSW)—Vostok window-scale atmospheric temperature deviations from calendar year 7962 BCE to 1989; (TDvSWm)—maximum TDvSW = 2.30 °C;[185] (TDvSWx)—mean DvSW = 0.07 °C;[186]

(TDgSW)—global window-scale atmospheric temperature deviations from calendar year 1929 to 2014; (TDvSWm)—maximum TDvSW = 0.93 °C;[187] and (Type I errors)—TDvSW > TDvSWm. Window-scale temperature deviations are mean temperature deviations obtained with a window of 50 years sliding in steps of 1 year, which are expressed in degrees Celsius (°C) relative to the Vostok base Kyr1 = 0.00 °C.

During 4,719 of the 9,951 observation years, the Vostok window-scale temperature deviations are higher than the Vostok mean window-scale temperature deviation.[188] Furthermore, during 518 of those 4,719 decision years, the Vostok window-scale temperature deviations (TDvSW) exceeded the maximum global window-scale temperature deviation of 0.93 °C (TDgSWm).[189] Hence, if the decision makers accept that the maximum global window-scale temperature deviation of 0.93 °C is so extreme that it indicates industrial global warming, then they also accept the probability that in the past 518 of their 4,719 decisions would have been incorrect. Therefore, they accept also that the probability of making a Type I error equals 0.11 (518 / 4719).[190]

(2) Thermal slopes

Table 9				
	Vostok Kyr10-Kyr1[191]		Global 1929-2014[192]	
	TDvLin °C/year	TDvLinR	TDgLin °C/year	TDgLinR
mean	0.01017	0.95	0.00561	0.98
median	0.00824	0.98	0.00612	0.99
STD	0.00754	0.06	0.00295	0.01
maximum	0.03172	1.00	0.00858	0.99
minimum	0.00100	0.81	0.00160	0.97
range	0.03072	0.19	0.00698	0.03
n	52	52	4	4

Table 9—uninterrupted increasing slopes of the 50-years window-scale temperatures.[193] This table shows the statistics for the following variables: (TDvLin)—linear function of TDvSW; (TDvSW (Dataset 16))—Vostok window-scale atmospheric temperature deviations from calendar year 7962 BCE to 1989; (TDvLinR)—explained variance of TDvLin; (TDgLin)—linear function of TDgSW; (TDgSW (Dataset 15))—global window-scale atmospheric temperature deviations from calendar year 1929 to 2014 (n = 86); (TDgLinR)—explained variance of TDgLin. Window-scale temperature deviations are mean temperature deviations obtained with a window of 50 years sliding in steps of 1 year, which are expressed in degrees Celsius (°C) relative to the Vostok base Kyr1 = 0.00 °C.

The high Vostok and global R-values show that linear functions explain adequately the variance in the data. The mean uninterrupted linear increase in Vostok window-scale temperature deviations during the last 10 Kyrs of 0.01017 °C/year is steeper than the maximum uninterrupted linear increase in global window-scale temperature deviations during the last 86 years of 0.00858 °C/year. Hence, the slope of the linear uninterrupted increase in global temperature is below the natural average and, therefore, it cannot support the hypothesis of industrial global warming.

In accord, the slopes of 26 of the 52 linear uninterrupted increases in the Vostok temperatures exceeded the global maximum slope of 0.00858 °C/year. Decision makers who accept that the maximum slope of the uninterrupted increases in global window-scale temperature deviations of 0.00858 °C/year is so extreme that it indicates industrial global warming, accept that the probability making a Type I error equals 0.50 (26 / 52).[194]

In addition, the linear increase in global window-scale temperature deviations, in degrees Celsius relative to the Vostok base Kyr1 = 0.00 °C, during the total global observation period from 1929 to 2014 (86 years), was evaluated.[195]

Table 10		
	TDvLin °C/year	TDvLinR
mean	0.00686	0.72
median	0.00487	0.86
STD	0.00616	0.31
maximum	0.03199	1.00
minimum	0.00005	0.00
range	0.03194	1.00
n	5095	5095

Table 10 (Dataset 16)—increasing slopes of the Vostok 50-years window-scale temperatures, computed for an observation window of 86 years, sliding across the last 10 Kyrs in steps of 1 year.[196] This table shows the statistics for the following variables: (TDvLin (Dataset 16))—linear function of TDvSW; (TDvSW (Dataset 16))—Vostok window-scale atmospheric temperature deviations from 7877

BCE to 1989; (TDvLinR (Dataset 16))—variance explained by TDvLin.

The linear increase in the global window-scale temperature deviations from 1929 to 2014 equals 0.00645 °C/year with an R-value of 0.98.[197] As shown in Table 10, this is less than the mean TDvLin of the Vostok window-scale temperature deviations of 0.00686 °C/year. Hence, the slope of the linear increase in global temperature is below the natural average and, therefore, it can neither support the hypothesis of industrial global warming.

In accord, 2,070 of the 5,095 linear increases in the Vostok window-scale temperature deviations, during the observation period from 7877 BCE to 1989, exceeded the global linear increase of 0.00645 °C/year during the observation period from 1929 to 2014.[198] Decision makers who accept that the linear increase in global window-scale temperature deviations of 0.00645 °C/year is so extreme that it indicates industrial global warming, accept that the probability of making a Type I error equals 0.41 (2070 / 5095).[199]

(3) Duration thermal increases

Table 11		
	Duration slopes in years	
	TDvd	TDgd
mean	97	20
median	91	21
STD	51	14
maximum	282	37
minimum	20	3
range	262	34
n	52	4

Table 11—duration uninterrupted increasing slopes of the 50-years window-scale temperatures.[200] This table shows the statistics for the following variables: (TDvd)—duration uninterrupted increasing slope of TDvSW; (TDvSW (Dataset

16))—Vostok window-scale atmospheric temperature deviations from calendar year 7962 BCE to 1989; (TDgd)—duration uninterrupted increasing of TDgSW; (TDgSW (Dataset 15))—global window-scale atmospheric temperature deviations from calendar year 1929 to 2014. Window-scale temperature deviations are mean temperature deviations obtained with a window of 50 years sliding in steps of 1 year, which are expressed in degrees Celsius (°C) relative to the Vostok base Kyr1 = 0.00 °C.

Table 11 (Dataset 16) shows that the maximum duration of the uninterrupted linear increases in global window-scale temperature deviations of 37 years is considerably shorter than the mean duration of the uninterrupted linear increases in the Vostok window-scale temperature deviations of 97 years. Hence, the maximum duration of the linear uninterrupted increase in global temperature is below the natural average and, therefore, it cannot support the hypothesis of industrial global warming.

In accord, 50 of the 52 uninterrupted increases in the Vostok window-scale temperatures lasted longer than the maximum uninterrupted increase in the global window-scale temperature of 37 years.[201] Decision makers who accept that the duration of the uninterrupted linear increase in global window-scale temperature deviations is so extreme that it indicates industrial global warming, accept that the probability of making a Type I error equals 0.96 (50 / 52).[202]

Ignoring the interruptions in the increase of global window-scale temperature deviations during the observation period of 86 years provides a linear increase of 0.00645 °C/year. This approach would change the probability of making a Type I error to 0.62 (32 / 52).[203] Nevertheless, the probability of a mistake remains extremely high. Furthermore, this last analysis is spurious, because it compares durations of uninterrupted Vostok thermal linear increases with the duration of interrupted global thermal linear increases.

Comments

The analyses of publically available data show that the uninterrupted slope of the increase, duration, and recent magnitude of the global atmospheric temperature has been within the natural limits during the last 135 years.[204] Hence, the recent increase in global temperature cannot support the hypothesis that industrial CO2 causes global warming. The results suggest that a reduction of the non-significant average global atmospheric temperature will have a non-significant effect on the climate. As shown in this section, decision makers who accept the hypothesis of artificial industrial global warming, face high probabilities of making a mistake.

Statisticians compute objectively the probability of accepting a false industrial global warming hypothesis (p). In contrast, decision makers set subjectively the probability of accepting the risk to make such an error (α). Hence, if $p < \alpha$, then the computed risk of making a mistake is less than the accepted risk of making a mistake. In that case, decision makers would accept the industrial global warming hypothesis.

Decisions about global warming are not about science or statistics. They are about you. As your life and future are at stake, you should be the ultimate decision maker. It is up to you whether you are willing to accept the objectively computed odds of being wrong.

For example, if you accept the hypothesis that the magnitude of the global temperature is caused by industrial pollution, then you accept an 11% probability of being wrong. The chance of being wrong is worse than playing Russian roulette with one bullet in a ten-shooter.

The data show that the slope of the current increase in global temperature is below the natural average. If you nevertheless accept the hypothesis that the uninterrupted linear increase in global temperature is caused by industrial pollution, then you accept a 50% probability of being wrong. The chance of being wrong is the same as playing Russian roulette with five bullets in a ten-shooter. Forget science and statistics, you could just flip a coin.

The data show that the duration of the current global thermal increase is below the natural average. Hence, there is no indication of industrial global warming. If you nevertheless accept the hypothesis that the duration of the uninterrupted linear increase in global temperature is caused by industrial pollution, then you accept a 96% probability of being wrong. The chance of being wrong is worse than playing Russian roulette with nine bullets in a ten-shooter. Such odds are alarming.

Despite the modifications and synchronization of the Vostok and global datasets, differences remain that might confound the statistical analyses and their interpretations. For example, the Vostok data provide information about the Antarctic climate, which might be incomparable to the global climate. The Vostok data are obtained in a single location, while the global data are the means of temperature deviations recorded at multiple locations. The data intervals of the original Vostok datasets might be too long for accurate predictions of annual global temperatures. Furthermore, the Vostok temperature deviations are reconstructed from the deuterium concentrations in the ice-core, while the global temperature deviations are based on direct atmospheric measurements. Nevertheless, the Vostok dataset is the most valid and reliable standard available for this study to evaluate the current global temperature.

Making decisions about climate without adequate information is like playing Russian roulette with six bullets in a six-shooter, while the survival of humanity is at stake. The consequences of accepting incorrectly the industrial global warming hypothesis are as disastrous as rejecting incorrectly the industrial global warming hypothesis. In the first scenario, humanity would not be prepared for the impending disaster. In the second one, humanity would waste precious resources on the prevention of an illusion.

VOSTOK CO2 / VOSTOK °C (423 KYRS)

VOSTOK CO2

Method

The aim of this section is to evaluate the atmospheric CO2 concentrations that were trapped in the Vostok ice during the last 423 Kyrs. The analyses in this section are based on Dataset 17.

Dataset 17 (Vostok423Kyrs CO2).[205] This dataset contains the following Vostok variables: (Gas age)—gas age from 414,085 to 2,342 years before calendar year 1989; and (CO2v)—Vostok atmospheric carbon dioxide concentrations, from 414,085 to 2,342 years before calendar year 1989, in parts per million (ppm). The data intervals are irregular.

Citation Dataset 17: *Trapped gas bubbles record the history of atmospheric CO2 concentrations for over 400,000 years.* IGBP PAGES/World Data Center for Paleoclimatology. NOAA/NGDC Paleoclimatology Program, Boulder CO, USA.

http://www.ncdc.noaa.gov/paleo/icecore/antarctica/vostok/vostok_data.html. Downloaded Friday, 24-Jun-2011 22:33:34 EDT. Last Updated Wednesday, 20-Aug-2008 11:24:22 EDT by paleo@noaa.gov. Please cite the original reference when using these data.

The dataset was derived from an ice core drilled by a French-Russian team at the Vostok station in Antarctica. Vostok is situated at North-bound latitude -78.47 * South-bound latitude -78.47; West-bound longitude 106.8 * East-bound longitude 106.8.

Results

During the observation period from 414,085 to 2,342 years before 1989, the statistics for the Vostok atmospheric CO2 concentrations in parts per million (ppm), comprising

Dataset 17 are: mean 234.0; median = 234.5; standard deviation = 28.7; maximum = 298.7; minimum = 182.2; and range = 116.5 ppm; n = 283.[206] Furthermore, the statistics for the data intervals concerning the Vostok atmospheric CO_2 comprising Dataset 17 are: mean 1,457; median = 1,111; standard deviation = 1,014; maximum = 5,441; minimum = 176; and range = 5,265 years; n = 281.[207]

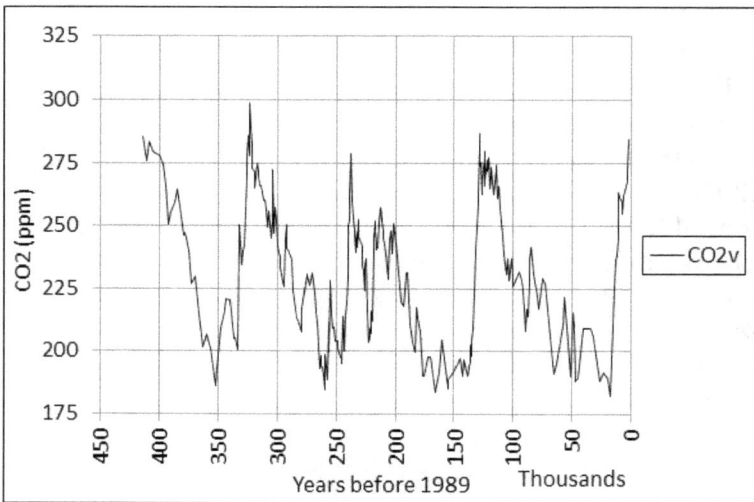

Figure 25 (Dataset 17)—Vostok CO2 423 Kyrs.[208] This graph shows: (CO2v)—Vostok atmospheric carbon dioxide concentrations, from 414,085 to 2,342 years before calendar year 1989, in parts per million (ppm). The data intervals are irregular.

Comments

The data intervals in Dataset 17 are so irregular (176 to 5441 years) that unequal regression towards the mean is likely to confound comparisons of CO_2 concentrations within this dataset. Regression towards the mean and the long data intervals make the minimum and maximum CO_2 values spurious. The last pre-industrial measurement of CO_2 dates back to about 350 BCE. No one knows what happened during that period or during any of the other long data intervals. Furthermore, the long duration of the data intervals prevent the time-scale CO2 concentrations to be converted into reliable millennial-scale CO2 concentrations.

VOSTOK SYNCHRONIZING CO2 AND °C

Method

The aim of this section is to synchronize the Vostok temperature deviations and CO2 concentrations as observed during the last 423 Kyrs. Dataset 2 and Dataset 17 were combined to create Dataset 18, which is the basis for the analyses in this section.

Dataset 2 (VostokTempDepth423Kyrs). This dataset contains the following Vostok variables: (Depth)—ice depth from 3,310 metres to 0 metre in steps of 1 metre; (Age)—ice age in years from 422,766 years before present (BP) to the present calendar year 1989; and (TDv)—depth-scale atmospheric temperature deviations in degrees Celsius (°C) relative to the Vostok base 1850-1989 = 0.00 °C.

To optimize the comparability of temperatures across this study, the base of the Vostok depth-scale temperature deviations comprising Dataset 2 (1850-1989 = 0.00 °C) was converted to (Kyr1 = 0.00 °C) by adding 0.46 °C to each Vostok depth-scale temperature deviation. This is the same correction as used to compute the Vostok millennial-scale temperature deviations.[209] The temperature deviations should not be detrended, because that would confound the relationship between temperature and CO2.

Dataset 17 (Vostok423Kyrs CO2). This dataset contains the following Vostok variables: (Gas age)—gas age from 414,085 to 2,342 years before calendar year 1989; and (CO2v)—Vostok atmospheric carbon dioxide concentrations, from 414,085 to 2,342 years before calendar year 1989, in parts per million (ppm). The data intervals are irregular.

Dataset 2 contains 3,311 depth-scale temperature deviations, while Dataset 17 contains only 283 CO_2 values. In addition, none of the depth-scale temperature deviations comprising Dataset 2 was measured at the same time as the CO_2 concentrations comprising Dataset 17. Therefore, the temperature deviation for each CO_2 value was computed by subjecting the two adjacent thermal values to linear interpolation.[210] This irregular-scale is not ideal, but it is the best method available to synchronize Dataset 2 and Dataset 17.

Dataset 18 (VostokTempCO2Synch412Kyrs).[211] This dataset contains the following Vostok variables: (Age)— from 414,085 to 2,342 years before calendar year 1989; (CO2v)—irregular-scale Vostok atmospheric carbon dioxide concentrations in parts per million (ppm); (TDv)—irregular-scale Vostok atmospheric temperature deviations in degrees Celsius (°C) relative to the Vostok base Kyr1 = 0.00 °C. This dataset is included in appendix: Dataset 18.

Results

Dataset 18 provides the following statistics for the Vostok atmospheric CO_2 concentrations in parts per million (ppm): mean 234.0; median = 234.5; standard deviation = 28.7; maximum = 298.7; minimum = 182.2; and range = 116.5 ppm; n = 283.[212] Furthermore, Dataset 18 provides the following statistics for the irregular-scale Vostok temperature deviations: mean -3.61; median = -4.05; standard deviation = 3.06; maximum = 3.64; minimum = -8.69; and range = 12.33 °C (Kyr1 = 0.00 °C); n = 283.[213]

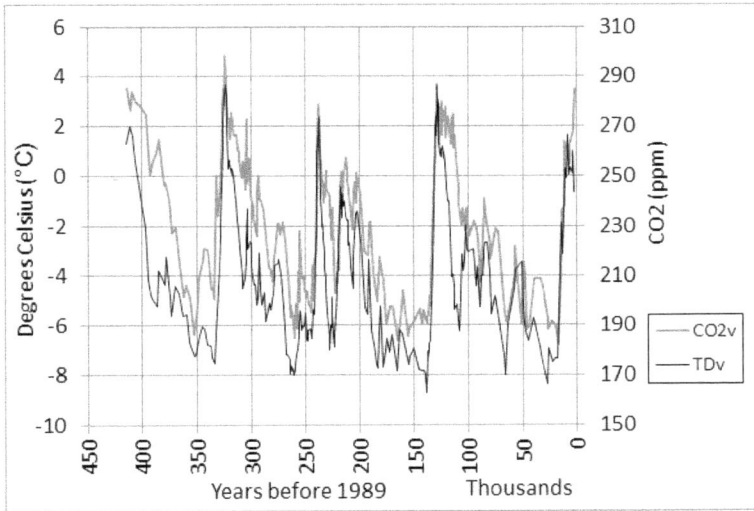

Figure 26 (Dataset 18)—Vostok CO2 and temperature 412 Kyrs.[214] This graph shows: (CO2v)—irregular-scale Vostok atmospheric carbon dioxide concentrations in parts per million (ppm); (TDv)—irregular-scale Vostok atmospheric temperature deviations in degrees Celsius (°C) relative to the Vostok base Kyr1 = 0.00 °C. The observation period ranges from 414,085 to 2,342 years before calendar year 1989.

During 1,292 years, from 3,634 to 2,342 years before 1989, the CO2 concentrations increased with 11.9 from 272.8 to 284.7 ppm, while the temperature decreased with 1.65 from 1.01 to -0.64 °C.[215]

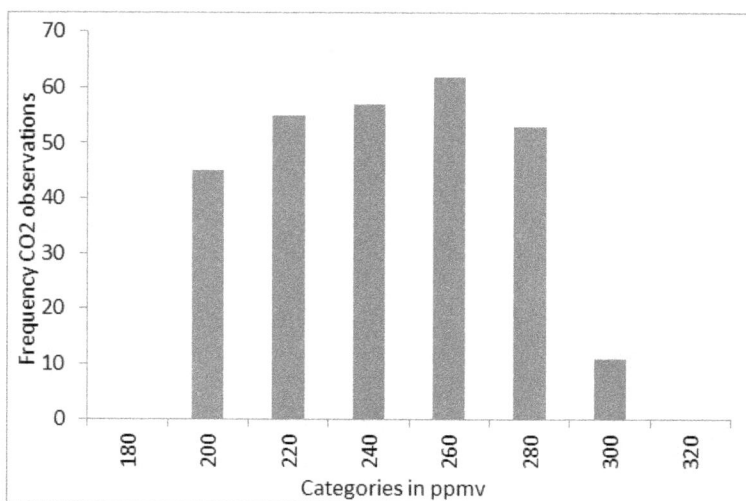

Figure 27 (Dataset 18)—histogram Vostok CO2 412 Kyrs.[216] This graph shows the frequency distribution of (CO2v)—Vostok atmospheric carbon dioxide concentrations, from 414,085 to 2,342 years before calendar year 1989, in parts per million (ppm). The data intervals are irregular.

The frequency distribution of the CO2 concentrations is rather flat and negatively skewed towards the lower values.

Comments

Compared to Dataset 2, the data resolution of Dataset 17 is low.[217] This is a weak point in the data analyses and further research should extract more CO2 data points from the Vostok ice. Figure 26 suggest that the patterns of the Vostok irregular-scale temperature deviations and CO2 concentrations are similar. However, the most recent changes in temperature and CO2, from 3,634 to 2,342 years before 1989, are in opposite directions and, therefore, contradict that similarity. The next section will evaluate the significance of the relationship between CO2 and temperature. Figure 27 suggest that the frequency distribution of the CO_2 concentrations comprising Dataset 18 is unsuitable for parametric statistical analyses.

VOSTOK CO2 VERSUS °C

Method

It is the aim of this section to analyse the correlation between the Vostok atmospheric CO_2 concentrations in parts per million (ppm) and the Vostok temperature deviations comprising Dataset 18.

Results

The statistics for the irregular-scale Vostok atmospheric CO_2 concentrations in parts per million (ppm), comprising Dataset 18 are: mean 234.0; median = 234.5; standard deviation = 28.7; maximum = 298.7; minimum = 182.2; and range = 116.5 ppm; n = 283.[218] Furthermore, the statistics for the Vostok irregular-scale temperature deviations comprising Dataset 18 are: mean -3.61; median = -4.05; standard deviation = 3.06; maximum = 3.64; minimum = -8.69; and range = 12.33 °C (Kyr1 = 0.00 °C); n = 283.[219]

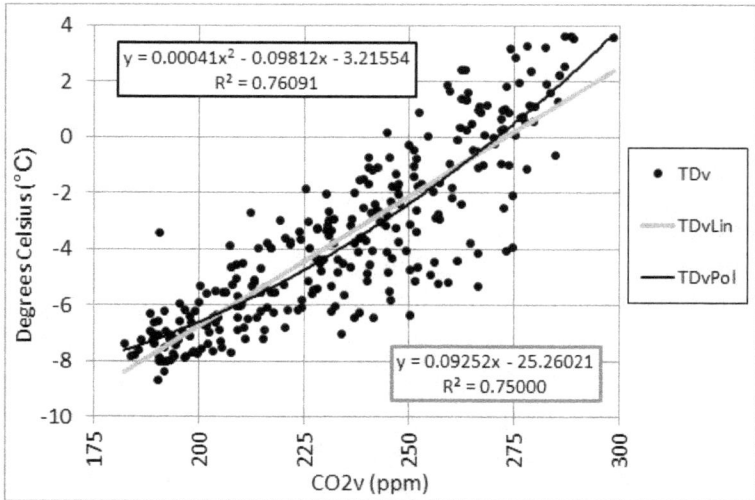

Figure 28 (Dataset 18)—Vostok temperature versus CO2 412 Kyrs.[220] This graph shows: (TDv)—irregular-scale Vostok atmospheric temperature deviations in degrees Celsius (°C) relative to the Vostok base Kyr1 = 0.00 °C; (CO2v)— irregular-scale Vostok atmospheric carbon dioxide concentrations in parts per million (ppm). The observation period ranges from 414,085 to 2,342 years before calendar year 1989, while the data intervals are irregular.

Figure 28 shows that a linear function explains the variance adequately (R = 0.75) and almost equals the explanatory strength of the second order polynomial function (R = 0.76). Increasing the polynomial function up to the sixth order did not improve significantly the explained variance.

Test 2 (Dataset 18)—randomized correlation TDv and CO2v (RCT).[221] In which: (TDv)—Vostok ranked irregular-scale temperature deviation; (CO2v)—Vostok ranked irregular-scale CO2 concentration; observation period from 414,085 to 2,342 before 1989. At each iteration, the correlation coefficient was computed after CO_2v was randomized relative to TDv. Using ranked data avoids the statistical problems associated with asymmetric frequency distributions, unequal variances, and outliers.

Test 2 provided the following statistics for the randomized correlation coefficients: mean = 0.00; median = 0.00; standard deviation = 0.06; maximum = 0.22; minimum = -0.20; and range = 0.42 random r; n = 283; original r = 0.87; R = 0.75; p = 0.0000; i = 10,000.[222]

Comments

The results reject the null hypothesis that the correlation between Vostok atmospheric temperature and CO_2 concentrations is a random effect. Therefore, the alternative hypothesis is accepted that the correlation between Vostok atmospheric temperature and CO_2 concentrations is a systematic effect.

Although a correlation coefficient is in itself objective, it is also one of the most abused statistics. A correlation coefficient provides information about the synchronization between the changes of two variables. It does not provide information about cause and effect.

A significant correlation between the hypothetical variables A and B could mean that A causes B, or alternatively, that B causes A. It could also mean that both variables cause each other. For example, low temperatures foster snow, but snow fosters also low temperatures because the white colour of snow reflects warm light.

Alternatively, A and B might not affect each other, but are both caused by one or more confounding variables. For example, there is a perfect correlation between the arrival of storks and lambs in Holland. Storks return each year from their migration at the same time that the lambs appear, while they leave when the lambs disappear. As you might have guessed, the arrival of the storks has nothing to do with the arrival of lambs and vice versa. Both events are caused by the upcoming spring. When the storks leave for warmer locations, the lambs have grown into sheep.

No matter how significant, a correlation coefficient cannot support the hypothesis that changes in atmospheric CO_2 concentrations cause proportional changes in atmospheric temperature or vice versa. The significant correlation reported in this study indicates only that the changes in CO_2 concentrations and temperature deviations are synchronized. Leaders beware—any claim that the significant correlation proves that rising CO_2 concentrations cause rising temperatures is unscientific nonsense.

HAWAII CO2 / HAWAII °C (1959-2014)

HAWAII °C CALIBRATED

Method

The aim of this section is to facilitate the comparison of temperatures across datasets by synchronizing the baselines of the Hawaiian and global thermal datasets. The global and Vostok temperatures are relative to the Vostok base Kyr1 = 0.00 °C. Therefore, the calibrated Hawaiian temperatures have to be relative to that base. Modification of Dataset 19 provided Dataset 20, which is the basis for the analyses in this section.

Dataset 19 (TemperaturesHawaiiAbsolute°C).[223] This dataset contains the following Hawaiian variables: (Th)—Hawaiian annual atmospheric temperatures from calendar year 1955 to 2014, in absolute degrees Celsius (°C).

The original dataset was provided by the Western Regional Climate Center and comprises the Monthly Average of Average Daily Temperature at MAUNA LOA SLOPE OBS 39, HI in Degrees Fahrenheit during the observation period from 1955 to 2015 (516198). It was last updated on Sep 13, 2015 and downloaded for this study on 13/09/2015 22:26 from http://www.wrcc.dri.edu/cgi-bin/cliMAIN.pl?hi6198.

Missing monthly values were estimated by linear interpolation of the adjacent monthly temperatures. The annual temperatures in Dataset 19 are the means of the reconstructed monthly temperatures. The monthly and annual temperatures were converted from absolute degrees Fahrenheit (°F) into absolute degrees Celsius (°C).

Equation 5 (Dataset 19)—converts degrees Fahrenheit into degrees Celsius: $°C = (°F - 32) * 5/9$. In which, (°C)—degrees Celsius; and (°F)—degrees Fahrenheit.[224]

Dataset 20 (TemperatureDeviationsHawaii).[225] This dataset contains the following Hawaiian variables: (Year)—calendar years from 1955 to 2014; (TDh)—Hawaiian annual atmospheric temperature deviations in degrees Celsius (°C) relative to the Vostok base Kyr1 = 0.00 °C.

The global temperature deviations (TDg) are calibrated to the Vostok base Kyr1 = 0.00 °C. Furthermore, the Hawaiian annual atmospheric temperature deviations (TDh) are calibrated to the mean of the global atmospheric temperature deviations from 1955 to 2014. Hence, the Hawaiian temperature deviations are also calibrated to the Vostok base Kyr1 = 0.00 °C.

It is emphasized that this recalibration of the Hawaiian temperature is an approximation, because the observation period of the Hawaiian temperature is 60 years, rather than the 1,000 years of the Vostok base. Furthermore, the reliability of the calibration relied on the accuracy of the previous calibration of the global temperatures. Nevertheless, it is the best what can be done with the available data.

Results

Dataset 19 provides the statistics of the Hawaiian temperatures in absolute degrees Celsius °C (Th), during the observation period from 1955 to 2014: mean = 7.30; median = 7.43; standard deviation = 0.75; maximum = 8.72; minimum = 5.69; and range = 3.03 °C; n = 60.[226] Dataset 14 provides the statistics of the global temperature deviations in degrees Celsius (°C), relative to the baseline Vostok Kyr1 = 0.00 °C (TDg), during the observation period from 1955 to 2014: mean = 0.88; median = 0.84; standard deviation = 0.26; maximum = 1.32; minimum = 0.45; and range = 0.87 °C; n = 60.[227]

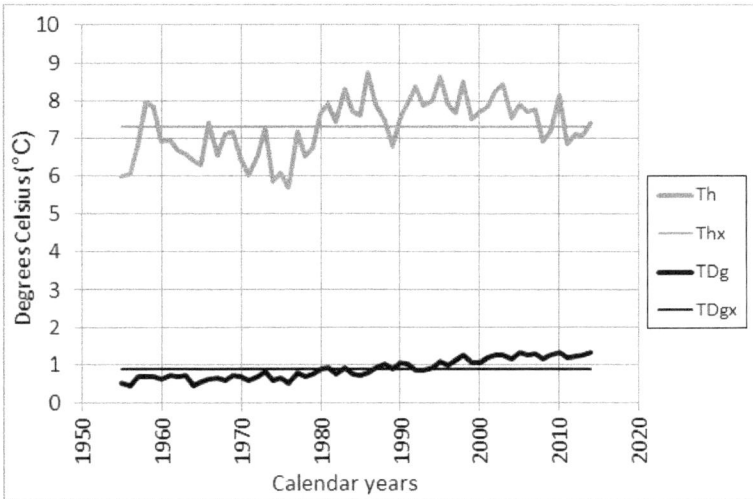

Figure 29—Hawaiian and global temperatures 1955-2014.[228] This graph shows: (Th (Dataset 19))—Hawaiian annual atmospheric temperatures in absolute degrees Celsius (°C); (Thx)—mean Th; (TDg (Dataset 14))—extended global annual atmospheric temperature deviations in degrees Celsius (°C) relative to the Vostok base Kyr1 = 0.00 °C; (TDgx)—mean TDg.

Recalibration Th

The Hawaiian annual temperatures in Figure 29 (Th), were recalibrated so that their mean equalled the mean of the global temperature deviations (TDg) of 0.88 °C relative to the Vostok base Kyr1 = 0.00 °C.

Dataset 20 provides the statistics of the resulting Hawaiian temperature deviations in degrees Celsius °C relative to the Vostok base Kyr1 = 0.00 °C (TDh), during the observation period from 1955 to 2014: mean = 0.88; median = 1.01; standard deviation = 0.75; maximum = 2.30; minimum = -0.73; and range = 3.03 °C; n = 60.[229]

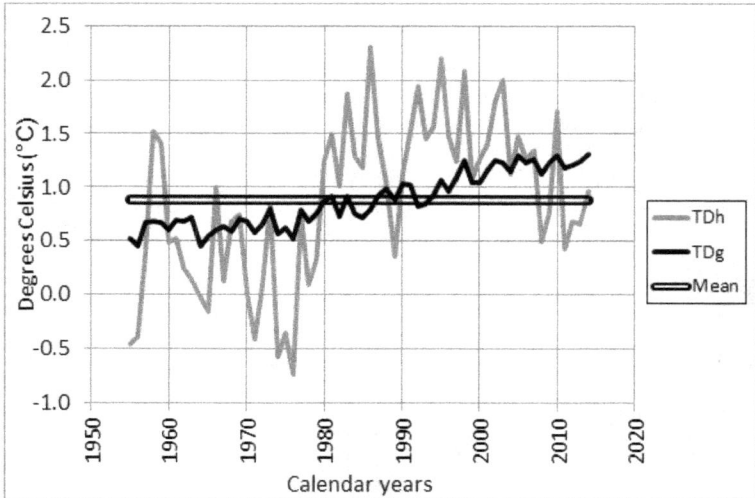

Figure 30—recalibrated Hawaiian and global temperatures 1955-2014.[230] This graph shows: (TDh (Dataset 20))—Hawaiian annual atmospheric temperature deviations in degrees Celsius (°C) relative to the Vostok base Kyr1 = 0.00 °C; (TDg (Dataset 14))—extended global annual atmospheric temperature deviations; TDh and TDg are expressed in degrees Celsius (°C) relative to the Vostok base Kyr1 = 0.00 °C.

Test 3—randomized correlation TDh and TDg.[231] In which: (TDh (Dataset 20))—Hawaii ranked annual temperature deviation; (TDg (Dataset 14))—Global ranked annual temperature deviation; observation period from 1955 to 2014. At each iteration, the correlation coefficient was computed after TDh was randomized relative to TDg. Using ranked data avoids the statistical problems associated with asymmetric frequency distributions, unequal variances, and outliers.

Test 3 provided the following statistics for the randomized correlation coefficients: mean = 0.00; median = 0.00; standard deviation = 0.13; maximum = 0.43; minimum = -

0.52; and range = 0.96 random r; n = 60; original r = 0.63; R = 0.40; p = 0.0000; i = 10,000.[232]

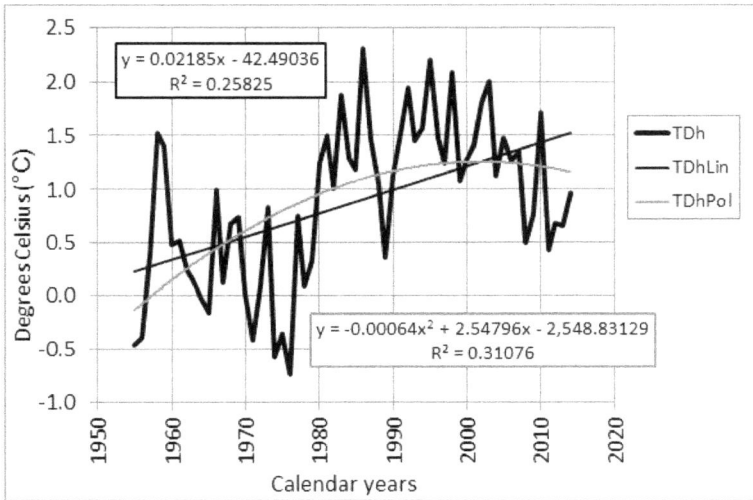

y = 0.02185x - 42.49036
R² = 0.25825

y = -0.00064x² + 2.54796x - 2,548.83129
R² = 0.31076

Figure 31 (Dataset 20)—Hawaiian temperatures 1955-2014.[233] This graph shows; (TDh)—Hawaiian annual atmospheric temperature deviations in degrees Celsius (°C) relative to the Vostok base Kyr1 = 0.00 °C; (TDhLin)—linear function of TDh; (TDhPol)—second order polynomial function of TDh.

The linear function (TDhLin) obtained an R-value of 0.26, while polynomial function (TDhPol) reached an R-value 0.31. The linear function explains 26% of the variance in the data, while the polynomial function explains 31%. Therefore, the polynomial function is a slightly better fit to the data than the linear function.

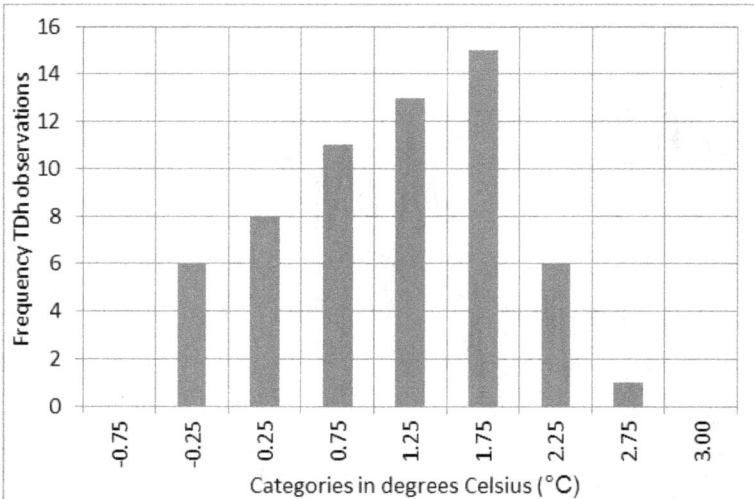

Figure 32 (Dataset 20)—histogram Hawaiian temperatures 1955-2014.[234] This histogram shows the frequency distribution of (TDh)—Hawaiian annual atmospheric temperature deviations in degrees Celsius (°C) relative to the Vostok base Kyr1 = 0.00 °C.

Comments

In this section, the Hawaiian temperature was calibrated to the global temperature, which is relative to the Vostok base Kyr1 = 0.00 °C. Therefore, the calibrated Hawaiian temperature is also relative to the Vostok base Kyr1 = 0.00 °C. The calibration is not ideal, but it is the best available option.

Hawaiian annual temperature deviations (TDh) are associated with one location, while the global annual temperature deviations (TDg) are the means of many locations. Therefore, regression towards the mean reduces the variation in the global annual temperature deviations (TDg) in comparison to the Hawaiian annual temperature deviations (TDh).

The significant correlation, between the Hawaiian temperatures deviations (TDh) and the global temperature deviations (TDg), suggests that the changes in Hawaiian temperature represent the changes in global temperature. However, the correlation coefficient leaves 60% of the variance in the data unexplained.

Figure 31 shows that the data are better explained by the polynomial function than by the linear function. This suggests that the Hawaiian temperature has reached a maximum around the year 2000. Since then, the decrease has been accelerating. However, this effect might be due to random variation around the mean. The observation period might be too short to account for the variance in the data. To increase the reliability of the analyses, the number of data points has to be increased or the variance in the data points has to be decreased.

The histogram in Figure 32 shows that the frequency distribution of the Hawaiian temperature deviations is negatively skewed towards the lower temperatures. This suggests that parametric statistical analyses are inappropriate for the evaluation of the Hawaiian temperature deviations in Dataset 20.

HAWAII CO2

Method

The aim of this section is to evaluate the Hawaiian atmospheric CO_2 concentrations as recorded by the Mauna Loa Observatory. The analyses in this section are based on Dataset 21.

Dataset 21 is based on data provided by the Mauna Loa Observatory (Scripps / NOAA / ESRL) and comprises the Monthly Mean CO_2 Concentrations (ppm). The file comprises Scripps data from March 1958 to April 1974 and NOAA-ESRL data from May 1974 onwards. This dataset was released by NOAA-ESRL on September 7, 2015 and downloaded for the study on 14/09/2015 13:47:19 from http://co2now.org/Current-CO2/CO2-Now/noaa-mauna-loa-co2-data.html.[235] Due to missing data points, the observation period of the dataset was truncated from 1958-2015 to 1959-2014.

Dataset 21 (Hawaii CO2).[236] This dataset contains the following Hawaiian variables: (Year)—calendar years from 1959 to 2014; (CO2h)—Hawaiian annual atmospheric carbon dioxide concentrations in part per million (ppm).

Results

Dataset 21 provides the statistics of the Hawaiian atmospheric CO2 concentrations in parts per million (ppm), during the observation period from 1959 to 2014: mean = 350.8; median = 348.3; standard deviation = 24.8; maximum = 398.6; minimum = 316.0; and range = 82.6 ppm; n = 56.[237]

Figure 33 (Dataset 21)—Hawaiian CO2 1959-2014.[238] This graph shows: (CO2h)—Hawaiian annual atmospheric carbon dioxide concentrations in part per million (ppm); (CO2hLin)—linear function of CO2h; and (CO2hPol)—second order polynomial function of CO2h.

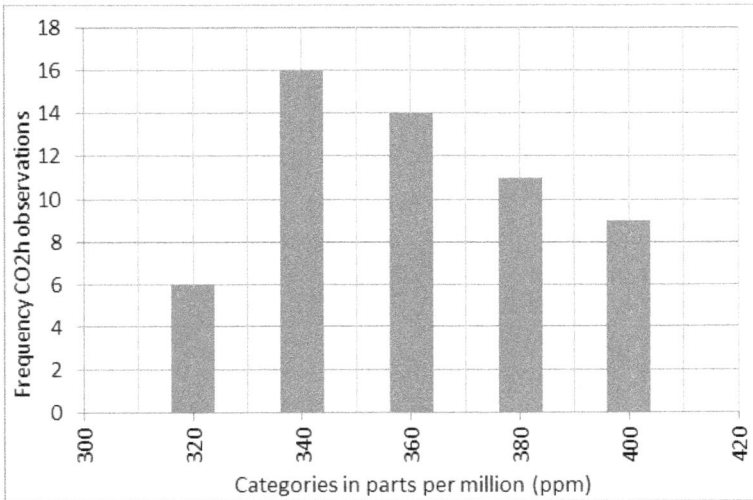

Figure 34 (Dataset 21)—histogram Hawaiian CO2 1959-2014.[239] This graph shows the frequency distribution of the Hawaiian annual atmospheric carbon dioxide concentrations in part per million (ppm).

Comments

The data from the Mauna Loa Observatory in Hawaii are important, because they provide the longest record of recent atmospheric CO2 concentrations.

Figure 33 shows that the Hawaiian atmospheric CO2 concentrations are increasing steadily over time and reached almost 400 ppm in 2014. Furthermore, it shows that the linear function (CO2hLin) explains 99% of the variation in CO2 concentrations. Nevertheless, the polynomial function (CO2hPol) is a perfect fit to the data. This would suggest that the increase in Hawaiian CO2 concentrations is slightly accelerating.

Figure 34 shows that the frequency distribution of the Hawaiian atmospheric CO2 concentrations is positively skewed towards the higher values. This suggests that parametric statistical analyses are inappropriate for the evaluation of the CO2 concentrations in Dataset 21.

HAWAII CO2 VERSUS °C

Method

The aim of this section is to evaluate the relationship between Hawaiian atmospheric temperatures and CO2 concentrations during the recent past. Dataset 20 and Dataset 21 were combined to create Dataset 22, which is the basis for the analyses in this section.

Dataset 20 (TemperatureDeviationsHawaii). This dataset contains the following Hawaiian variables: (Year)—calendar years from 1955 to 2014; (TDh)—Hawaiian annual atmospheric temperature deviations in degrees Celsius (°C) relative to the Vostok base Kyr1 = 0.00 °C.

The global temperature deviations (TDg) are calibrated to the Vostok base Kyr1 = 0.00 °C. Furthermore, the Hawaiian annual atmospheric temperature deviations (TDh) are calibrated to the mean of the global atmospheric temperature deviations from 1955 to 2014. Hence, the Hawaiian temperature deviations are also calibrated to the Vostok base Kyr1 = 0.00 °C.

Dataset 21 (Hawaii CO2). This dataset contains the following Hawaiian variables: (Year)—calendar years from 1959 to 2014; (CO2h)—Hawaiian annual atmospheric carbon dioxide concentrations in part per million (ppm).

Dataset 22 (Hawaii temperature and CO2).[240] This dataset contains the following Hawaiian variables: (Year)—calendar years from 1959 to 2014; (TDh)—Hawaiian atmospheric temperature deviations in degrees Celsius (°C) relative to the Vostok base Kyr1 = 0.00 °C; and (CO2h)—Hawaiian atmospheric CO2 concentrations in parts per million (ppm). This dataset is included in appendix: Dataset 22.

Results

Dataset 22 provides the statistics of the Hawaiian atmospheric temperature deviations in degrees Celsius (°C) relative to the Vostok base Kyr1 = 0.00 °C, during the observation period from 1959 to 2014: mean = 0.92; median 1.04; standard deviation = 0.73; maximum = 2.30; minimum = -0.73; and range = 3.03 °C; n = 56.[241] Furthermore, it provides the statistics of the Hawaiian atmospheric CO_2 concentrations in parts per million (ppm), during the observation period from 1959 to 2014: mean = 350.8; median = 348.3; standard deviation = 24.8; maximum = 398.6; minimum = 316.0; and range = 82.6 ppm; n = 56.[242]

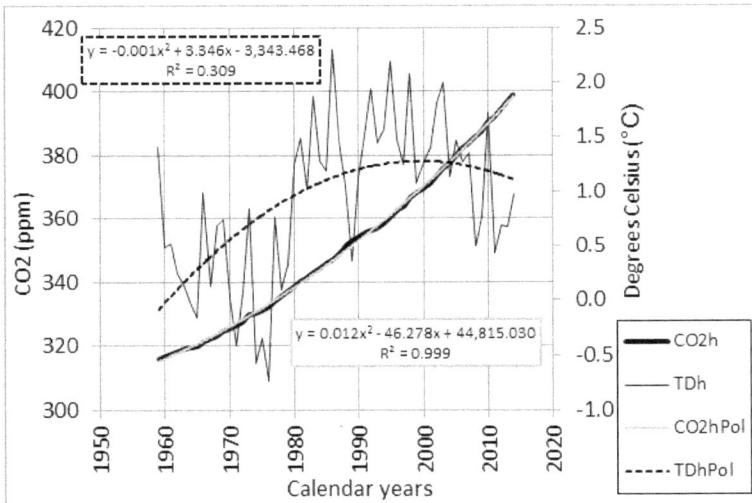

Figure 35 (Dataset 22)—Hawaiian temperatures and CO_2 1959-2014.[243] This graph shows: (CO2h)—Hawaiian atmospheric CO_2 concentrations in parts per million (ppm); (TDh)—Hawaiian atmospheric temperature deviations in degrees Celsius (°C) relative to the Vostok base Kyr1 = 0.00 °C; (TDhPol)—second order polynomial function of TDh; and (CO2hPol)—second order polynomial function of CO2h.

Test 4 (Dataset 22)—randomized correlation CO2h and TDh.[244] In which: (CO2h)—Hawaii ranked annual CO2 concentration; (TDh)—Hawaii ranked annual temperature deviation; observation period from 1959 to 2014. At each iteration, the correlation coefficient was computed after CO2h was randomized relative to TDh. Using ranked data avoids the statistical problems associated with asymmetric frequency distributions, unequal variances, and outliers.

Test 4 provided the following statistics for the randomized correlation coefficients: mean = 0.00; median = 0.00; standard deviation = 0.13; maximum = 0.48; minimum = - 0.52; and range = 1.00 random r; n = 56; original r = 0.48; R = 0.23; p = 0.0000; i = 10,000.[245]

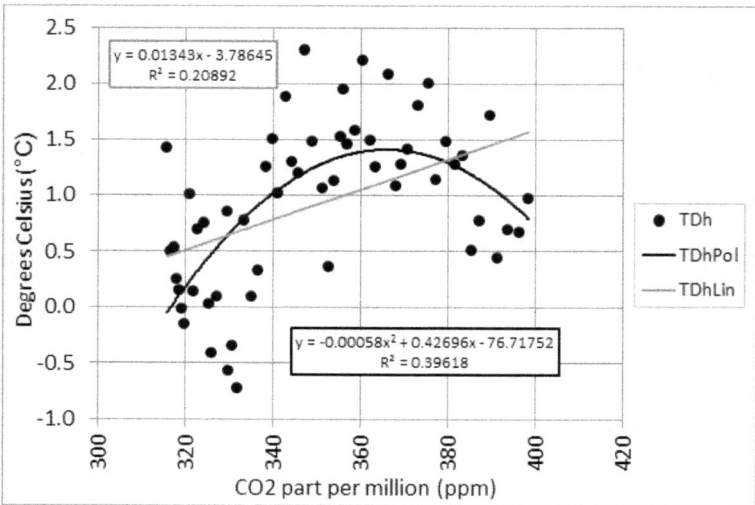

Figure 36 (Dataset 22)—scatterplot Hawaiian temperatures versus CO2 1959-2014.[246] This graph shows: (TDh)—Hawaiian atmospheric temperature deviations in degrees Celsius (°C) relative to the Vostok base Kyr1 = 0.00 °C; as a function of (CO2h)—Hawaiian atmospheric CO2 concentrations in parts per million (ppm); (TDhPol)—second

order polynomial function of TDh; and (TDhLin)—linear function of TDh.

Figure 36 shows that the linear function (TDhLin) obtained an R-value of 0.21, while polynomial function (TDhPol) reached an R value 0.40. Hence, the polynomial function is a slightly better fit to the data than the linear function.

Comments

The polynomial functions presented in Figure 35 suggest that the increase in the Hawaiian atmospheric CO_2 concentrations is steadily accelerating since 1959, while the decrease in Hawaiian atmospheric temperature is accelerating since 2000. Hence, these contradictory changes cannot support the hypothesis that an increase in atmospheric CO_2 will cause an increase in atmospheric temperature. However, the effect might be caused by the large random variation around the mean. This variation in the Hawaiian temperature could be reduced by using the mean annual temperatures concerning multiple global locations.

The polynomial function presented in Figure 36 is a better fit to the data than the linear function. This function suggests that the decrease in temperature accelerates after reaching a maximum CO_2 concentration of about 360 ppm. Hence, the results reject the hypothesis that the temperature increases with increasing CO_2 concentrations. Alternatively, if that hypothesis would be true, then the result suggests that another variable decreases the temperature at the higher CO_2 concentrations.

The polynomial function of the Hawaiian temperatures in Figure 35 explains 31% of the variance in those data. To increase the explained variance, the next section will evaluate the relationship between the mean global atmospheric temperatures and the Hawaiian CO_2 atmospheric concentrations.

HAWAII CO2 / GLOBAL °C (1959-2014)

CORRELATION CO2 AND °C

Method

It is the aim of this section to analyse the relationship between the Hawaiian annual atmospheric CO_2 concentrations and the global annual atmospheric temperature deviations from 1959 to 2014. Dataset 14 and Dataset 23 were combined to create Dataset 24, which is the basis for the analyses in this section.

Dataset 14 (VostokGlobalTempSynchronized). This dataset contains the following variables: (Year)—calendar years from 989 to 2014; (TDv)—Vostok semi-annual atmospheric temperature deviations from 989 to 1989; (TDg)—extended global annual atmospheric temperature deviations from calendar year 1850 to 2014; (Syn)—means of TDv and TDg during synchronization period from calendar year 1850 to 1989. TDg and TDv are expressed in degrees Celsius (°C) relative to the Vostok base Kyr1 = 0.00 °C.

Dataset 23 (CO2Hawaii1959-2014).[247] This dataset contains the following Hawaiian variables: Year; CO_2 Concentration Posted by NOAA; Uncertainty Posted by NOAA; and CO_2 Concentration Calculated.

Dataset 23 was downloaded for this study on the 2nd of September 2105 from the website http://co2now.org/Current-CO2/CO2-Now/noaa-mauna-loa-co2-data.html "At CO2Now.org, data for March 1958 - April 1974 was obtained by Charles David Keeling of the Scripps Institution of Oceanography (Scripps). Data for CO2 since May 1974 was obtained by the National Oceanic and Atmospheric Administration (NOAA). The Scripps Institution of Oceanography also maintains a CO2 monitoring program at the Mauna Loa Observatory."

Dataset 24 (GlobalTempHawaiiCO2-1959-2014).[248] This dataset contains the following variables: (Year)—calendar years from 1959 to 2014; (TDg)—global annual atmospheric temperature deviations in degrees Celsius (°C) relative to the Vostok base Kyr1 = 0.00 °C; (CO2h)—Hawaiian atmospheric carbon dioxide concentrations in parts per million (ppm).

Results

The statistics for the global annual atmospheric temperature deviations from 1959 to 2014, comprising Dataset 24 are: (TDg) mean 0.90; median = 0.87; standard deviation = 0.25; maximum = 1.32; minimum = 0.45; and range = 0.87 °C; n = 56.[249] Furthermore, the statistics for the Hawaiian annual atmospheric CO2 concentrations from 1959 to 2014, comprising Dataset 24 are: (CO2h) mean 350.8; median = 348.3; standard deviation = 24.8; maximum = 398.6; minimum = 316.0; and range = 82.6 ppm; n = 56.[250] Regression towards the mean reduced the maximum CO2h with 12.0% from 398.6 to 350.8 ppm.[251]

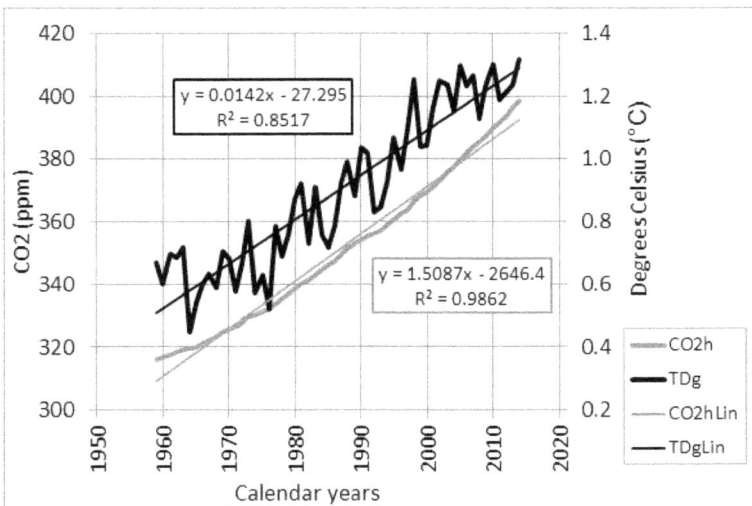

Figure 37 (Dataset 24)—global temperatures and Hawaiian CO2 1959-2014.[252] This graph shows: (CO2h)—Hawaiian atmospheric carbon dioxide concentrations in parts per million (ppm); (TDg)—global annual atmospheric temperature deviations in degrees Celsius (°C) relative to the Vostok base Kyr1 = 0.00 °C; (CO2hLin)—linear function of CO2h; and (TDgLin)—linear function of TDg.

Figure 37 shows that TDgLin explains 85% of the variance in the TDg values, while CO2Lin explains 99% of the variance in the CO2h values. Hence, both linear functions represent adequately those atmospheric data.

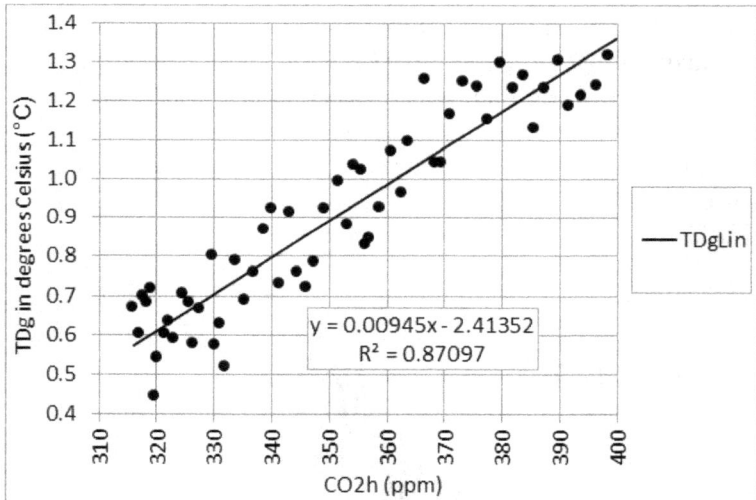

$$y = 0.00945x - 2.41352$$
$$R^2 = 0.87097$$

Figure 38 (Dataset 24)—global temperatures versus Hawaiian CO2 1959-2014.[253] This graph shows: (TDg)—global annual atmospheric temperature deviations in degrees Celsius (°C) relative to the Vostok base Kyr1 = 0.00 °C; as a function of (CO2h)—Hawaiian atmospheric carbon dioxide concentrations in parts per million (ppm); and (TDgLin)—linear function of TDg.

Test 5 (Dataset 24)—randomized correlation TDg and CO2h.[254] In which: (TDg)—Global ranked annual temperature deviation; (CO2h)—Hawaii ranked annual CO2 concentration; observation period from 1959 to 2014. At each iteration, the correlation coefficient was computed after TDg was randomized relative to CO2h. Using ranked data avoids the statistical problems associated with asymmetric frequency distributions, unequal variances, and outliers.

Test 5 provided the following statistics for the randomized correlation coefficients: mean = 0.00; median = 0.00; standard deviation = 0.13; maximum = 0.51; minimum = -0.50; and range = 1.01 random r; n = 56; original r = 0.92; R = 0.84; p = 0.0000; i = 10,000.[255]

Comments

The interpretation of statistics can be misleading. During the observation period from 1959 to 2014, the CO2 concentrations at Hawaii reached a maximum of 398.6 ppm in 2014. That is the highest concentration recorded during the last 423 Kyrs. However, that does not mean that it is the highest CO2 concentration that occurred during the last 423 Kyrs.

The Hawaiian maximum CO2 concentration of 398.6 ppm is an annual mean obtained during the 56 years observation period from 1959 to 2014. In contrast, the mean CO2 concentration during that same observation period was 350.8 ppm. Hence, regression towards the mean has a strong effect on the maximum and reduced it with 12.0%.[256]

From 353 BCE until 1959, the atmospheric CO2 concentrations are unknown.[257] Furthermore, during the observation period from 414,085 to 2,342 years before 1989, the maximum data interval concerning the Vostok atmospheric CO2 was 5,441 years.[258] During that long period, the maximum CO2 concentration is unknown. Hence, one can only guess what regression towards the mean has done to the maxima of the Vostok CO2 concentrations.

What is known is that regression towards the mean reduced that unknown maximum to 209.1 ppm.[259] During those 5,441 years, the maximum CO_2 concentrations might have been much higher than the 398.6 ppm recorded in 2014. At this moment, detailed data are not available. Consequently, any historical interpretation of the 398.6 ppm is non-conclusive and misleading. Leaders beware—any claim that the high CO_2 concentrations in 2014 are an all-time high is unscientific nonsense. Unfortunately, scientists do not know.

The correlation between the Hawaiian CO_2 and global temperature during the observation period from 1959 to 2014 ($r = 0.92$) is slightly higher than the correlation between the CO_2 and temperature in Vostok during the last 423 Kyrs ($r = 0.87$). Furthermore, both correlations are highly significant ($p = 0.0000$). This suggests that the datasets are valid and reliable. The difference might be attributed to the higher resolution provided by the annual means of the recent datasets.

As mentioned previously, despite the significant correlation, a correlation analysis cannot evaluate whether rising CO_2 concentrations cause rising temperatures. Leaders beware—any claim that the significant correlation proves that rising CO_2 concentrations cause rising temperatures is unscientific nonsense. A more specific statistical analysis is required than a correlation analysis.

CO2 PREDICTS TEMPERATURE

Method

The aim of this section is to test the hypothesis that changes in the atmospheric CO_2 concentrations will change proportionally the atmospheric temperature. Therefore, the following predictive algorithms will be evaluated:

CO2 predicts temperature
If TDp > 0, then TDc > 0
If TDp < 0, then TDc < 0

In which, CO2p = predictor of the change in the global annual temperature deviation (CO2hz - TDgz); TDc = actual change in the global annual temperature deviation (TDgz1 - TDgz); Furthermore, CO2hz = z-score of the Hawaiian annual CO2 concentration during the year under examination; TDgz = z-score of the global annual temperature deviation during the year under examination; and TDgz1 = z-score of the global annual temperature deviation during the following year. Therefore:

CO2 predicts temperature (z-scores)
If (CO2hz - TDgz) > 0, then (TDgz1 - TDgz) > 0
If (CO2hz - TDgz) < 0, then (TDgz1 - TDgz) < 0

The advantage of using z-scores, or standard-scores, is that they facilitate comparisons across variables with different units of measurements. In this case, between CO2 concentrations in parts per million (ppm) and temperatures in degrees Celsius (°C). After the original units are transformed into z-scores, they are directly comparable and can be subjected to arithmetical operations. The analyses in this section are based on Dataset 24.

Dataset 24 (GlobalTempHawaiiCO2-1959-2014). This dataset contains the following variables: (Year)—calendar years from 1959 to 2014; (TDg)—global annual atmospheric temperature deviations in degrees Celsius (°C) relative to the Vostok base Kyr1 = 0.00 °C; (CO2h)—Hawaiian atmospheric carbon dioxide concentrations in parts per million (ppm).

Results

Dataset 24 provides the statistics for the global annual temperature deviations from 1959 to 2014 (TDg): mean 0.90;

median = 0.87; standard deviation = 0.25; maximum = 1.32; minimum = 0.45; and range = 0.87 °C; n = 56.[260] Furthermore, Dataset 24 provides the statistics for the Hawaiian annual CO2 concentrations from 1959 to 2014 (CO2h): mean 350.8; median = 348.3; standard deviation = 24.8; maximum = 398.6; minimum = 316.0; and range = 82.6 ppm; n = 56.[261]

Table 12

	CO2hz	TDgz	CO2p	TDc
mean	0.00	0.00	0.00	0.05
median	-0.10	-0.10	-0.01	0.13
stdev	1.00	1.00	0.37	0.47
maximum	1.93	1.66	0.76	1.07
minimum	-1.40	-1.81	-0.77	-1.08
range	3.33	3.47	1.53	2.15
n	56	56	55	55

Table 12—statistics of CO2hz, TDgz, CO2p, and TDc.[262] This table shows the statistics for the following variables: (CO2hz)—z-score of CO2h; (CO2h (Dataset 24))—Hawaiian atmospheric carbon dioxide concentrations, from 1959 to 2014, in parts per million (ppm); (TDgz)—z-score of TDg; (TDg (Dataset 24))—global annual atmospheric temperature deviations, from 1959 to 2014, in degrees Celsius (°C) relative to the Vostok base Kyr1 = 0.00 °C; (CO2p)—predicts TDc (CO2p = CO2hz − TDgz); (TDc)—change in next year's temperature deviation (TDc = TDgz1 − TDgz); and TDgz1—z-score of next year's temperature deviation.

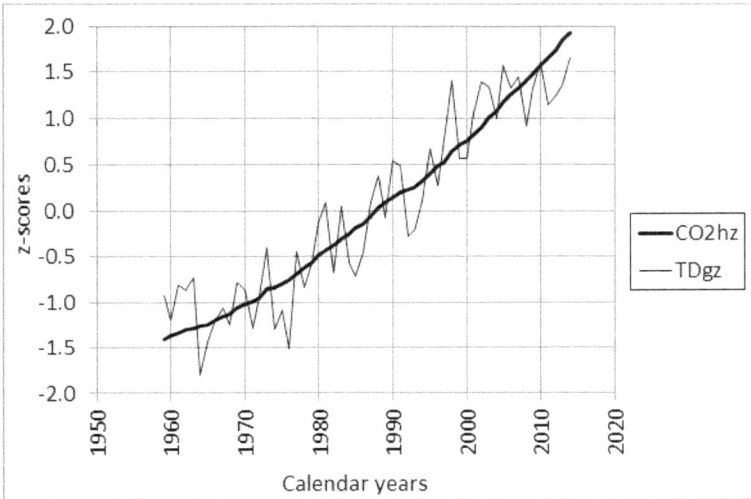

Figure 39 (Dataset 24) z-scores Hawaiian CO2 and global temperatures 1959-2014.[263] This graph shows: (CO2hz)—z-scores Hawaiian atmospheric carbon dioxide concentrations in parts per million (ppm); and (TDgz)—z-scores global annual atmospheric temperature deviations in degrees Celsius (°C) relative to the Vostok base Kyr1 = 0.00 °C.

The CO2hz values show a steady increase over time, while the TDgz values show variation around the CO2hz values.

Positive CO2pz-values predicted retrospectively 25 of the 27 actual increases in annual TDgz-values (93%). Negative CO2pz-values values predicted retrospectively 20 of the 28 actual decreases in annual TDgz-values (71%). In total, 45 of the 55 CO2p-values predicted retrospectively the actual changes in the annual TDgz-values (82%).[264]

Test 6 (Dataset 24)—randomized correlation CO2p and TDc.[265] In which: (CO2p)—predictor of annual temperature deviations (CO2p = CO2hz – TDgz); (CO2hz)—Hawaii z-score annual CO2 concentration; (TDgz)—Global z-score annual temperature deviation; (TDc)—change in annual temperature deviations (TDc = TDgz next year – TDgz current year); observation period from 1959 to 2014. At each iteration, the correlation coefficient was computed after CO_2p was randomized relative to TDc. Using z-scores avoids the statistical problems associated with different units of measurement.

Test 6 provided the following statistics for the randomized correlation coefficients: mean = 0.00; median = 0.00; standard deviation = 0.14; maximum = 0.52; minimum = -0.51; and range = 1.02 random r; n = 55; original r = 0.63; R = 0.40; p = 0.0000; i = 10,000.[266]

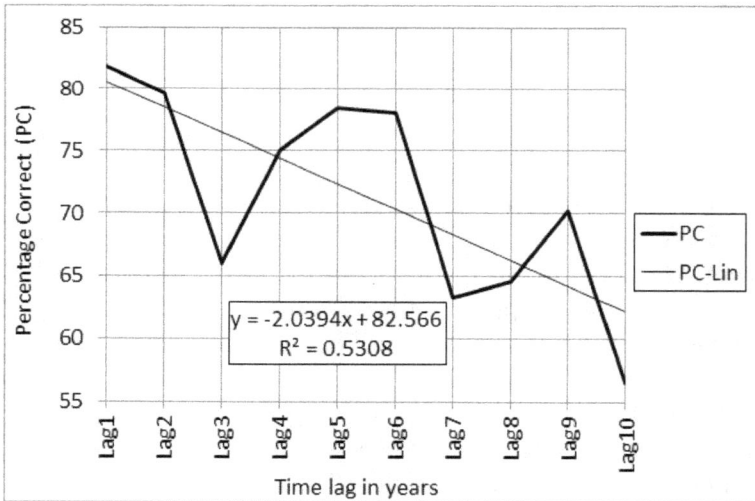

Figure 40 (Dataset 24)—CO2 predicts temperatures.[267] This graph shows: (PC)—percentage correct of retrospective predictions obtained with the predictor of change in global temperature (CO2p) from 1 to 10 years into the future; and (PC-Lin)—linear function of PC.

Test 7 (Dataset 24)—randomized probability test evaluated the significance of the percentage correctly predicted TDc by CO2p one year into the future.[268] In which: (CO2p)—predictor of annual temperature deviations (CO2p = CO2hz – TDgz); (CO2hz)—Hawaii z-score annual

CO_2 concentration; (TDgz)—Global z-score annual temperature deviation; (TDc)—change in annual temperature deviations (TDc = (TDgz at year + 1) − (TDgz at year)); observation period from 1959 to 2014. At each iteration, the percentage correctly predicted TDc by CO2p was computed after randomization. Using z-scores avoids the statistical problems associated with different units of measurement.

Test 7 provided the following statistics: original prediction = 82; mean 67; median = 67; standard deviation = 6; maximum = 93; minimum = 40; and range = 53 % correct; n = 55; i = 10,000; and p = 0.0026.[269]

The probability (p) is the likelihood that the percentage of correct predictions obtained with the randomized global annual temperatures is higher than the percentage of correct predictions obtained with the non-randomized global annual temperatures.

Test 8 (Dataset 24)—randomized probability test evaluated the significance of the percentage correctly predicted TDc by CO2p ten year into the future.[270] In which: (CO2p)—predictor of annual temperature deviations (CO2p = CO2hz − TDgz); (CO2hz)—Hawaii z-score annual CO_2 concentration; (TDgz)—Global z-score annual temperature deviation; (TDc)—change in annual temperature deviations (TDc = (TDgz at year + 10) − (TDgz at year)); observation period from 1959 to 2014. At each iteration, the percentage correctly predicted TDc by CO2p was computed after randomization. Using z-scores avoids the statistical problems associated with different units of measurement.

Test 8 provided the following statistics: original prediction = 57; mean 67; median = 67; standard deviation = 6; maximum = 87; minimum = 41; and range = 46 %; n = 46; i = 10,000; and p = 0.9189.[271]

Comments

The results suggest that the Hawaii annual CO_2 concentrations predict with statistical significance 82% of the changes in global annual temperature one year into the future. In addition, the results suggest that Hawaii annual CO_2 concentrations failed to predict the changes in the global annual temperatures ten years into the future. Furthermore, the significant correlation coefficient, between the CO_2 predictor of changes in atmospheric temperature and the actual changes in atmospheric temperature ($r = 0.63$), supports the hypothesis that the atmospheric CO_2 concentration drives the atmospheric temperature.

The linear function of the percentage correct predictions (PC-Lin) shows that the predictive power of CO_2 decreases with about 2% per year. If the predictive power was the result of the similar trends in CO_2 and temperature, then one would expect that the predictive power of CO_2 would remain constant across time lags. Hence, the results suggest that CO_2 is a predictor of future temperatures.

The results provide tentative support for the hypothesis that changes in the atmospheric CO_2 concentrations change proportionally the atmospheric temperature. However, both variables show a trend and the observation period is only 56 years. Therefore, more data are required urgently.

BILATERAL CLIMATE-CHANGE

VOSTOK CO2 PREDICTS VOSTOK TEMPERATURE

Method

The aim of this section is to develop a mathematical equation that predicts Vostok atmospheric temperatures based on Vostok atmospheric CO2 concentrations. To minimize the possible confounding effect of human activity during the current interglacial, Dataset 25 was created by truncating the observation period of Dataset 18.[272] The analyses in this section are based on Dataset 25.

Dataset 18 (VostokTempCO2Synch412Kyrs). This dataset contains the following Vostok variables: (Age)— from 414,085 to 2,342 years before calendar year 1989; (CO2v)—irregular-scale Vostok atmospheric carbon dioxide concentrations in parts per million (ppm); (TDv)—irregular-scale Vostok atmospheric temperature deviations in degrees Celsius (°C) relative to the Vostok base Kyr1 = 0.00 °C.

Dataset 25 (VostokTempCO2Synch403Kyrs).[273] This dataset contains the following Vostok variables: (Age)— from 414,085 to 11,013 years before calendar year 1989; (CO2v)—irregular-scale Vostok atmospheric carbon dioxide concentrations in parts per million (ppm); (TDv)—irregular-scale Vostok atmospheric temperature deviations in degrees Celsius (°C) relative to the Vostok base Kyr1 = 0.00 °C; (TDvCO2v)—irregular-scale Vostok atmospheric temperature deviation in degrees Celsius (°C) relative to the Vostok base Kyr1 = 0.00 °C, as predicted by the Vostok atmospheric CO2 concentrations substituted in Equation 6.

Results

Dataset 25 contains: (CO2v)—irregular-scale Vostok atmospheric carbon dioxide concentrations in parts per million (ppm); for the observation period from 414,085 to

11,013 years before calendar year 1989. The statistics are: mean 233.1; median = 233.2; standard deviation = 28.5; maximum = 298.7; minimum = 182.2; and range = 116.5 ppm; n = 276.[274] Furthermore, Dataset 25 contains: (TDv)—irregular-scale Vostok atmospheric temperature deviations in degrees Celsius (°C) relative to the Vostok base Kyr1 = 0.00 °C; for the observation period from 414,085 to 11,013 years before calendar year 1989. The statistics are: mean -3.71; median = -4.08; standard deviation = 3.03; maximum = 3.64; minimum = -8.69; and range = 12.33 °C; n = 276.[275]

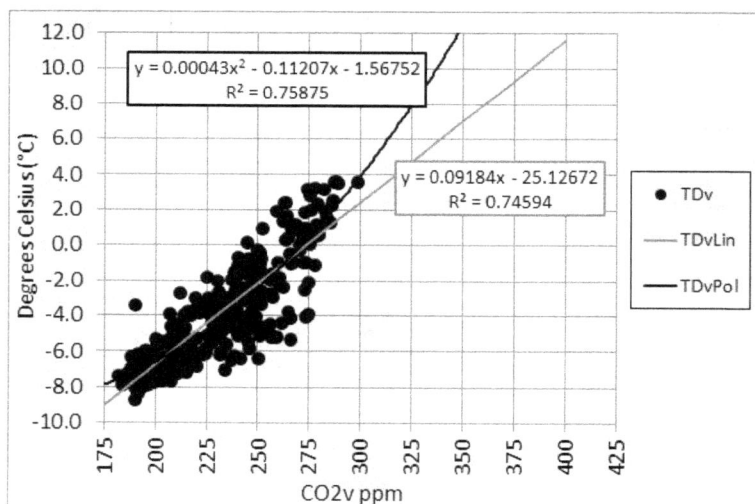

Figure 41 (Dataset 25)—Vostok temperature as a function of Vostok CO2.[276] This graph shows: (TDv)—irregular-scale Vostok atmospheric temperature deviations in degrees Celsius (°C) relative to the Vostok base Kyr1 = 0.00 °C; (CO2v)—irregular-scale Vostok atmospheric carbon dioxide concentrations in parts per million (ppm); (TDvLin)—linear function of TDv; and (TDvPol)—second order polynomial function of TDv. The observation period ranges from 412,096 to 9,024 BCE.[277]

The high R-values (0.76 and 0.75) support the notion that both functions explain equally well the variance in the data.

The linear function in Figure 41 was transformed into Equation 6 in order to predict the Vostok atmospheric temperatures with the Vostok atmospheric CO2 concentrations.

Equation 6 (Dataset 25)—estimates the long-term natural Vostok temperature for a given Vostok CO2 concentration: TDvCO2v = (0.0918422886772854 * CO2v) - 25.1267163418446.[278] In which: (TDvCO2v)—estimated Vostok atmospheric temperature deviation in degrees Celsius (°C) relative to the Vostok base Kyr1 = 0.00 °C; and (CO2v)—observed Vostok atmospheric carbon dioxide concentrations in parts per million (ppm).

At a CO2 concentration of 0.0 ppm, Equation 6 predicts that the temperature will be -25.13 °C. On the other hand, at a CO2 concentration of 400.0 ppm, Equation 6 predicts that the temperature will be 11.61 °C. Hence, the potential range of temperature affected by CO2 would be 36.74 °C.[279]

Equation 6 was transformed into Equation 7.

Equation 7 (Dataset 25)—estimates the long-term natural Vostok CO2 concentration for a given Vostok temperature: CO2vTDv = (TDv + 25.1267163418446) / 0.0918422886772854.[280] In which: (CO2vTDv)—estimated Vostok atmospheric carbon dioxide concentration in parts per million (ppm); (TDv)—observed Vostok atmospheric temperature deviation in degrees Celsius (°C) relative to the Vostok base Kyr1 = 0.00 °C.

Dataset 25 contains: (TDvCO2C)—Vostok atmospheric temperature deviation in degrees Celsius (°C) relative to the Vostok base Kyr1 = 0.00 °C, as predicted by Equation 6. The statistics are: mean = -3.71; median = -4.07; standard deviation = 2.64; maximum = 3.74; minimum = -7.56; and range = 11.30 °C; n = 276.[281]

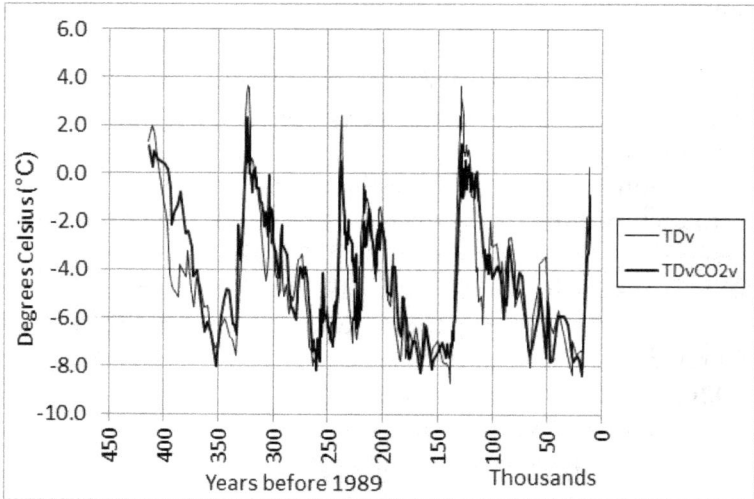

Figure 42 (Dataset 25)—predicted Vostok temperature.[282] This graph shows: (TDv)—irregular-scale Vostok atmospheric temperature deviations in degrees Celsius (°C) relative to the Vostok base Kyr1 = 0.00 °C; (TDvCO2v)— TDv predicted by substituting CO2v in Equation 6. The observation period ranges from 414,085 to 11,013 years before calendar year 1989.

Test 9 (Dataset 25)—randomized correlation TDv and TDvCO2v.[283] In which: (TDv)—Vostok ranked irregular-scale atmospheric temperature deviations in degrees Celsius (°C) relative to the Vostok base Kyr1 = 0.00 °C; (TDvCO2v)—Vostok ranked irregular-scale atmospheric temperature deviation in degrees Celsius (°C) relative to the Vostok base Kyr1 = 0.00 °C, as predicted by the Vostok atmospheric CO2 concentrations substituted in Equation 6; observation period from 414,085 to 11,013 before 1989. At each iteration, the correlation coefficient was computed after TDv was randomized relative to TDvCO2v. Using ranked data avoids the statistical problems associated with asymmetric frequency distributions, unequal variances, and outliers.

Test 9 provided the following statistics for the randomized correlation coefficients: mean = 0.00; median = 0.00; standard deviation = 0.06; maximum = 0.23; minimum = -0.23; and range = 0.46 random r; n = 276; original r = 0.86; R = 0.75; p = 0.0000; i = 10,000.[284]

Comments

The high R-value of the linear function of the Vostok atmospheric temperatures, and the significant correlation between the Vostok atmospheric temperatures and the Vostok atmospheric CO_2 concentrations, support the notion that the Vostok long-term Equation 6 is a reliable predictor of the atmospheric temperature. This equation supports the notion that CO_2 controls an atmospheric temperature range of more than 37.00 °C.[285] This range is huge in comparison to the alleged global warming of 1.32 °C in 2014.[286] Before we consider any adjustment of the atmospheric CO_2 concentration, we should carefully evaluate the possible consequences. The likelihood of increasing climate change towards its extremes is substantial. Further research in this direction is urgent and vital.

HAWAIIAN CO2 PREDICTS GLOBAL TEMPERATURE

Method

The aim of this section is to develop a mathematical equation that predicts global atmospheric temperatures based on Hawaiian atmospheric CO_2 concentrations. Dataset 26 was created by adding the predicted global atmospheric temperature deviations (TDgCO2h) to Dataset 24. The analyses in this section are based on Dataset 25.

Dataset 24 (GlobalTempHawaiiCO2-1959-2014). This dataset contains the following variables: (Year)—calendar years from 1959 to 2014; (TDg)—global annual atmospheric temperature deviations in degrees Celsius (°C) relative to the Vostok base Kyr1 = 0.00 °C; (CO2h)—Hawaiian atmospheric carbon dioxide concentrations in parts per million (ppm).

Dataset 26 (HawaiiCO2PredictGlobal°C).[287] This dataset contains the following variables: (Year)—calendar years from 1959 to 2014; (TDg)—global annual atmospheric temperature deviations in degrees Celsius (°C) relative to the Vostok base Kyr1 = 0.00 °C; (CO2h)—Hawaiian annual atmospheric carbon dioxide concentrations in parts per million (ppm); (TDgCO2h)—global atmospheric temperature deviations in degrees Celsius (°C) relative to the Vostok base Kyr1 = 0.00 °C, as predicted by the Hawaiian atmospheric CO2 concentrations substituted in Equation 8.

Results

Dataset 26 contains: (TDg)—global annual atmospheric temperature deviations in degrees Celsius (°C) relative to the Vostok base Kyr1 = 0.00 °C; from calendar year 1959 to 2014. The statistics are: mean 350.8; median = 348.3; standard deviation = 24.8; maximum = 398.6; minimum = 316.0; and range = 82.6 ppm; n = 56.[288] Furthermore, Dataset 26 contains: (CO2h)—Hawaiian atmospheric carbon dioxide concentrations in parts per million (ppm); from calendar year 1959 to 2014. The statistics are: mean 0.90; median = 0.87; standard deviation = 0.25; maximum = 1.32; minimum = 0.45; and range = 0.87 °C; n = 56.[289]

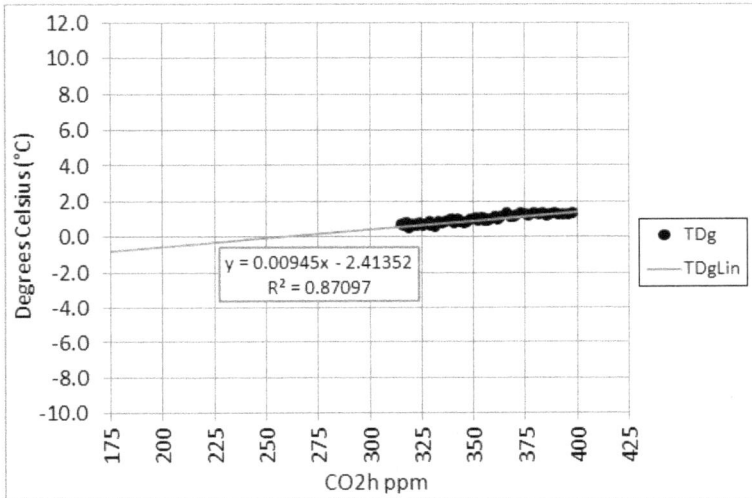

Figure 43 (Dataset 26)—global temperature as a function of Hawaiian CO2.[290] This graph shows: (TDg)—global annual atmospheric temperature deviations in degrees Celsius (°C) relative to the Vostok base Kyr1 = 0.00 °C; (CO2h)—Hawaiian atmospheric carbon dioxide concentrations in parts per million (ppm); (TDgLin)—linear function of TDg. The observation period ranges from calendar year 1959 to 2014.

The high R-value of 0.87 supports the notion that the linear function explains adequately the variance in the data.

The linear function in Figure 43 was transformed into Equation 8 in order to predict global atmospheric temperatures with the Hawaiian atmospheric CO2 concentrations.

Equation 8 (Dataset 26)—estimates the short-term artificial global temperature for a given Hawaiian CO2 concentration: TDgCO2h = (0.00944709321227046 * CO2h) - 2.41351639838839.[291] In which: (TDgCO2h)—estimated global atmospheric temperature deviation in degrees Celsius (°C) relative to the Vostok base Kyr1 = 0.00 °C; (CO2h)—

observed Hawaiian atmospheric carbon dioxide
concentration in parts per million (ppm).

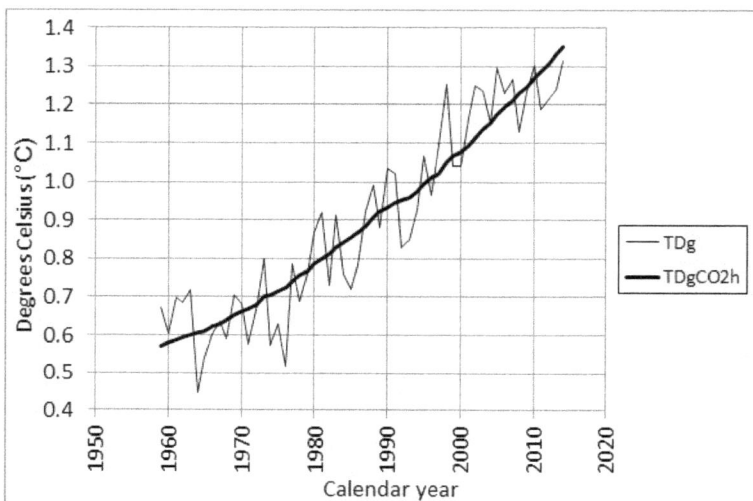

Figure 44: (Dataset 26)—predicted global temperature.[292]
This graph shows: (TDg)—global annual atmospheric
temperature deviations in degrees Celsius (°C) relative to the
Vostok base Kyr1 = 0.00 °C; (TDgCO2h)—global
atmospheric temperature deviation in degrees Celsius (°C)
relative to the Vostok base Kyr1 = 0.00 °C, as predicted by
the Hawaiian atmospheric CO2 concentrations substituted in
Equation 8. The observation period ranges from calendar
year 1959 to 2014.

Test 10 (Dataset 26)—randomized correlation TDg and
TDgCO2h.[293] In which: (TDg)—global ranked annual
atmospheric temperature deviations in degrees Celsius (°C)
relative to the Vostok base Kyr1 = 0.00 °C; (TDgCO2h)—
global ranked annual atmospheric temperature deviations in
degrees Celsius (°C) relative to the Vostok base Kyr1 = 0.00
°C, as predicted by the Hawaiian atmospheric CO2
concentrations substituted in Equation 8; observation period
from 1959 to 2014. At each iteration, the correlation

coefficient was computed after TDg was randomized relative to TDgCO2h. Using ranked data avoids the statistical problems associated with asymmetric frequency distributions, unequal variances, and outliers.

Test 10 provided the following statistics for the randomized correlation coefficients: mean = 0.00; median = 0.00; standard deviation = 0.14; maximum = 0.46; minimum = -0.49; and range = 0.95 random r; n = 56; original r = 0.92; R = 0.84; p = 0.0000; i = 10,000.[294]

The significant correlation coefficient and high R-value support the notion that substituting the recorded Hawaiian annual CO_2 concentration in the linear function of Figure 43 provides an adequate predictor of the associated global temperature.

Comments

The high R-value of the linear function of the global atmospheric temperatures, and the significant correlation between the global atmospheric temperatures and the Hawaiian atmospheric CO_2 concentrations, support the notion that Equation 8 is a reliable predictor of the global atmospheric temperature.

THERMAL GAP

Method

The aim of this section is to evaluate the difference between predictions of the atmospheric temperature based on the Vostok thermal linear function (Equation 6) versus the global thermal linear function (Equation 8). Dataset 24 and Dataset 25 were combined to create Dataset 27, which is the basis for the analyses in this section.

Dataset 24 (GlobalTempHawaiiCO2-1959-2014). This dataset contains the following variables: (Year)—calendar years from 1959 to 2014; (TDg)—global annual atmospheric temperature deviations in degrees Celsius (°C) relative to the

Vostok base Kyr1 = 0.00 °C; (CO2h)—Hawaiian atmospheric carbon dioxide concentrations in parts per million (ppm).

Dataset 25 (VostokTempCO2Synch403Kyrs). This dataset contains the following Vostok variables: (Age)—from 414,085 to 11,013 years before calendar year 1989; (CO2v)—irregular-scale Vostok atmospheric carbon dioxide concentrations in parts per million (ppm); (TDv)—irregular-scale Vostok atmospheric temperature deviations in degrees Celsius (°C) relative to the Vostok base Kyr1 = 0.00 °C; (TDvCO2v)—irregular-scale Vostok atmospheric temperature deviation in degrees Celsius (°C) relative to the Vostok base Kyr1 = 0.00 °C, as predicted by the Vostok atmospheric CO2 concentrations substituted in Equation 6.

Dataset 27 (ThermalGap).[295] This dataset contains the following variables: (Year)—calendar years from 412,096 BCE to 2014; (TDg)—global annual atmospheric temperature deviations in degrees Celsius (°C) relative to the Vostok base Kyr1 = 0.00 °C, and (CO2h)—Hawaiian atmospheric carbon dioxide concentrations in parts per million (ppm), from 1959 to 2014; (CO2v)—irregular-scale Vostok atmospheric carbon dioxide concentrations in parts per million, and (TDv)—irregular-scale Vostok atmospheric temperature deviations in degrees Celsius (°C) relative to The Vostok base Kyr1 = 0.00 °C, from 412,096 to 9,024 BCE. This dataset is included in appendix: Dataset 27.

Results

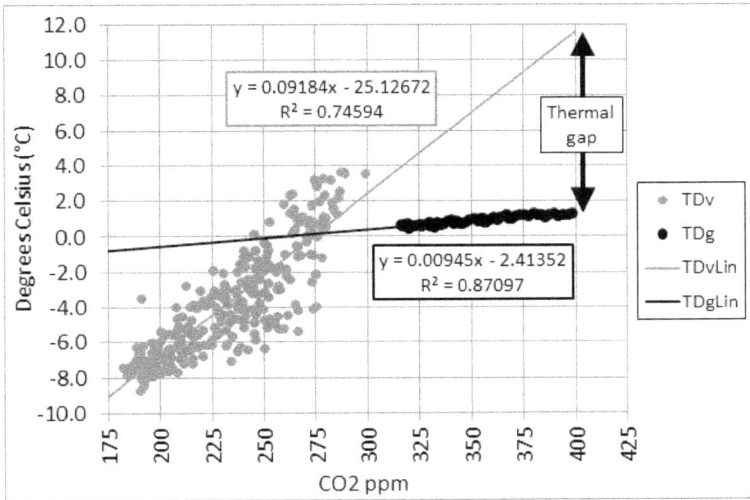

Figure 45 (Dataset 27)—thermal gap.[296] This graph shows: (TDv)—irregular-scale atmospheric temperature deviations, from 412,096 to 9,024 BCE, in degrees Celsius (°C) relative to The Vostok base Kyr1 = 0.00 °C; (TDg)—global annual atmospheric temperature deviations, from 1959 to 2014, in degrees Celsius (°C) relative to the Vostok base Kyr1 = 0.00 °C; (TDvLin)—linear function of TDv; (TDgLin)—linear function of TDg; and (CO2 ppm)—atmospheric carbon dioxide concentrations in parts per million.

Figure 45 shows that the Vostok temperature deviations as a linear function of CO2 concentrations (TDvLin = 0.09184 °C/ppm) is ten times steeper than the slope of the global temperature deviations as a linear function of CO2 concentrations (TDgLin = 0.00945 °C/ppm).[297] This difference was tested with a Sliding Window Test (SWT).

Test 11 (Dataset 27)—sliding window test (SWT), comprising 56 data pairs shifting in steps of 1 pair, evaluated whether the Vostok natural long-term linear function (TDvLin = 0.09184 °C/ppm) is significantly larger than the global artificial short-term linear function of (TDgLin = 0.00945 °/ppm).[298] In which: (TDvLin)—linear function of TDv as a function of CO2v; (TDv)—irregular-scale Vostok atmospheric temperature deviations in degrees Celsius (°C) relative to the Vostok base Kyr1 = 0.00 °C; (CO2v)— irregular-scale Vostok atmospheric carbon dioxide concentrations in parts per million; and observation period TDv and CO2v from 412,096 to 9,024 BCE; versus (TDgLin)—linear function of TDg as a function of CO2h; (TDg)—global annual atmospheric temperature deviations in degrees Celsius (°C) relative to the Vostok base Kyr1 = 0.00 °C; (CO2h)—Hawaiian atmospheric carbon dioxide concentrations in parts per million (ppm); and observation period TDg and CO2h from 1959 to 2014.

Test 11 provided the following statistics for the obtained TDvLin values: mean = 0.09511; median = 0.09626; standard deviation = 0.01322; maximum = 0.11713; minimum = 0.05669; and range = 0.06044 °C/ppm; n = 221; p = 0.000.[299]

Hence, the results of Test 11 suggest that the slope of the TDvLin is significantly steeper than the slope of the TDgLin.

In 2014, the annual atmospheric CO2 concentration at Hawaii reached 398.6 ppm.[300] At that CO2 concentration, the global linear function (TDgLin) predicts an atmospheric temperature of 1.37 °C relative to the Vostok base Kyr1 = 0.00 °C.[301] On the other hand, the Vostok linear function (TDvLin) predicts for the same CO2 concentration of 398.6 ppm an atmospheric temperature of 11.48 °C relative to the Vostok base Kyr1 = 0.00 °C.[302] Consequently, the thermal gap between the two predictions equals 11.48 - 1.35 = 10.13 °C.[303]

Test 12 (Dataset 25)—randomized correlation TDv and CO2v.[304] In which: (TDv)—Vostok ranked irregular-scale atmospheric temperature deviations in degrees Celsius (°C) relative to the Vostok base Kyr1 = 0.00 °C; (CO2v)—Vostok ranked irregular-scale atmospheric carbon dioxide concentrations in parts per million (ppm); observation period from 414,085 to 11,013 years before 1989. At each iteration, the correlation coefficient was computed after CO2v was randomized relative to TDv. Using ranked data avoids the statistical problems associated with asymmetric frequency distributions, unequal variances, and outliers.

Test 12 provided the following statistics for the randomized correlation coefficients: mean = 0.00; median = 0.00; standard deviation = 0.06; maximum = 0.23; minimum = -0.21; and range = 0.44 random r; n = 276; original r = 0.86; R = 0.75; p = 0.0000; i = 10,000.[305]

Test 13 (Dataset 26)—randomized correlation TDg and CO2h.[306] In which: (TDg)—global ranked annual atmospheric temperature deviations in degrees Celsius (°C) relative to the Vostok base Kyr1 = 0.00 °C; (CO2h)— Hawaiian ranked annual atmospheric carbon dioxide concentrations in parts per million (ppm); observation period from 1959 to 2014. At each iteration, the correlation coefficient was computed after TDg was randomized relative to CO2h. Using ranked data avoids the statistical problems associated with asymmetric frequency distributions, unequal variances, and outliers.

Test 13 provided the following statistics for the randomized correlation coefficients: mean = 0.00; median = 0.00; standard deviation = 0.13; maximum = 0.46; minimum = -0.50; and range = 0.95 random r; n = 56; original r = 0.92; R = 0.84; p = 0.0000; i = 10,000.[307]

Thermal outliers are temperature deviations that are both higher than the global maximum of 1.32 °C in 2014 and associated with CO2 concentrations below the Hawaiian maximum of 398.6 ppm in 2014.[308] Thermal outliers obtained the following statistics: mean 2.50; median = 2.40; standard deviation = 0.79; maximum = 3.64; minimum = 1.32; and range = 2.32 °C; n = 21.[309] The statistics for the associated CO2 concentrations are: mean 277.0; median = 278.2; standard deviation = 11.0; maximum = 298.7; minimum = 259.0; and range = 39.7 ppm; n = 21.[310] The statistics for the associated Age are: mean 230,600; median =

237,831; standard deviation = 106,933; maximum = 410,831; minimum = 126,475; and range = 284,356 years before 1989; n = 21.[311]

In order to reduce the recent global temperature of 1.32 °C, President Obama plans to decrease the CO_2 concentrations with about 30% (120 ppm) by 2025.[312] The thermal effect of Obama's plan depends on the function between CO_2 and temperature. Equation 8 represents the confounded short-term linear global function (TDgLin), which predicts for a CO_2 decrease of 120 ppm in CO_2 a decrease of 1.13 °C in temperature.[313] Hence, this result would roughly restore the Vostok base Kyr1 = 0.00 °C. On the other hand, Equation 6 represents the long-term linear Vostok function (TDvLin), which predicts for a CO_2 decrease of 120 ppm in CO_2 a decrease of 11.02 °C in temperature.[314]

Comments

The thermal gap is the difference between the estimated and actual annual global atmospheric temperature deviations. The conventional unilateral climate-change hypothesis holds that industrial greenhouse gasses, such as carbon dioxide (CO_2), increase the temperature by trapping heat in the atmosphere.[315] However, this hypothesis cannot explain adequately the huge thermal gap. Therefore, I propose the bilateral climate-change hypothesis, which holds that the effect of artificial global warming compensates for the effect of natural global cooling. Low glacial temperatures are increased by solar heat that is trapped by domestic and industrial atmospheric greenhouse gasses such as carbon dioxide (CO_2). In 2014, the thermal effect of artificial global warming did overcompensate the thermal effect of natural global cooling.

ONSET THERMAL GAP

Method

The aim of this section is to evaluate the onset of the previously mentioned thermal gap. The thermal gap was defined as the difference between the estimated and actual annual global atmospheric temperature deviations. The analyses in this section are based on Dataset 18.

Dataset 18 (VostokTempCO2Synch412Kyrs). This dataset contains the following Vostok variables: (Age)— from 414,085 to 2,342 years before calendar year 1989; (CO2v)—irregular-scale Vostok atmospheric carbon dioxide concentrations in parts per million (ppm); (TDv)—irregular-scale Vostok atmospheric temperature deviations in degrees Celsius (°C) relative to the Vostok base Kyr1 = 0.00 °C.

Results

Vostok CO2 and temperature (8 Kyrs)

Dataset 18 provides the statistics for the Vostok CO2 concentrations in ppm (CO2v) from 8,134 to 353 BCE: mean 266.2; median = 262.2; standard deviation = 10.0; maximum = 284.7; minimum = 254.6; and range = 30.1 ppm; n = 7.[316] Furthermore, Dataset 18 provides the statistics for the associated Vostok temperature deviations in degrees Celsius (TDv) from 8,134 to 353 BCE: mean 0.34; median = 0.09; standard deviation = 0.76; maximum = 1.64; minimum = -0.64; and range = 2.28 °C; n = 7.[317]

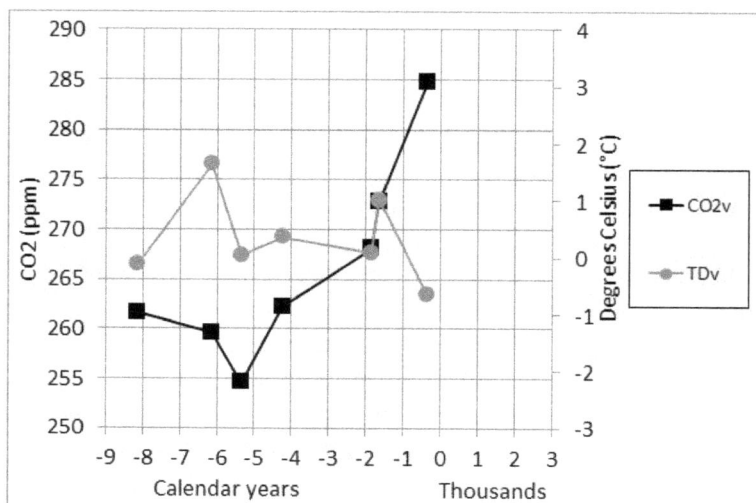

Figure 46 (Dataset 18)—Vostok CO2 and temperature (10 Kyrs).[318] This graph shows, for the observation period from 8,134 to 353 BCE: (CO2v)—irregular-scale Vostok atmospheric carbon dioxide concentrations in parts per million (ppm); (TDv)—irregular-scale Vostok atmospheric temperature deviations in degrees Celsius (°C) relative to the Vostok base Kyr1 = 0.00 °C.

To predict the natural temperatures (TDvCO2v) during the observation period from 8,134 BCE to 2014 CE, the Vostok CO2 concentrations (COv) in Dataset 18 were substituted in the Vostok long-term Equation 6. This provided the following statistics for TDvCO2v: mean -0.68; median = -1.05; standard deviation = 0.92; maximum = 1.02; minimum = -1.74; and range = 2.76 °C; n = 7.[319] However, those statistics might be spurious, because the data intervals of the CO2 reconstructions in Dataset 18 are irregular, while the total number of data pairs of CO2 and associated temperatures is rather small for the intended purpose. Nevertheless, it is as good as it gets. The equations describing the relationship between CO2 concentrations and temperatures will be computed next.

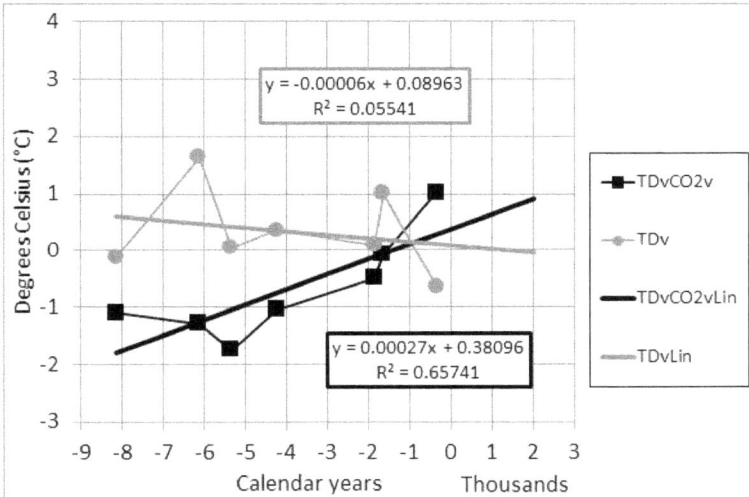

$y = -0.00006x + 0.08963$
$R^2 = 0.05541$

$y = 0.00027x + 0.38096$
$R^2 = 0.65741$

Figure 47 (Dataset 18)—Vostok CO2 and temperature (10 Kyrs).[320] This graph shows, for the observation period from 8,134 to 353 BCE: (TDvCO2v)—irregular-scale Vostok atmospheric temperature deviation in degrees Celsius (°C) relative to the Vostok base Kyr1 = 0.00 °C, as predicted by the Vostok atmospheric CO2 concentrations substituted in Equation 6; (TDv)—irregular-scale Vostok atmospheric temperature deviations in degrees Celsius (°C) relative to the Vostok base Kyr1 = 0.00 °C; (TDvCO2vLin)—linear function of TDvCO2v; (TDvLin)—linear function of TDv.

Although linear functions are preferable for their simplicity, Figure 47 shows that the linear function TDvCO2vLin explains about 66% of the variance in the data, while the linear function TDvLin explains only 6%. To increase those R-values, both linear functions were changed into second order polynomial functions.

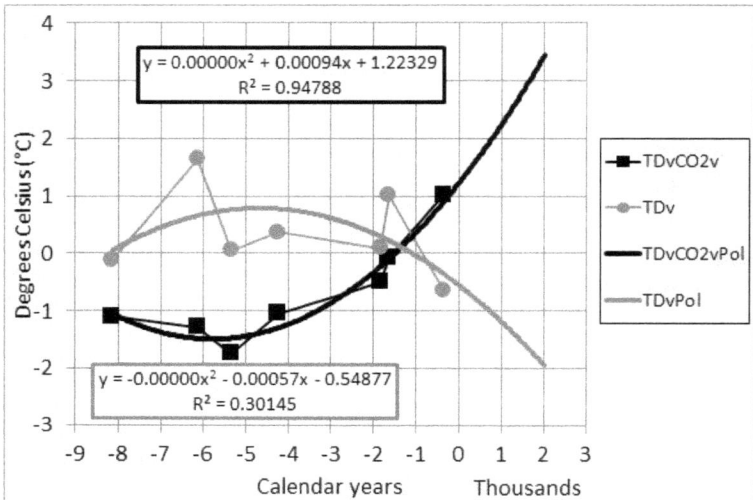

Figure 48 (Dataset 18)—Vostok CO2 and temperature (10 Kyrs).[321] This graph shows, for the observation period from 8,134 to 353 BCE: (TDvCO2v)—irregular-scale Vostok atmospheric temperature deviation in degrees Celsius (°C) relative to the Vostok base Kyr1 = 0.00 °C, as predicted by the Vostok atmospheric CO2 concentrations substituted in Equation 6; (TDv)—irregular-scale Vostok atmospheric temperature deviations in degrees Celsius (°C) relative to the Vostok base Kyr1 = 0.00 °C; (TDvCO2vLin)—second order polynomial function of TDvCO2v; (TDvLin)—second order polynomial function of TDv.

Figure 48 shows that the second order polynomial function TDvCO2vPol explains about 95% of the variance in the data, while the linear function TDvPol explains 30%. Although the fit of TDvPol to the data is better than the fit of TDvLin, the 30% explained variance is still low. Further research should test the tentative results presented in this study. More CO2 concentrations, concerning the observation period from 8,134 to 353 BCE, should be extracted from the Vostok ice as soon as possible.

Figure 48 shows that the polynomial functions TDvCO2vLin and TDvLin cross each other at about 1,500 BCE. After that time, the temperature predictions based on the Vostok long-term CO2 concentrations are higher than the temperature predictions based on the actual temperatures. This support the notion that the thermal gap started at about 1,500 BCE. The polynomial functions in Figure 48 predict that the thermal gap would reach 5.39 °C in 2014.[322] The last data pair included in this analysis is dated 353 BCE and, therefore, this thermal gap cannot be explained by recent industrial activity. Hence, the onset of this thermal gap supports the notion that domestic CO2 has affected the climate for thousands of years.

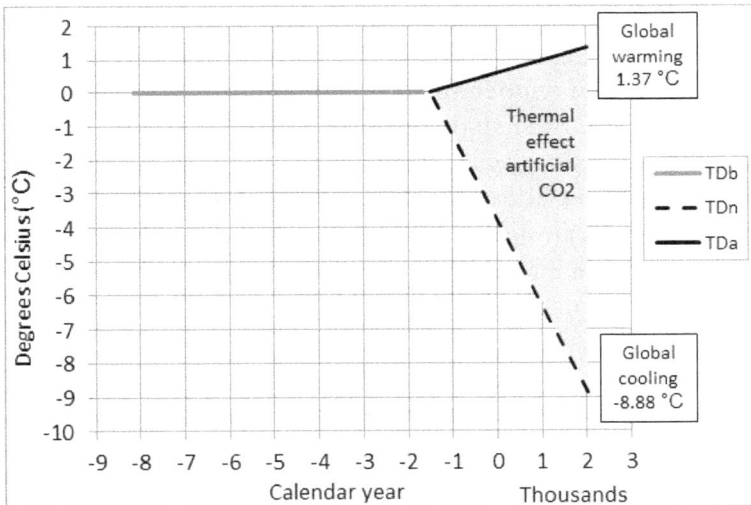

Figure 49 (Dataset 28)—thermal effect of artificial CO2 during the current interglacial.[323] This graph shows: (TDb)—Vostok base Kyr1 = 0.00 degrees Celsius (°C); (TDn)—hypothesized natural glacial decline of atmospheric temperature deviations in degrees Celsius (°C) relative to the Vostok base Kyr1 = 0.00 °C; (TDa)—artificial atmospheric temperature deviations, as estimated by Equation 8, in

degrees Celsius (°C) relative to the Vostok base Kyr1 = 0.00 °C.

Comment

Onset

Figure 47 shows an uninterrupted increase in Vostok atmospheric CO_2 since 5,338 BCE. This increase might indicate the release of artificial domestic CO_2 and the onset of the hypothesized thermal gap. Furthermore, Figure 48 shows that the polynomial functions of the Vostok actual temperatures, and the temperatures predicted by the Vostok natural long-term Equation 6, cross at about 1,500 BCE. Hence, the data support the notion that the thermal gap started sometime between 5,338 and 1,500 BCE.

Figure 49 illustrates the idea that the observed temperature has been considerably decreased by the hypothesized natural glacial decline in temperature. If that compensation would be confirmed by further study, then humanity exists on a knives' edge. Without the glacial decline, the current temperature would have been 11.61 °C, which is much higher than the alleged global warming of 1.32 °C as observed in 2014. The results of such a high temperature would drive humanity to the edge of extinction.

The results suggest that the hypothesized glacial decline has almost reached the all-time glacial low. When that decline stops, the predicted thermal effect of CO_2 will switch from the global confounded short-term Equation 8 to the Vostok natural long-term Equation 6. This switch would multiply the thermal effect of any further increase in artificial CO_2 by a magnitude of ten. Furthermore, Figure 2 shows that the glacial temperature is unstable. Therefore, the possibility is not excluded that the decrease in natural glacial temperature will turnaround in the near future. Such an increase, in synchrony with the currently increasing artificial CO_2 concentration, could push the actual temperature beyond all-time highs. Unfortunately, the data show that this devastation could start at any time, without further warning, and reach critical levels within a few years.

The estimates about the onset of the thermal gap are based on only 7 data pairs of natural CO_2 concentrations and temperatures, which are unequally divided over the current interglacial period of about 10,000 years. The ensuing long and irregular data intervals are inadequate for a valid and reliable statistical analysis. Nevertheless, if these alarming results are true, then reliable data are crucial for the understanding of climate change. It is imperative that research is directed urgently towards the extraction of additional CO_2 concentrations from the Vostok ice.

It could be argued that the thermal gap is the result of unreliable linear functions. However, the slope of the Vostok temperature deviations as a linear function of the Vostok CO_2 concentrations (0.09184 °C/ppm) is significant and ten times steeper than the slope of the recent global temperature deviations as a function of the Hawaiian CO_2 concentrations (0.00945 °C/ppm).[324] The high R-values of these linear functions show that each one explains adequately the variance in the data.

The global temperature deviation reached 1.32 °C in 2014.[325] In the same year, the CO_2 concentration at Hawaii reached 398.6 ppm.[326] Based on that concentration, the historical Vostok linear function estimates an atmospheric temperature of 11.48 °C in 2014.[327] Consequently, the linear functions predict retrospectively for 2014 a thermal gap of 10.16 °C (11.48 – 1.32 °C).[328]

Compared to the alleged global warming of 1.32 °C, the magnitude of the thermal gap is astonishing. Nevertheless, the second order polynomial function of the Vostok temperature deviations as a function of CO_2 concentrations estimated for 2014 an even higher temperature of 22.82 °C.[329] This suggests that the linear estimate is a conservative minimum.

It could be argued that the recent global temperature of 1.32 °C shows that the prediction with the Vostok linear function of 11.48 °C in 2014 is incorrect.[330] However, the Vostok ice-core datasets are widely accepted and the resulting linear function is based on 276 data pairs across an observation period of more than 400,000 years.[331] That linear function is based on pre-industrial natural conditions spanning several glacial and interglacial periods. In contrast, the global linear function is based on only 56 annual data pairs relating exclusively to interglacial conditions confounded by artificial domestic and industrial variables. Consequently, the Vostok linear function is at least as reliable as the global linear function.

Compared with the steep slope of the Vostok linear function, the almost flat global linear function could indicate that CO_2 has reached a thermal ceiling effect. In that case, the dangers of global warming would be limited. However, the polynomial function of the Vostok temperature deviations suggests that the increase in temperature accelerates in response to increasing CO_2 values.[332] This acceleration suggests that that the recent relatively low global temperature of 1.32 °C is not caused by a thermal ceiling of CO_2.

Furthermore, twenty-one thermal outliers provide valid information about the natural relationship between temperature and CO_2, because they are spread over the four previous interglacials.[333] These outliers show that temperatures above the recent maximum of 1.32 °C have occurred naturally below the Vostok maximum CO_2 concentration of 298.7 ppm.[334] Nevertheless, this Vostok concentration is 25% lower than the recent maximum CO_2 concentration in Hawaii of 398.6 ppm.[335] Hence, the results do not support the notion of a CO_2 thermal ceiling.

The thermal gap could also be explained by a delayed thermal response to the increasing atmospheric CO_2 concentration. However, atmospheric temperatures differ so much between day and night, and between summer and winter, that it seems to be highly unlikely that it would take more than 56 years for the atmospheric temperature to adjust to the slowly increasing CO_2 concentration.

It could also be argued that global warming is melting the polar ice-sheets and glaciers. That melting ice could work as a gigantic heat-sink that cools down the atmosphere by transferring energy from the atmosphere to the oceans. That heat-sink could explain the flat slope of the global linear function. However, the Vostok data show that during each interglacial, temperatures have exceeded the current global temperature without changing the Vostok linear function between CO_2 and temperature. Consequently, the data reject the notion that a natural heat-sink could explain the thermal gap.

An explanation for the thermal gap could be that CO_2 does not affect the temperature significantly. If true, then the unilateral- and bilateral hypotheses should be rejected. However, the results suggest that in 82% of the cases, the CO_2 concentration of any particular year predicts correctly the temperature of the following year.[336] Furthermore, the results show that the correlation coefficients between the temperature deviations and CO_2 concentrations for both the historical Vostok dataset and the recent global-Hawaiian dataset are positive, high, and significant. The correlation between CO_2 and temperature remained about the same, while the slope of their function decreased significantly. Hence, the results support the notion that a new variable affects the recent thermal CO_2 function.

Alternatively, the thermal gap could be explained by a difference between the thermal effect of natural CO_2 and industrial CO_2. If the thermal effect of the recent industrial CO_2 would be considerably smaller than the thermal effect of the Vostok long-term natural CO_2, then one would expect a lower current temperature than predicted by the long-term Vostok linear function. However, as far as I know, scientists have found no significant difference between the thermal effects of natural and artificial CO2. Hence, a difference between the characteristics of artificial and natural CO2 is unlikely to explain the thermal gap.

The thermal gap could also be explained by confounding variables such as volcanic and industrial pollutants. Those substances could block a part of solar energy from entering the atmosphere and, therefore, reduce the greenhouse effect. However, natural confounding variables make a weak argument, because they are likely to have affected equally the long-term Vostok data and the short-term global data. Although industrial confounding variables could provide an alternative explanation for the thermal gap, industrial pollution started about 150 to 200 years ago. Hence, industrial variables could not explain the significant long duration and thermal stability of the current interglacial. Neither could they account for the onset of the thermal gap about 3,500 years ago. Hence, the proposed bilateral climate-change hypothesis provides a better explanation for the thermal gap. Nevertheless, further research is required urgently to examine the possibility of short-term confounding variables.

THERMAL EFFECT CO2 REDUCTION

In 2014, the global annual temperature reached 1.32 degrees Celsius (°C) relative to the Vostok base $Kyr1 = 0.00$ °C, at a CO2 concentration of 398.6 parts per million (ppm).[337]

President Obama plans to decrease the CO2 concentrations by 2025 with 30% from 400 to 280 ppm.[338] Unfortunately, his plan could backfire. For example, if 100% of the CO2 would be removed from the atmosphere, then the long-term Vostok Equation 6 predicts a temperature of -25.13 °C.[339] Hence, this suggests that Obama's reduction of 30% has the potential to create a glacial period or ice-age.

The global short-term Equation 8 predicts that Obama's target to reach a CO2 concentration of 280 ppm will produce a temperature of 0.23 °C.[340]   On the other hand, the Vostok long-term Equation 6 predicts that a CO_2 concentration of 280 ppm will produce a temperature of 0.59 °C.[341] Apparently, both equations support President Obama's decision. Nevertheless, this study suggests that his plan is unlikely to restore the Vostok base Kyr1 = 0.00 °C.

At a CO_2 concentration of 400 ppm, the short-term Equation 8 predicts a global temperature of 1.37 °C, which is in accord with the observed global temperature in 2014 of 1.32 °C.[342]   In contrast, the long-term Equation 6 predicts for the same CO_2 concentration a temperature of 11.61 °C. Hence, for a CO_2 concentration of 280 ppm, Equation 8 predicts a decrease in temperature of 1.13 °C (1.37 – 0.23), while Equation 6 predicts a decrease of 11.02 °C (11.61 – 0.59).[343]   This large difference shows that those equations cannot be applied to the same situation. At least, one equation is invalid for President Obama's plan.

Method

It is the aim of this section to evaluate the reduction in temperature resulting from a reduction in artificial CO_2. The notion will be evaluated that the long-term Vostok Equation 6 and the short-term global Equation 8 are both invalid for President Obama's plan. The analyses in this section are based on Dataset 28.

Dataset 28 (CO_2 forged thermal decrease).[344] This dataset contains the following Vostok variables: (CO_2)—atmospheric carbon dioxide concentrations in parts per million (ppm); (TDvLin)—linear function of the Vostok atmospheric temperature deviations, from 414,085 to 11,013 years before calendar year 1989, in degrees Celsius (°C) relative to the Vostok base Kyr1 = 0.00 °C; (TDgLin)—linear function of the global annual atmospheric temperature deviations, from 1959 to 2014, in degrees Celsius (°C) relative to the Vostok base Kyr1 = 0.00 °C; (TDa)—artificial atmospheric temperature deviation, as predicted by Equation 8, in degrees

Celsius (°C) relative to the Vostok base Kyr1 = 0.00 °C, at an atmospheric CO2 concentration of 400 parts per million (ppm) including artificial CO2; (TDn)—natural atmospheric temperature deviation in degrees Celsius (°C) relative to the Vostok base Kyr1 = 0.00 °C (TDn = TDa—thermal gap); and (TDdLin)—atmospheric temperature deviations in degrees Celsius (°C) relative to the Vostok base Kyr1 = 0.00 °C as a linear function of CO2 concentrations between TDa and TDn.

Results

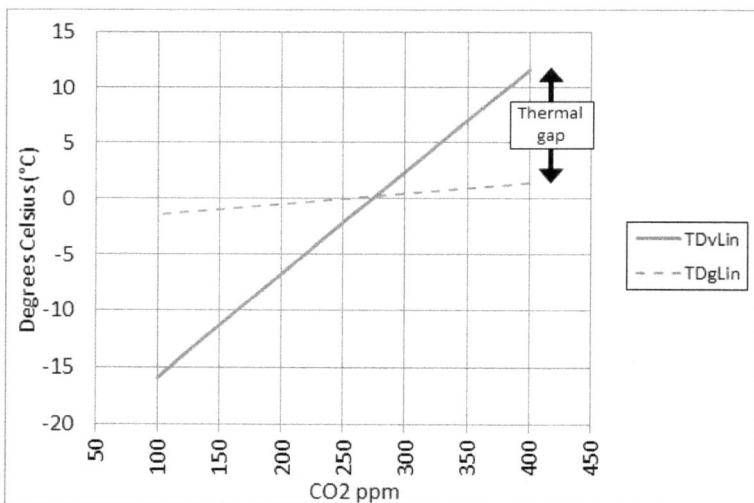

Figure 50 (Dataset 28)—predicted thermal gap.[345] This graph shows: (TDvLin)—linear function of the Vostok atmospheric temperature deviations, from 414,085 to 11,013 years before calendar year 1989, in degrees Celsius (°C) relative to the Vostok base Kyr1 = 0.00 °C; and (TDgLin)— linear function of the global annual atmospheric temperature deviations, from 1959 to 2014, in degrees Celsius (°C) relative to the Vostok base Kyr1 = 0.00 °C.

At a CO2 concentration of 400 ppm, the Vostok Equation 6 (TDvLin) predicts a natural temperature of 11.61 °C.[346] On the other hand, the global Equation 8 (TDgLin) predicts an artificial temperature of 1.37 °C.[347]   However, that temperature is confounded, because artificial CO2 has been released in the atmosphere affecting Equation 8. In accord, the thermal gap equals 10.24 °C (11.61 - 1.37) at the current CO2 concentration of 400 ppm.[348]   Hence, the data support the notion that a confounding variable has decreased the current natural temperature with 10.24 °C. The data suggest that this variable is not related to CO2. Therefore, it is proposed that the hidden decrease in the current natural temperature is due to natural glacial conditions.

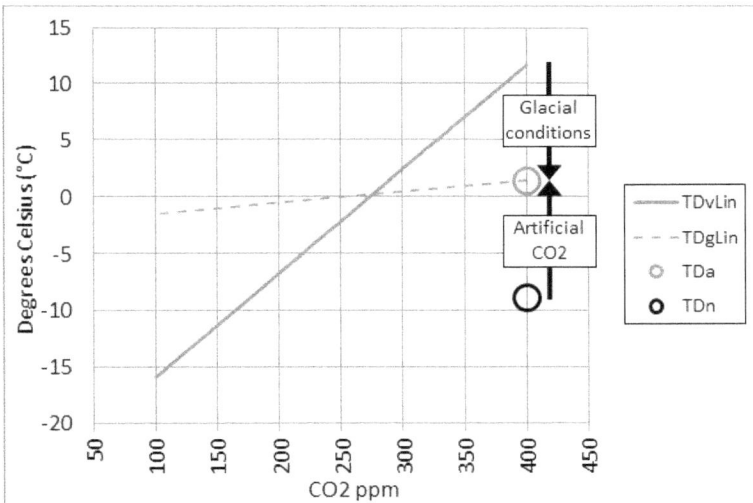

Figure 51 (Dataset 28)—thermal action equals reaction.[349] This graph shows: (TDvLin)—linear function of the Vostok atmospheric temperature deviations, from 414,085 to 11,013 years before calendar year 1989, in degrees Celsius (°C) relative to the Vostok base Kyr1 = 0.00 °C; (TDgLin)—linear function of the global annual atmospheric temperature deviations, from 1959 to 2014, in degrees Celsius (°C) relative to the Vostok base Kyr1 = 0.00 °C; (TDa)—artificial atmospheric temperature deviation, as predicted by Equation

8, in degrees Celsius (°C) relative to the Vostok base Kyr1 = 0.00 °C, at an atmospheric CO2 concentration of 400 parts per million (ppm) including artificial CO2; and (TDn)—natural atmospheric temperature deviation in degrees Celsius (°C) relative to the Vostok base Kyr1 = 0.00 °C (TDn = TDa—thermal gap).

Action equals reaction. If glacial conditions have decreased the current natural temperature (TDn) with 10.24 °C, then the hidden natural temperature would be 10.24 °C lower than the observable current artificial temperature (TDa) of 1.37 °C. Hence, the data suggest that the current natural temperature would be -8.88 °C (1.37 - 10.24).[350] It is proposed that the natural glacial temperature is compensated for by the thermal effect of artificial CO2 concentrations. Hence, the observed current temperature of 1.37 °C is the precarious balance between huge natural and artificial forces.

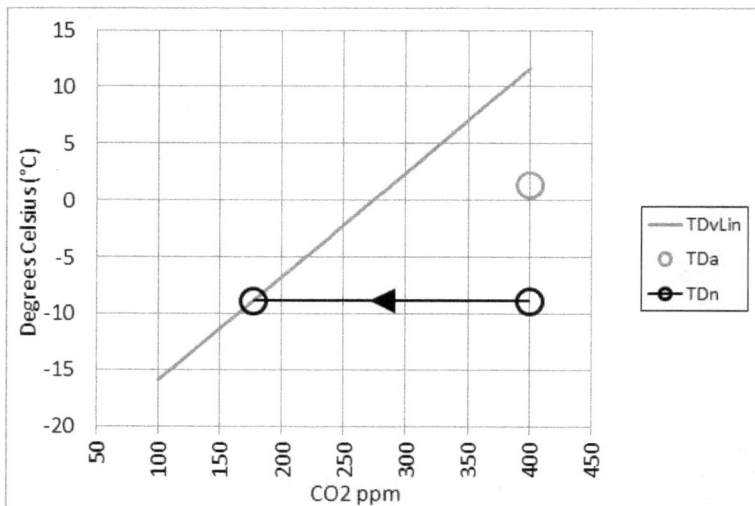

Figure 52 (Dataset 28)—natural CO2 concentration at the predicted natural glacial temperature.[351] This graph shows: (TDvLin)—linear function of the Vostok atmospheric temperature deviations, from 414,085 to 11,013 years before

calendar year 1989, in degrees Celsius (°C) relative to the Vostok base Kyr1 = 0.00 °C; (TDa)—artificial atmospheric temperature deviation, as predicted by Equation 8, in degrees Celsius (°C) relative to the Vostok base Kyr1 = 0.00 °C, at an atmospheric CO2 concentration of 400 parts per million (ppm) including artificial CO2; and (TDn)—natural atmospheric temperature deviation in degrees Celsius (°C) relative to the Vostok base Kyr1 = 0.00 °C (TDn = TDa—thermal gap).

Equation 7 predicts for the hidden current natural temperature (TDn) of -8.88 °C, a natural CO2 concentration of 176.9 ppm.[352]

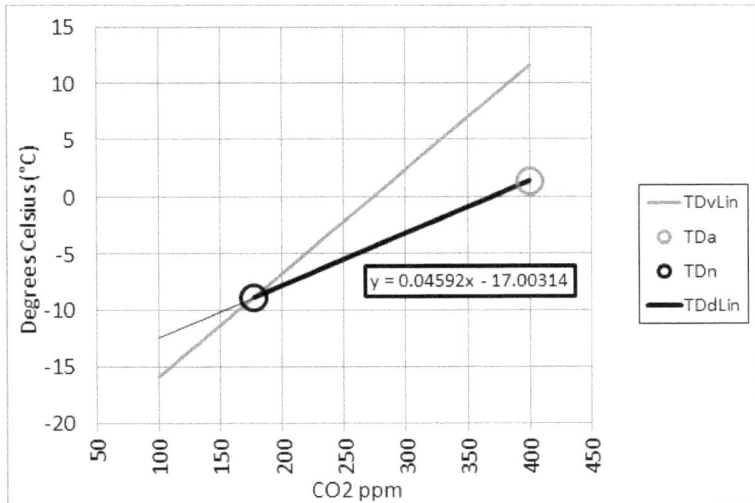

Figure 53 (Dataset 28)—linear equation thermal decrease towards natural balance.[353] This graph shows: (TDvLin)—linear function of the Vostok atmospheric temperature deviations, from 414,085 to 11,013 years before calendar year 1989, in degrees Celsius (°C) relative to the Vostok base Kyr1 = 0.00 °C; (TDa)—artificial atmospheric temperature deviation, as predicted by Equation 8, in degrees Celsius (°C) relative to the Vostok base Kyr1 = 0.00 °C, at an atmospheric

CO_2 concentration of 400 parts per million (ppm) including artificial CO_2; (TDn)—natural atmospheric temperature deviation in degrees Celsius (°C) relative to the Vostok base Kyr1 = 0.00 °C (TDn = TDa—thermal gap); and (TDdLin)—atmospheric temperature deviations in degrees Celsius (°C) relative to the Vostok base Kyr1 = 0.00 °C as a linear function of CO_2 concentrations between TDa and TDn.

The straight line shown in Figure 53, which connects the artificial temperature (TDa) and the natural temperature (TDn), represents the decrease in temperature resulting from a decrease in artificial CO_2. As shown in the graph, this decrease equates the linear function (TDdLin): y = 0.0459211443386427x - 17.0031368489373.

TDdLin was transformed into Equation 9.

Equation 9 (Dataset 28)—estimates the temperature for a given CO_2 concentration within the range from TDa to TDn: TDdCO2d = (0.0459211443386428 * CO2d) - 17.0031368489373.[354] In which: (TDdCO2d)—estimated decreased atmospheric temperature deviation in degrees Celsius (°C) relative to the Vostok base Kyr1 = 0.00 °C; (CO2d)—decreased atmospheric carbon dioxide concentration in parts per million (ppm); (TDa)—artificial atmospheric temperature deviation, as estimated by Equation 8, in degrees Celsius (°C) relative to the Vostok base Kyr1 = 0.00 °C; (TDn)—estimated natural atmospheric temperature deviation in degrees Celsius (°C) relative to the Vostok base Kyr1 = 0.00 °C (TDn = TDa - thermal gap).

Equation 9 was transformed into Equation 10.

Equation 10 (Dataset 28)—estimates the CO_2 concentration for a given temperature within the range from TDa to TDn: CO2dTDd = (TDd + 17.0031368489373) / 0.0459211443386427.[355] In which: (CO2dTDd)—estimated decreased atmospheric carbon dioxide concentration in parts per million (ppm); (TDd)—decreased atmospheric temperature deviation in degrees Celsius (°C) relative to the Vostok base Kyr1 = 0.00 °C. (TDa) —artificial atmospheric temperature deviation, as estimated by Equation 8, in degrees Celsius (°C) relative to the Vostok base Kyr1 = 0.00 °C; (TDn)—natural atmospheric temperature deviation in degrees Celsius (°C) relative to the Vostok base Kyr1 = 0.00 °C (TDn = TDa - thermal gap).

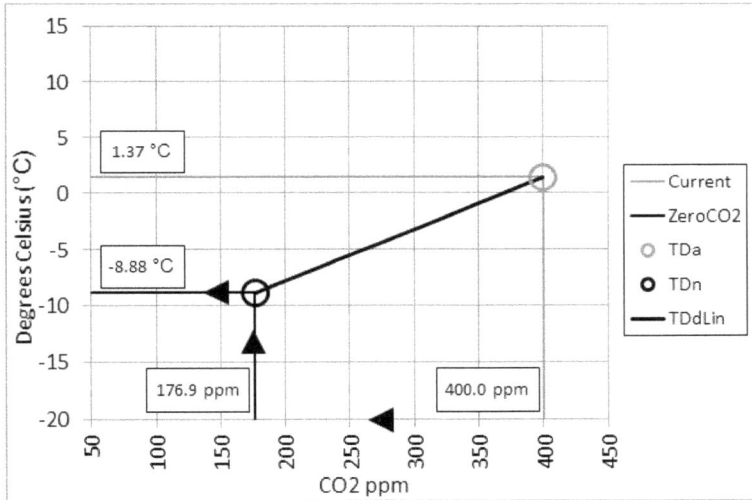

Figure 54 (Dataset 28)—temperature at zero artificial CO_2.[356] This graph shows: (Current)—current atmospheric temperature of 1.37 °C at an artificial atmospheric CO_2 concentration of 400.0 ppm; (ZeroCO2)—atmospheric temperature of -8.88 °C at a natural CO_2 concentration of 176.9 ppm, which equals an artificial CO_2 concentration of zero; (TDa)—artificial atmospheric temperature deviation, as predicted by Equation 8, in degrees Celsius (°C) relative to the Vostok base Kyr1 = 0.00 °C, at an atmospheric CO_2 concentration of 400 parts per million (ppm) including artificial CO_2; (TDn)—natural atmospheric temperature deviation in degrees Celsius (°C) relative to the Vostok base Kyr1 = 0.00 °C (TDn = TDa—thermal gap); and (TDdLin)—atmospheric temperature deviations in degrees Celsius (°C) relative to the Vostok base Kyr1 = 0.00 °C as a

linear function of CO_2 concentrations between TDa and TDn.

Equation 9 predicts that a removal of all artificial atmospheric CO_2 will result in a natural balance comprising a CO_2 concentration of 176.9 ppm and a temperature deviation of -8.88 °C.[357] In accord, Figure 54 illustrates that the current artificial CO_2 concentration equals 223.1 ppm (400.0 – 176.3 ppm).[358] Hence, the results of this study suggest that the artificial CO_2 equals 56% of the atmospheric CO_2.[359]

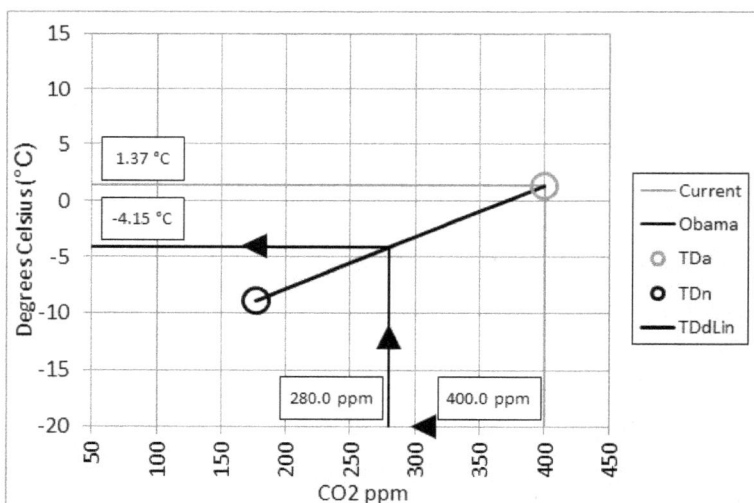

Figure 55 (Dataset 28)—Obama's CO_2 reduction to 280 ppm.[360] This graph shows: (Current)—current atmospheric temperature of 1.37 °C at an artificial atmospheric CO_2 concentration of 400.0 ppm; (Obama)—artificial atmospheric CO_2 concentration of 280.0 ppm resulting in an atmospheric temperature of -4.15 °C in 2025; (TDa)—artificial atmospheric temperature deviation, as predicted by Equation 8, in degrees Celsius (°C) relative to the Vostok base Kyr1 = 0.00 °C, at an atmospheric CO_2 concentration of 400 parts per million (ppm) including artificial CO_2; (TDn)—natural

atmospheric temperature deviation in degrees Celsius (°C) relative to the Vostok base Kyr1 = 0.00 °C (TDn = TDa— thermal gap); and (TDdLin)—atmospheric temperature deviations in degrees Celsius (°C) relative to the Vostok base Kyr1 = 0.00 °C as a linear function of CO_2 concentrations between TDa and TDn.

Equation 9 predicts that Obama's 30% reduction of CO_2 with 120.0 ppm, from 400.0 to 280.0, will reduce the temperature with 5.51 °C, from 1.37 to -4.15 °C.[361] Hence, the data suggest that Obama's plan would overshoot the target and create devastating glacial conditions.

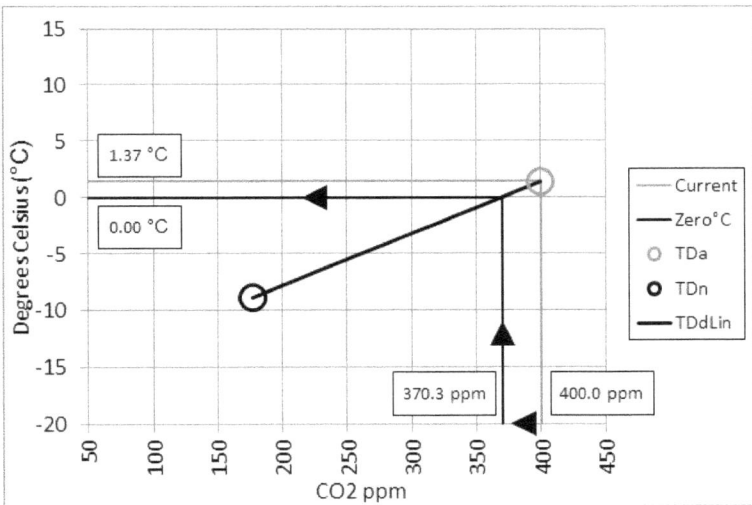

Figure 56 (Dataset 28)—restoring Vostok base Kyr1 = 0.00 °C.[362] This graph shows: (Current)—current atmospheric temperature of 1.37 °C at an artificial atmospheric CO_2 concentration of 400.0 ppm; (Zero°C)— artificial atmospheric CO_2 concentration of 370.3 ppm resulting in the atmospheric temperature of the Vostok base Kyr1 = 0.00 °C; (TDa)—artificial atmospheric temperature deviation, as predicted by Equation 8, in degrees Celsius (°C) relative to the Vostok base Kyr1 = 0.00 °C, at an atmospheric

CO2 concentration of 400 parts per million (ppm) including artificial CO2; (TDn)—natural atmospheric temperature deviation in degrees Celsius (°C) relative to the Vostok base Kyr1 = 0.00 °C (TDn = TDa—thermal gap); and (TDdLin)—atmospheric temperature deviations in degrees Celsius (°C) relative to the Vostok base Kyr1 = 0.00 °C as a linear function of CO2 concentrations between TDa and TDn.

Equation 10 predicts that a relatively small reduction in CO2 of 29.7 ppm, from 400.0 to 370.3 ppm, will reduce the temperature deviation from 1.37 °C to the Vostok base Kyr1 = 0.00 °C.[363] The thermal gap suggests that the hidden natural temperature has decreased. Natural glacial conditions will drag the observed global temperature down faster than it has been increasing. Therefore, it is proposed that the relatively small decrease in CO2 will decrease the global temperature deviation to 0.00 °C. The results of this study suggest that the short-term global linear function TDgCO2h of Equation 8 is confounded by the opposing effects of artificial CO2 and glacial conditions. Therefore, the long-term linear function TDdCO2d of Equation 9 is a more valid and reliable predictor of the decrease in temperature resulting from a decrease in artificial CO2.

Comments

Removing all the artificial atmospheric CO2 is likely to stop global warming and restore the natural balance between CO2 and temperature. However, be careful what you wish for, because this decrease could create devastating glacial conditions. The data support the notion that the temperature deviation would drop to -8.88 degrees Celsius (°C).[364] This is likely to kill all vegetation and destroy the global food supply. Hence, artificial CO2 is not just a hazardous substance that allegedly threatens our survival with global warming. Equally important, it is a vital substance for the survival of humanity, because it keeps the current global temperature around the save Vostok base Kyr1 = 0.00 °C and it maintains our food production. Therefore, we should be extremely careful about decreasing the artificial CO2 concentration.

The data of this study support the notion that a decrease of 29.7 ppm in atmospheric CO2, from 400.0 to 370.3 ppm, would decrease the average atmospheric global temperature, from 1.37 °C to the 'normal' temperature of Vostok base Kyr1 = 0.00 °C.[365]

In contrast, President Obama intents to decrease the atmospheric CO2 with 120 ppm, from 400.0 to 280.0 ppm, by 2025.[366] Unfortunately, this study suggests that the intended reduction in CO2 could decrease the atmospheric temperature with 5.51 °C, from 1.37 to -4.15 °C, by 2025.[367] Once set in motion, such political developments are difficult to turn around. Consequently, Obama's plan could backfire and create devastating glacial conditions that could force humanity to the edge of extinction by 2025.

The data show that the uninterrupted increase and duration of the global atmospheric temperature has been within the natural limits.[368] Therefore, reducing the non-significant average temperature to normal is likely to have a non-significant effect on the effects of climate change.

The real threat to our survival is the hidden thermal gap. This gap suggests that huge artificial and natural thermal forces oppose each other. The nemonik question is whether we have enough CO2 to maintain the precarious balance between those forces.

APPENDICES

BIBLIOGRAPHY

GISTEMP Team, N. G. (2015). GISS Surface Temperature Analysis (GISTEMP). *Dataset accessed 2015-05-03 at http://data.giss.nasa.gov/gistemp/*.

Gore, A. (2006). *An Inconvenient Truth*. Amazon.

Hansen, J. R. (2010). Global surface temperature change. *Rev. Geophys., 48, RG4004, doi:10.1029/2010RG000345*.

Moore, G. P., & McCabe, D. S. (2003). *Introduction to the Practice of Statistics*. New York: W. H. Freeman and Company.

NOAA/NGDC Paleoclimatology Program, B. C. (n.d.).

Petit, J. R. (2001). *Vostok Ice Core Data for 420,000 Years*. Boulder CO, USA: IGBP PAGES/World Data Center for Paleoclimatology Data Contribution Series #2001-076. NOAA/NGDC Paleoclimatology Program.

Petit, J. R., Jouzel, J., Raynaud, D., Barkov, N. I., M, B. J., Basile, I., et al. (1999). Climate and Atmospheric History of the Past 420,000 years from the Vostok Ice Core, Antarctica. *Nature, 399*, 429-436.

Schade, A. (2015). bioPAD: Nemonik Thinking (PowerPoint). Dunedin: nemonik-thinking.org.

Schade, A. (2015). *Climate-Change and Dunedin (video)*. Dunedin: nemonik-thinking.org.

Schade, A. (2016). *Dictionary Nemonik Thinking*. nemonik-thinking.org.

Schade, A. (2016). *Glossary Nemonik Thinking*. nemonik-thinking.org.

Schade, A. (2016). *Think Smarter with Nemonik Thinking*. nemonik-thinking.org.

Schade, A. (planned 2017). *Lao Zi's Dao De Jing*. nemonik-thinking.org.

Schade, A. (planned 2017). *Sun Zi's The Art of War*. nemonik-thinking.org.

Schade, A. (planned 2017). *The Unreal Reality*. nemonik-thinking.org.

GLOSSARY

Affecters—mental signals are generated by subconscious affectorial thinking, which influence the conscious without explanations. Affecters do not rely on conscious reasoning or facts, and therefore, they are by definition non-rational and illogical. Affecters include beliefs, common sense, desires, discoveries, emotions, fantasies, habits, heuristics, ideas, impulses, innovations, insights, inspirations, intuitions, inventions, novelties, reactions, reflexes, routines, sensibility, skills, etc.

Affectorial thinking—subconscious part of nemonik thinking that deals with the unpredictable disorder of reality by generating affecters.

Alpha (α)—probability that a decision maker accepts to make a mistake. In statistics, an α-value would range traditionally from lax (0.05) to strict (0.001).

Alternative hypothesis (Ha)—statement that the effect is systematic, rather than random. See Hypothesis and Null hypothesis (Ho).

Antithesis—description of reality that contradicts a thesis. See Hegel.

Artificial global warming—comprises domestic and industrial global warming.

Artificial—manmade.

Bilateral climate-change hypothesis—holds that the effect of artificial global warming compensates for the effect of natural global cooling. Low glacial temperatures are increased by solar heat that is trapped by domestic and industrial atmospheric greenhouse gasses such as carbon dioxide (CO_2).

Carbon dioxide (CO_2)—greenhouse gas that allegedly traps heat in the atmosphere.

Climate change—long-term and fundamental change in the global climate. Climate change comprises long-term natural cycles of global warming and cooling, and recently artificial domestic and industrial global warming.

Climate control—voluntary and involuntary artificial ways to control the natural climate.

Climatology—study of the climate.

CO2—See carbon dioxide.

Confounding variables—variables not included in the statistical analyses and are not eliminated by experimental design, which disturb the results.

Criterion—See Alpha.

Data point—numerical single observation, fact, measurement, or score in a dataset. For example: 1.0 is a data point in the dataset (2.0, 3.0, -2.0, **1.0**, and -1.0).

Data snooping—selecting a part of the dataset under examination in order to find a significant result. Data snooping breaches the process of random selection in which each data point has an equal likelihood to be selected. Consequently, the selected part might not represent the population and could cause spurious results. Data snooping is used as an exploratory tool to create new hypotheses.

Dataset—collection of data points. For example (2.0, 3.0, -2.0, 1.0, and -1.0).

Descriptive statistics—branch of statistics concerned with computing numbers (statistics) that describe the centricity and spread of data points comprising a dataset. Descriptive statistics include the mean (m), standard deviation (STD), maximum (max), minimum (min), range, and the number of data points (n).

Domestic global warming—hypothesis that humanity has unknowingly affected the climate by releasing domestic greenhouse gasses. Domestic greenhouse gasses are likely to have increased with decreasing atmospheric temperatures and increasing global population. See Global warming.

Facts—testable descriptions of reality that are supported adequately by sensory perception and rational thinking. See Hypothesis.

Glacial decline—thermal decline at the onset of a glacial period. It stretches from the interglacial thermal peak to an arbitrary limit of -4.00 degrees Celsius (°C) relative to the Vostok base Kyr1 = 0.00 °C.

Glacial period—long-term climatological period with mean millennial temperature deviations below the mean temperature deviation of the Vostok base Kyr1 = 0.00 °C.

Global cooling—long-term and fundamental decrease in the global temperature. Global cooling comprises natural and artificial decreases in global temperature. See Climate change.

Global warming—long-term and fundamental increase in the global temperature. Global warming comprises natural and artificial increases in global temperature. See Climate change.

Greenhouse gasses—gasses trapping heat in the atmosphere. See carbon dioxide.

Ha—See Alternative hypothesis.

Hegel, Frederich (1770-1831)—German philosopher who introduced a dialectic that describes the progress of knowledge. The dialectic holds that each thesis evokes an antithesis, which is followed by a synthesis. This synthesis becomes the new thesis. Hegel's dialectic is a process of successive approximation.

Ho—See Null hypothesis.

Hypothesis—untested, but testable description of reality. See Null hypothesis (Ho), Alternative hypothesis (Ha), and Facts.

Industrial global warming—hypothesis that humanity's industry has pollutes the climate by releasing too much greenhouse gasses. See Climate change.

Interglacial period—long-term climatological period with mean temperature deviations per Kyr equal or above the mean temperature deviation of the Vostok base Kyr1 = 0.00 °C. The mean duration of past interglacial periods is about 4.6 Kyrs.

Kyr—millennium or period of 1,000 years.

Maximum (max)—data point with the largest magnitude in a dataset. For the dataset (2.0, **3.0**, -2.0, 1.0, and -1.0), the maximum = 3.0.

Mean or average (m)—statistic that measures the central value of a dataset, which is the average of the data points. The mean is computed with the equation: $m = (x_1 + x_2 + ...x_n) / n$. In which, x_1, x_2, etc. are data points, and (n) is the number of data points. For the dataset (2.0, 3.0, -2.0, 1.0, and -1.0), the mean equals: $(2.0 + 3.0 + -2.0 + 1.0 + -1.0) / 5 = 3.0 / 5 = 0.6$.

Median—statistic that measures the central value of a dataset, which is the number in the middle of that dataset sorted by magnitude. For example, the sorted dataset (2.0, 3.0, -2.0, 1.0, and -1.0) would be (3.0, 2.0, **1.0**, -1.0, -2.0) and the number in the middle = 1.0.

Millennium—1000 years or Kyr.

Minimum (min)—data point with the smallest magnitude in a dataset. For the dataset (2.0, 3.0, -**2.0**, 1.0, and -1.0), the minimum = -2.0.

Natural global cooling—part of the natural climatological cycle characterized by long-term decreasing temperatures. See Global warming.

Natural global warming—part of the natural climatological cycle characterized by long-term increasing temperatures. See Global warming.

Natural—not manmade or artificial.

Nemonik thinking—exhaustive cognitive system that synthesizes rational, creative, and intuitive thinking. It maximizes success by evaluating 17 nemoniks, which are memorized keywords describing all the perceived aspects of the mind, reality, and their interaction. As defined by Lao Tzu—"Success is obtaining what you seek and escaping what you suffer."

Nemonik-accelerator—increases the speed of thinking by fostering agreement during disagreement, while fostering disagreement during agreement. This nemonik tool is based on Hegel's dialectic.

Nemoniks—17 memorized mnemonics describing all the perceived aspects of the mind, reality, and their interaction. The four mental nemoniks are objective, collective, creative, and reactive—each one representing a mindmode. In addition, the thirteen operational nemoniks are advance, stay, retreat, accumulate, preserve, dispose, act, wait, prepare, accept, reject, reveal, and conceal. Nemoniks guide the process of thinking, improve the memory with defragmentation, prompt the memory to release associated information, and help the creative mindmode to generate ideas, while the constant readiness reduces stress levels. The word nemonik is a phonetic notation of the Greek word mnemonic meaning memory aid. However, nemoniks are more than mnemonics. Nemoniks are tools for consciously managing the larger subconscious, guiding the process of thinking, improving the memory by defragmentation and prompting, and activating the creative mindmode to generate ideas. Best of all, you do not have to worry about all those improvements because they will happen automatically in the reactive mindmode.

n—number of data points in a dataset. For the dataset (2.0, 3.0, -2.0, 1.0, and -1.0), n = 5.

Normal or Gaussian frequency distribution—dataset described by a symmetric, unimodal, and bell shaped frequency curve. The mean and the median of a normal frequency distribution are the same. In parametric statistics, a normal population distribution is a basis for evaluating of data points and samples.

Null hypothesis (Ho)—statement that the effect is random, rather than systematic. Null refers to the lack of a systematic effect. See Hypothesis and Alternative hypothesis (Ha).

One-tailed p-test—evaluates the probability (p) that a data point is significant higher than the mean, or alternatively, that the data point is significant lower than the mean. For example, the evaluation of global warming (only warming) would require a one-tailed p-test, while climate change (warming and cooling) requires a two-tailed p-test. The criterion for a one-tailed test equals (α). Hence, the unilateral and bilateral hypotheses require different tests. See Two-tailed p-test.

Parameters—numbers describing the population distribution. Parameters include the mean (m), standard deviation (STD), maximum (max), minimum (min), range, and the number of data points (n). See Statistics (descriptive).

Parametric statistics—branch of statistics concerned with evaluating the probability that an effect is either systematic or random, by comparing the statistics of a sample with the parameters of a normal dataset. For that reason, the probabilities of making incorrect decisions (Type I and II errors) are computed. The main weakness of parametric statistics is the assumption that the data point or sample under examination is part of a population with a normal frequency distribution.

ppm—unit to measure gas meaning Parts Per Million. Alternatively, ppm is referred to as ppmv.

ppmv—unit to measure gas meaning Parts Per Million Volume. Alternatively, ppmv is referred to as ppm.

Probability (p)—computed probability of making a Type I error. See Alpha.

p—See Probability.

Random effect—effect attributed to random variation around the mean. Opposite: Systematic effect.

Randomization test—non-parametric test that computes the significance of statistics by comparing the original statistics of the sample with iterated statistics obtained from the randomized sample. This test creates its own population distribution, which is based on the data points of the sample. The randomization test is one of the most versatile tests in statistics. It can be used to test the significance of means, medians, correlation coefficients, linear functions, etc.

Range—is the maximum of the data points minus the minimum of the data points. For the dataset (2.0, **3.0**, **-2.0**, 1.0, and -1.0), the range = 3.0 – -2.0 = 3.0 + 2.0 = 5.0.

Rational thinking—conscious part of nemonik thinking that deals with the predictable order of reality by submitting facts to reason, which creates new facts.

Significant—computed probability of making an error (p)is smaller than the accepted probability of making an error (α). A significant effect is considered systematic, rather than random.

Standard deviation (STD)—measures the spread or variation of the data points around their mean. The STD is computed with the equation: STD = $\sqrt{(s(x_i - m)^2 / (n - 1))}$. In which, $\sqrt{}$ = square root; s = sum or total; x_i = all data points in the dataset from x_i to x_n; m = mean of the dataset; n = number of data points in the dataset. For dataset (2.0, 3.0, -2.0, 1.0, and -1.0); the mean = 0.6; and n = 5.[369] In accord, STD = $\sqrt{(((2.0 - 0.6)^2 + (3.0 - 0.6)^2 + (-2.0 - 0.6)^2 + (1.0 - 0.6)^2)(-1.0 - 0.6)^2) / (5 - 1))}$ = $\sqrt{(((1.4)^2}$

$+ (2.4)^2 + (-2.6)^2 + (0.4)^2 + (-1.6)^2) / 4) = \sqrt{(((1.96 + 5.76 + 6.76 + 0.16 + 2.56)) / 4)} = \sqrt{(17.20 / 4)} = \sqrt{4.30} = 2.1.$

Standard-score—See z-score.

Statistic (descriptive)—number that describes the centricity, spread, or size of a dataset sampled from a population.

Statistics—field of study concerned with collecting, organizing, describing, analysing, and evaluating numerical facts called data.

Synthesis—merger of separate parts of a phenomenon into an entirety. In the context of Hegel's dialectic, a synthesis is a description of reality that merges a thesis and an antithesis into a new thesis. A synthesis might create synergy. See Hegel.

Systematic effect—effect attributed to a specific variable. Opposite Random effect.

Temperature deviations—temperatures expressed as differences from a mean temperature during a base period. Deviations are also indicated by the Greek letter delta. Temperature deviations could be expressed as degrees Celsius, Fahrenheit, or Kelvin. See Absolute temperatures.

Thermal gap—difference between the estimated and actual annual global atmospheric temperature deviations.

Thermal outliers—temperature deviations that are both higher than the global maximum of 1.32 °C in 2014 and associated with CO_2 concentrations below the Hawaiian maximum of 398.6 ppm in 2014.[370]

Thermal—refers to temperature.

Thesis—a description of reality. See Hegel.

Trend—a one directional change in the data across time. See also Cycles.

Two-tailed p-test—evaluates the probability (p)that a data point is either significant higher or lower than the mean. For example, the evaluation of global warming (only warming) would require a one-tailed p-test, while climate change (warming and cooling) requires a two-tailed p-test. The criterion for a two-tailed test equals α / 2. See One-tailed p-test.

Type I error—incorrect decision to reject a true null hypothesis (Ho) and, therefore, accept the false alternative hypothesis (Ha). This error leads to the incorrect conclusion that a random effect is systematic. Opposite Type II error.

Type II error—incorrect decision to accept a false null hypothesis (Ho) and, therefore, reject the alternative hypothesis (Ha). This error leads to the incorrect conclusion that a systematic effect is random. Opposite Type I error.

Unilateral climate-change hypothesis—holds that industrial greenhouse gasses, such as carbon dioxide (CO_2), increase the global temperature by trapping heat in the atmosphere.

z-score or standard-score—difference between a data point and the mean, expressed in standard deviations. The z-score is computed with the equation: $z = (x - m)$ / STD. In which, x = the data point under examination; m = mean of the dataset; and STD = standard deviation of the dataset. For dataset (2.0, **3.0**, -2.0, 1.0, and -1.0), $m = 0.6$; STD = 2.07; and the data point to be tested = 3.0. The z-score of 3.0 = (3.0 − 0.6) / 2.07 = 2.4 / 2.07 = 1.16. Based on tables representing a normal distribution, the p-value for the z-score of 1.16 = 0.1230 (Moore & McCabe, 2003). Hence, p (0.1230) > α (0.05) and, therefore, the data point (3.0) is not significantly higher than the mean of the dataset (0.6). The difference between that data point and the mean has to be attributed to a random effect. In contrast, a data point of 6.0 would obtain a z-score = (6.0 − 0.6) / 2.07 = 5.4 / 2.07 = 2.60.[371] The p-value for the z-score of 2.60 = 0.0047. Hence, p (0.0047) < α (0.05) and, therefore, the data point (6.0) is significantly higher than the mean of the dataset (0.6). The difference between that data point and the mean has to be attributed to a systematic effect.

α (alpha)—See Alpha.

LISTS

LIST OF FIGURES

Figure 1 (Dataset 2)—Vostok depth-scale temperatures. This graph shows: (TDv)—depth-scale atmospheric temperature deviations from 422,766 years before present (BP) to the present calendar year 1989, in degrees Celsius (°C) relative to the Vostok base 1850-1989 = 0.00 °C; (TDvMax)—maximum TDv = 3.23 °C; (Age)—ice age from 422,766 years BP to the present calendar year 1989. 39

Figure 2 (Dataset 3)—Vostok millennial-scale temperatures. This graph shows: (TDv)—millennial-scale atmospheric temperature deviations from Kyr423 to Kyr1, in degrees Celsius (°C) relative to the Vostok base 1850-1989 = 0.00 °C; and (TDvMax)—maximum TDv = 3.03 °C. 42

Figure 3 (Dataset 4)—Vostok temperatures base Kyr1. This graph shows: (TDv)—millennial-scale atmospheric temperature deviations from Kyr423 to Kyr1, in degrees Celsius (°C) relative to the Vostok base Kyr1 = 0.00 °C; (TDvMax)—maximum TDv = 3.49 °C. 44

Figure 4 (Dataset 4)—Vostok thermal trend. This graph shows: (TDv)—millennial-scale atmospheric temperature deviations from Kyr423 to Kyr1, in degrees Celsius (°C) relative to the Vostok base Kyr1 = 0.00 °C; (TDvMax)—maximum TDv = 3.49 °C; and (TDvLin)—linear trend of TDv = 0.00508 °C/Kyr. 46

Figure 5 (Dataset 5)—Vostok detrended temperatures. This graph shows: (TDv)—detrended millennial-scale atmospheric temperature deviations from Kyr423 to Kyr1, in degrees Celsius (°C) relative to the Vostok base Kyr1 = 0.00 °C; (TDvMax)—maximum TDv = 2.30 °C; and (TDvLin)—residual linear trend of TDv = 0.00000 °C/Kyr. 48

Figure 6 (Dataset 5)—histogram Vostok detrended temperatures. This graph shows the frequency distribution of the Vostok detrended millennial temperature deviations from Kyr423 to Kyr1, in degrees Celsius (°C) relative to the Vostok base Kyr1 = 0.00 °C. 49

Figure 7 (Dataset 5)—Vostok glacial declines. This graph shows: (TDv)—detrended millennial-scale atmospheric temperature deviations from Kyr423 to Kyr1, in degrees Celsius (°C) relative to the Vostok base Kyr1 = 0.00 °C; (Decline)—thermal glacial decline; and (Limit)—arbitrary end of thermal decline = -4.00 °C. 58

Figure 8 (Dataset 5)—Vostok glacial decline window. This graph shows: (TDv)—detrended millennial-scale atmospheric temperature deviations from year 10,000 before present (BP) to the present calendar year 1989, in degrees Celsius (°C) relative to the Vostok base Kyr1 = 0.00 °C,; (CIGstart)—start current interglacial period at 9,926 years BP; (PIGmin)—minimum duration previous interglacials = 1,032 years; (PIGmean)—mean duration previous interglacials = 3,575 years; (PIGmax)—maximum duration previous interglacials = 7,682 years; (DWul)—upper-limit of the decline window = -0.30 °C/Kyr; (DWmean)—mean decline = -0.52 °C/Kyr; and (DWll)—lower-limit of the decline window = -1.00 °C/Kyr (DWll). 60

Figure 9 (Dataset 5)—Vostok thermal stability index (TSI). This graph shows: (TSI-CIG)—thermal stability index during the current interglacial of 10 Kyrs; and (TSI-SW)—thermal stability indexes during the observation period from Kyr414 to Kyr11, as obtained with a sliding window of 10 Kyrs. 63

Figure 10 (Dataset 5)—Vostok thermal stability index versus temperature. This graph shows: (TSI-SW)—Vostok thermal stability indexes from Kyr414 to Kyr11, as obtained with a sliding window of 10 Kyrs; and (TDx-SW)—Vostok mean temperature deviations as obtained with a sliding window of 10 Kyrs from Kyr414 to Kyr11............. 64

Figure 11 (Dataset 6)—Vostok annual temperature deviations 10 Kyrs. This graph shows: (TDv)—Vostok semi-annual atmospheric temperature deviations from 10,000 years before present (BP) to the present calendar year 1989, in degrees Celsius (°C) relative to the Vostok base 1850-1989 = 0.00 °C,; (TDvMax)—maximum TDv = 2.06 °C. 67

Figure 12 (Dataset 7)—Vostok detrended annual temperatures 10 Kyrs. This graph shows: (TDv)—Vostok detrended semi-annual atmospheric temperature deviations from 10,000 years before present (BP) to the present calendar year 1989, in degrees Celsius (°C) relative to the Vostok base Kyr1 = 0.00 °C; (TDvMax)—maximum TDv = 2.48 °C; and (TDvMin)—minimum TDv = -1.52 °C. . 70

Figure 13(Dataset 7)—histogram Vostok detrended annual temperatures 10 Kyrs. This graph shows the frequency distribution of the detrended Vostok semi-annual atmospheric temperature deviations from 10,000 years before present (BP) to the present calendar year 1989, in degrees Celsius (°C) relative to the Vostok base Kyr1 = 0.00 °C... 71

Figure 14 (Dataset 8)—Vostok window-scale temperature deviations 10 Kyrs. This graph shows: (TDv)—semi-annual Vostok atmospheric temperature deviations; and (TDvSW)—Vostok window-scale atmospheric temperature deviations, which are the mean temperature deviations obtained with a sliding window of 50 years. The observation period ranges from 10,000 years before the present (BP) to the present calendar year 1989, while the temperature deviations are expressed in degrees Celsius (°C) relative to Vostok base Kyr1 = 0.00 °C. 74

Figure 15 (Dataset 8)—histogram Vostok window-scale temperature deviations 10 Kyrs. This graph shows the frequency distribution of: (TDvSW)—Vostok window-scale atmospheric temperature deviations, which are the mean temperature deviations obtained with a sliding window of 50 years. The observation period ranges from 10,000 years before the present (BP) to the present calendar year 1989, while the temperature deviations are expressed in degrees Celsius (°C) relative to Vostok base Kyr1 = 0.00 °C. 75

Figure 16 (Dataset 10)—global temperatures base 1951-1980. This graph shows: (TDg)—global annual atmospheric temperature deviations from calendar year 1880 to 2014, in degrees Celsius (°C) relative to the global base 1951-1980 = 0.00 °C; and (TDgBase)—mean TDg from 1951-1980 = 0.00 °C. 77

Figure 17 (Dataset 10)—histogram global temperatures base 1951-1980. This graph shows the frequency distribution of the global annual atmospheric temperature deviations (TDg) from calendar year 1880 to 2014, in degrees Celsius (°C) relative to the global base 1951-1980 = 0.00 °C. 78

Figure 18—Vostok temperatures. This graph shows: (TDv (Dataset 11))—Vostok detrended semi-annual atmospheric temperature deviations from 1,000 years before present (BP) to the present calendar year 1989, in degrees Celsius (°C) relative to the Vostok base Kyr1 = 0.00 °C; (V-sync)—Vostok mean temperature deviation from calendar year 1850 to 1989 = 0.46 °C; (TDg (Dataset 10))—global annual atmospheric temperature deviations from calendar year 1880 to 2014, in degrees Celsius (°C) relative to TDgBase; and (TDgBase)—global mean temperature deviation from calendar year 1951 to 1980, in degrees Celsius (°C) = 0.00 °C. 80

Figure 19 (Dataset 12)—extended global temperatures. This graph shows: (TDg)—global annual atmospheric temperature deviations from calendar year 1880 to 2014, in degrees Celsius (°C) relative to the global base 1951-1980 = 0.00 °C; (ExtPol)—TDgPol extended to 1850; (ExtLin)—

TDgLin extended to 1850; (LinPol)—backwards extension of TDg from 1850 to 1879, comprising the means of ExtPol and ExtLin; (TDgPol)—second order polynomial function of TDg; and (TDgLin)—linear function of TDg.83

Figure 20 (Dataset 13)—Vostok and global synchronization periods. This graph shows: (TDv)—Vostok detrended semi-annual atmospheric temperature deviations from 989 to 1989, in degrees Celsius (°C) relative to the Vostok base Kyr1 = 0.00 °C; (TDvSyn)—mean TDv from 1850 to 1989 = 0.46 °C; (TDgExt)—extended global annual atmospheric temperature deviations from 1850 to 2014, in degrees Celsius (°C) relative to the global base 1951-1980 = 0.00 °C; and (TDgSyn)—mean TDgExt from 1850 to 1989 = -0.18 °C. 86

Figure 21 (Dataset 14)—Vostok and global temperatures synchronized. This graph shows: (TDv)—Vostok semi-annual atmospheric temperature deviations from 989 to 1989; (TDg)—extended global annual atmospheric temperature deviations from calendar year 1850 to 2014; (Syn)—means of TDv and TDg during synchronization period from calendar year 1850 to 1989 = 0.46 °C; maximum global temperature deviation in 2014 = 1.32 °C. TDg and TDv are expressed in degrees Celsius (°C) relative to the Vostok base Kyr1 = 0.00 °C. 89

Figure 22 (Dataset 15)—global annual- and window-scale temperatures. This graph shows: (TDg)—global annual-scale atmospheric temperature deviations from calendar year 1880 to 2014; and (TDgSW)—global window-scale atmospheric temperature deviations from calendar year 1929 to 2014, which are the mean temperature deviations obtained with a window of 50 years sliding in steps of 1 year. TDg and TDgSW are in degrees Celsius (°C) relative to the Vostok base Kyr1 = 0.00 °C. 92

Figure 23 (Dataset 15)—histogram global window-scale temperatures. This graph shows: (TDgSW)—frequency distribution of the global window-scale temperature deviations from calendar year 1929 to 2014, in degrees Celsius (°C) relative to the Vostok base Kyr1 = 0.00 °C. . 93

Figure 24 (Dataset 16)—Vostok and global window-scale temperatures. This graph shows: (TDvSW)—Vostok window-scale atmospheric temperature deviations from calendar year 7962 BCE to 1989; (TDvSWm)—maximum TDvSW = 2.30 °C; (TDvSWx)—mean DvSW = 0.07 °C; (TDgSW)—global window-scale atmospheric temperature deviations from calendar year 1929 to 2014; (TDvSWm)—maximum TDvSW = 0.93 °C; and (Type I errors)—TDvSW > TDvSWm. Window-scale temperature deviations are mean temperature deviations obtained with a window of 50 years sliding in steps of 1 year, which are expressed in degrees Celsius (°C) relative to the Vostok base Kyr1 = 0.00 °C. 97

Figure 25 (Dataset 17)—Vostok CO2 423 Kyrs. This graph shows: (CO2v)—Vostok atmospheric carbon dioxide concentrations, from 414,085 to 2,342 years before calendar year 1989, in parts per million (ppm). The data intervals are irregular. 106

Figure 26 (Dataset 18)—Vostok CO2 and temperature 412 Kyrs. This graph shows: (CO2v)—irregular-scale Vostok atmospheric carbon dioxide concentrations in parts per million (ppm); (TDv)—irregular-scale Vostok atmospheric temperature deviations in degrees Celsius (°C) relative to the Vostok base Kyr1 = 0.00 °C. The observation period ranges from 414,085 to 2,342 years before calendar year 1989. 109

Figure 27 (Dataset 18)—histogram Vostok CO2 412 Kyrs. This graph shows the frequency distribution of (CO2v)—Vostok atmospheric carbon dioxide concentrations, from 414,085 to 2,342 years before calendar year 1989, in parts per million (ppm). The data intervals are irregular. 110

Figure 28 (Dataset 18)—Vostok temperature versus CO2 412 Kyrs. This graph shows: (TDv)—irregular-scale Vostok atmospheric temperature deviations in degrees Celsius (°C) relative to the Vostok base Kyr1 = 0.00 °C; (CO2v)— irregular-scale Vostok atmospheric carbon dioxide concentrations in parts per million (ppm). The observation period ranges from 414,085 to 2,342 years before calendar year 1989, while the data intervals are irregular................. 112

Figure 29—Hawaiian and global temperatures 1955-2014. This graph shows: (Th (Dataset 19))—Hawaiian annual atmospheric temperatures in absolute degrees Celsius (°C); (Thx)—mean Th; (TDg (Dataset 14))—extended global annual atmospheric temperature deviations in degrees Celsius (°C) relative to the Vostok base Kyr1 = 0.00 °C; (TDgx)—mean TDg. 117

Figure 30—recalibrated Hawaiian and global temperatures 1955-2014. This graph shows: (TDh (Dataset 20))— Hawaiian annual atmospheric temperature deviations in degrees Celsius (°C) relative to the Vostok base Kyr1 = 0.00 °C; (TDg (Dataset 14))—extended global annual atmospheric temperature deviations; TDh and TDg are expressed in degrees Celsius (°C) relative to the Vostok base Kyr1 = 0.00 °C.. 118

Figure 31 (Dataset 20)—Hawaiian temperatures 1955-2014. This graph shows; (TDh)—Hawaiian annual atmospheric temperature deviations in degrees Celsius (°C) relative to the Vostok base Kyr1 = 0.00 °C; (TDhLin)—linear function of TDh; (TDhPol)—second order polynomial function of TDh. 119

Figure 32 (Dataset 20)—histogram Hawaiian temperatures 1955-2014. This histogram shows the frequency distribution of (TDh)—Hawaiian annual atmospheric temperature deviations in degrees Celsius (°C) relative to the Vostok base Kyr1 = 0.00 °C....................................... 120

Figure 33 (Dataset 21)—Hawaiian CO2 1959-2014. This graph shows: (CO2h)—Hawaiian annual atmospheric carbon dioxide concentrations in part per million (ppm); (CO2hLin)—linear function of CO2h; and (CO2hPol)— second order polynomial function of CO2h.....................122

Figure 34 (Dataset 21)—histogram Hawaiian CO2 1959-2014. This graph shows the frequency distribution of the Hawaiian annual atmospheric carbon dioxide concentrations in part per million (ppm)..............................123

Figure 35 (Dataset 22)—Hawaiian temperatures and CO2 1959-2014. This graph shows: (CO2h)—Hawaiian atmospheric CO2 concentrations in parts per million (ppm); (TDh)—Hawaiian atmospheric temperature deviations in degrees Celsius (°C) relative to the Vostok base Kyr1 = 0.00 °C; (TDhPol)—second order polynomial function of TDh; and (CO2hPol)—second order polynomial function of CO2h.................................125

Figure 36 (Dataset 22)—scatterplot Hawaiian temperatures versus CO2 1959-2014. This graph shows: (TDh)— Hawaiian atmospheric temperature deviations in degrees Celsius (°C) relative to the Vostok base Kyr1 = 0.00 °C; as a function of (CO2h)—Hawaiian atmospheric CO2 concentrations in parts per million (ppm); (TDhPol)— second order polynomial function of TDh; and (TDhLin)—linear function of TDh..................................126

Figure 37 (Dataset 24)—global temperatures and Hawaiian CO2 1959-2014. This graph shows: (CO2h)—Hawaiian atmospheric carbon dioxide concentrations in parts per million (ppm); (TDg)—global annual atmospheric temperature deviations in degrees Celsius (°C) relative to the Vostok base Kyr1 = 0.00 °C; (CO2hLin)—linear function of CO2h; and (TDgLin)—linear function of TDg. 130

Figure 38 (Dataset 24)—global temperatures versus Hawaiian CO2 1959-2014. This graph shows: (TDg)—global annual atmospheric temperature deviations in degrees Celsius (°C) relative to the Vostok base Kyr1 = 0.00 °C; as a function of (CO2h)—Hawaiian atmospheric carbon dioxide concentrations in parts per million (ppm); and (TDgLin)— linear function of TDg. 130

Figure 39 (Dataset 24) z-scores Hawaiian CO2 and global temperatures 1959-2014. This graph shows: (CO2hz)—z-scores Hawaiian atmospheric carbon dioxide concentrations in parts per million (ppm); and (TDgz)—z-scores global annual atmospheric temperature deviations in degrees Celsius (°C) relative to the Vostok base Kyr1 = 0.00 °C. 135

Figure 40 (Dataset 24)—CO2 predicts temperatures. This graph shows: (PC)—percentage correct of retrospective predictions obtained with the predictor of change in global temperature (CO2p) from 1 to 10 years into the future; and (PC-Lin)—linear function of PC. 136

Figure 41 (Dataset 25)—Vostok temperature as a function of Vostok CO2. This graph shows: (TDv)—irregular-scale Vostok atmospheric temperature deviations in degrees Celsius (°C) relative to the Vostok base Kyr1 = 0.00 °C; (CO2v)—irregular-scale Vostok atmospheric carbon dioxide concentrations in parts per million (ppm); (TDvLin)—linear function of TDv; and (TDvPol)—second order polynomial function of TDv. The observation period ranges from 412,096 to 9,024 BCE. 140

Figure 42 (Dataset 25)—predicted Vostok temperature. This graph shows: (TDv)—irregular-scale Vostok atmospheric temperature deviations in degrees Celsius (°C) relative to the Vostok base Kyr1 = 0.00 °C; (TDvCO2v)—TDv predicted by substituting CO2v in Equation 6. The observation period ranges from 414,085 to 11,013 years before calendar year 1989. 142

Figure 43 (Dataset 26)—global temperature as a function of Hawaiian CO2. This graph shows: (TDg)—global annual atmospheric temperature deviations in degrees Celsius (°C)

relative to the Vostok base Kyr1 = 0.00 °C; (CO2h)—
Hawaiian atmospheric carbon dioxide concentrations in
parts per million (ppm); (TDgLin)—linear function of TDg.
The observation period ranges from calendar year 1959 to
2014. 145

Figure 44: (Dataset 26)—predicted global temperature. This
graph shows: (TDg)—global annual atmospheric
temperature deviations in degrees Celsius (°C) relative to
the Vostok base Kyr1 = 0.00 °C; (TDgCO2h)—global
atmospheric temperature deviation in degrees Celsius (°C)
relative to the Vostok base Kyr1 = 0.00 °C, as predicted by
the Hawaiian atmospheric CO2 concentrations substituted
in Equation 8. The observation period ranges from
calendar year 1959 to 2014...146

Figure 45 (Dataset 27)—thermal gap. This graph shows:
(TDv)—irregular-scale atmospheric temperature deviations,
from 412,096 to 9,024 BCE, in degrees Celsius (°C) relative
to The Vostok base Kyr1 = 0.00 °C; (TDg)—global annual
atmospheric temperature deviations, from 1959 to 2014, in
degrees Celsius (°C) relative to the Vostok base Kyr1 =
0.00 °C; (TDvLin)—linear function of TDv; (TDgLin)—
linear function of TDg; and (CO2 ppm)—atmospheric
carbon dioxide concentrations in parts per million. 149

Figure 46 (Dataset 18)—Vostok CO2 and temperature (10
Kyrs). This graph shows, for the observation period from
8,134 to 353 BCE: (CO2v)—irregular-scale Vostok
atmospheric carbon dioxide concentrations in parts per
million (ppm); (TDv)—irregular-scale Vostok atmospheric
temperature deviations in degrees Celsius (°C) relative to
the Vostok base Kyr1 = 0.00 °C...154

Figure 47 (Dataset 18)—Vostok CO2 and temperature (10
Kyrs). This graph shows, for the observation period from
8,134 to 353 BCE: (TDvCO2v)—irregular-scale Vostok
atmospheric temperature deviation in degrees Celsius (°C)
relative to the Vostok base Kyr1 = 0.00 °C, as predicted by

the Vostok atmospheric CO2 concentrations substituted in
Equation 6; (TDv)—irregular-scale Vostok atmospheric
temperature deviations in degrees Celsius (°C) relative to
the Vostok base Kyr1 = 0.00 °C; (TDvCO2vLin)—linear
function of TDvCO2v; (TDvLin)—linear function of TDv.
.... 155

Figure 48 (Dataset 18)—Vostok CO2 and temperature (10
Kyrs). This graph shows, for the observation period from
8,134 to 353 BCE: (TDvCO2v)—irregular-scale Vostok
atmospheric temperature deviation in degrees Celsius (°C)
relative to the Vostok base Kyr1 = 0.00 °C, as predicted by
the Vostok atmospheric CO2 concentrations substituted in
Equation 6; (TDv)—irregular-scale Vostok atmospheric
temperature deviations in degrees Celsius (°C) relative to
the Vostok base Kyr1 = 0.00 °C; (TDvCO2vLin)—second
order polynomial function of TDvCO2v; (TDvLin)—
second order polynomial function of TDv...................... 156

Figure 49 (Dataset 28)—thermal effect of artificial CO2
during the current interglacial. This graph shows: (TDb)—
Vostok base Kyr1 = 0.00 degrees Celsius (°C); (TDn)—
hypothesized natural glacial decline of atmospheric
temperature deviations in degrees Celsius (°C) relative to
the Vostok base Kyr1 = 0.00 °C; (TDa)—artificial
atmospheric temperature deviations, as estimated by
Equation 8, in degrees Celsius (°C) relative to the Vostok
base Kyr1 = 0.00 °C.. 157

Figure 50 (Dataset 28)—predicted thermal gap. This graph
shows: (TDvLin)—linear function of the Vostok
atmospheric temperature deviations, from 414,085 to
11,013 years before calendar year 1989, in degrees Celsius
(°C) relative to the Vostok base Kyr1 = 0.00 °C; and
(TDgLin)—linear function of the global annual
atmospheric temperature deviations, from 1959 to 2014, in
degrees Celsius (°C) relative to the Vostok base Kyr1 =
0.00 °C.. 164

Figure 51 (Dataset 28)—thermal action equals reaction. This graph shows: (TDvLin)—linear function of the Vostok atmospheric temperature deviations, from 414,085 to 11,013 years before calendar year 1989, in degrees Celsius (°C) relative to the Vostok base Kyr1 = 0.00 °C; (TDgLin)—linear function of the global annual atmospheric temperature deviations, from 1959 to 2014, in degrees Celsius (°C) relative to the Vostok base Kyr1 = 0.00 °C; (TDa)—artificial atmospheric temperature deviation, as predicted by Equation 8, in degrees Celsius (°C) relative to the Vostok base Kyr1 = 0.00 °C, at an atmospheric CO2 concentration of 400 parts per million (ppm) including artificial CO2; and (TDn)—natural atmospheric temperature deviation in degrees Celsius (°C) relative to the Vostok base Kyr1 = 0.00 °C (TDn = TDa— thermal gap)..165

Figure 52 (Dataset 28)—natural CO2 concentration at the predicted natural glacial temperature. This graph shows: (TDvLin)—linear function of the Vostok atmospheric temperature deviations, from 414,085 to 11,013 years before calendar year 1989, in degrees Celsius (°C) relative to the Vostok base Kyr1 = 0.00 °C; (TDa)—artificial atmospheric temperature deviation, as predicted by Equation 8, in degrees Celsius (°C) relative to the Vostok base Kyr1 = 0.00 °C, at an atmospheric CO2 concentration of 400 parts per million (ppm) including artificial CO2; and (TDn)—natural atmospheric temperature deviation in degrees Celsius (°C) relative to the Vostok base Kyr1 = 0.00 °C (TDn = TDa—thermal gap). 166

Figure 53 (Dataset 28)—linear equation thermal decrease towards natural balance. This graph shows: (TDvLin)— linear function of the Vostok atmospheric temperature deviations, from 414,085 to 11,013 years before calendar year 1989, in degrees Celsius (°C) relative to the Vostok base Kyr1 = 0.00 °C; (TDa)—artificial atmospheric temperature deviation, as predicted by Equation 8, in degrees Celsius (°C) relative to the Vostok base Kyr1 = 0.00 °C, at an atmospheric CO2 concentration of 400 parts per million (ppm) including artificial CO2; (TDn)—natural atmospheric temperature deviation in degrees Celsius (°C) relative to the Vostok base Kyr1 = 0.00 °C (TDn = TDa—thermal gap); and (TDdLin)—atmospheric temperature deviations in degrees Celsius (°C) relative to the Vostok base Kyr1 = 0.00 °C as a linear function of CO2 concentrations between TDa and TDn.............................. 167

Figure 54 (Dataset 28)—temperature at zero artificial CO2. This graph shows: (Current)—current atmospheric temperature of 1.37 °C at an artificial atmospheric CO2 concentration of 400.0 ppm; (ZeroCO2)—atmospheric temperature of -8.88 °C at a natural CO2 concentration of 176.9 ppm, which equals an artificial CO2 concentration of zero; (TDa)—artificial atmospheric temperature deviation, as predicted by Equation 8, in degrees Celsius (°C) relative to the Vostok base Kyr1 = 0.00 °C, at an atmospheric CO2 concentration of 400 parts per million (ppm) including artificial CO2; (TDn)—natural atmospheric temperature deviation in degrees Celsius (°C) relative to the Vostok base Kyr1 = 0.00 °C (TDn = TDa—thermal gap); and (TDdLin)—atmospheric temperature deviations in degrees Celsius (°C) relative to the Vostok base Kyr1 = 0.00 °C as a linear function of CO2 concentrations between TDa and TDn. 169

Figure 55 (Dataset 28)—Obama's CO2 reduction to 280 ppm. This graph shows: (Current)—current atmospheric temperature of 1.37 °C at an artificial atmospheric CO2 concentration of 400.0 ppm; (Obama)—artificial atmospheric CO2 concentration of 280.0 ppm resulting in an atmospheric temperature of -4.15 °C in 2025; (TDa)—artificial atmospheric temperature deviation, as predicted by Equation 8, in degrees Celsius (°C) relative to the Vostok base Kyr1 = 0.00 °C, at an atmospheric CO2 concentration of 400 parts per million (ppm) including artificial CO2;

(TDn)—natural atmospheric temperature deviation in degrees Celsius (°C) relative to the Vostok base Kyr1 = 0.00 °C (TDn = TDa—thermal gap); and (TDdLin)—atmospheric temperature deviations in degrees Celsius (°C) relative to the Vostok base Kyr1 = 0.00 °C as a linear function of CO2 concentrations between TDa and TDn. 170

Figure 56 (Dataset 28)—restoring Vostok base Kyr1 = 0.00 °C. This graph shows: (Current)—current atmospheric temperature of 1.37 °C at an artificial atmospheric CO2 concentration of 400.0 ppm; (Zero°C)—artificial atmospheric CO2 concentration of 370.3 ppm resulting in the atmospheric temperature of the Vostok base Kyr1 = 0.00 °C; (TDa)—artificial atmospheric temperature deviation, as predicted by Equation 8, in degrees Celsius (°C) relative to the Vostok base Kyr1 = 0.00 °C, at an atmospheric CO2 concentration of 400 parts per million (ppm) including artificial CO2; (TDn)—natural atmospheric temperature deviation in degrees Celsius (°C) relative to the Vostok base Kyr1 = 0.00 °C (TDn = TDa—thermal gap); and (TDdLin)—atmospheric temperature deviations in degrees Celsius (°C) relative to the Vostok base Kyr1 = 0.00 °C as a linear function of CO2 concentrations between TDa and TDn. 171

LIST OF TABLES

Table 1—modification Vostok temperature deviations during 423 Kyrs. This table shows the statistics for the following variables: (TDvd (Dataset 2))—depth-scale atmospheric temperature deviations, during the last 423 Kyrs, in degrees Celsius (°C) relative to the Vostok base 1850-1989 = 0.00 °C; (TDvm (Dataset 3))—millennial-scale atmospheric temperature deviations, during the last 423 Kyrs, in degrees Celsius (°C) relative to the Vostok base 1850-1989 = 0.00 °C; (TDvmb (Dataset 4))—millennial-scale atmospheric temperature deviations, during the last 423 Kyrs, in degrees Celsius (°C) relative to the Vostok base Kyr1 = 0.00 °C; (TDv (Dataset 5))—detrended millennial-scale atmospheric temperature deviations, during the last 423 Kyrs, in degrees Celsius (°C) relative to the Vostok base Kyr1 = 0.00 °C. . 50

Table 2 (Dataset 5)—durations of glacials during 423 Kyrs. This table shows the start and finish of glacial periods in years before the present calendar year 1989 (YBP); their duration in years; and the statistics of those glacials. 52

Table 3 (Dataset 5)—durations of interglacials during 423 Kyrs. This table shows the start and finish of interglacial periods in years before the present calendar year 1989 (YBP); their duration in years; and the statistics of those interglacials. 54

Table 4 (Dataset 5)—duration of the current interglacial. This table shows the start and finish of the current interglacial period in years before the present calendar year 1989 (YBP); and its duration in years. Furthermore, it shows the maximum duration of the previous interglacials; critical interglacial duration; z-score of the current interglacial duration; p-value of the current interglacial duration; and the number of previous interglacials. 55

Table 5 (Dataset 5)—statistics of the glacial decline window. This table shows the statistics for the following variables:

(TD-CIG)—mean TDv last 10 Kyrs; (TDv)—detrended millennial-scale atmospheric temperature deviations in degrees Celsius (°C) relative to the Vostok base Kyr1 = 0.00 °C; (D-PIG)—durations of the previous interglacials in years; (GD-PIG) glacial declines ending the previous interglacials in °C/Kyr...59

Table 6 (Dataset 5)—sliding window of thermal stability. This table shows the statistics for the following variables: (TD-CIG)—detrended millennial-scale atmospheric temperature deviations, during the current interglacial of 10 Kyrs, in degrees Celsius (°C) relative to the Vostok base Kyr1 = 0.00 °C; (TSI-SW)—thermal stability indexes as obtained with a sliding window of 10 Kyrs. This window shifted from Kyr414 to Kyr11 in steps of 1 Kyr; (TSI-CIG)—thermal stability index of the current interglacial during the last 10 Kyrs...62

Table 7—modification Vostok temperature deviations during 10 Kyrs. This table shows the statistics for the following variables: (TDvd (Dataset 2))—Vostok depth-scale atmospheric temperature deviations, during the last 10 Kyrs, in degrees Celsius (°C) relative to the Vostok base 1850-1989 = 0.00 °C; (TDva (Dataset 6))—Vostok semi-annual atmospheric temperature deviations, during the last 10 Kyrs, in degrees Celsius (°C) relative to the Vostok base 1850-1989 = 0.00 °C; (TDvab)—Vostok semi-annual temperature deviations, during the last 10 Kyrs, in degrees Celsius (°C) relative to the Vostok base Kyr1 = 0.00 °C; (TDv)—detrended semi-annual atmospheric temperature deviations, during the last 10 Kyrs, in degrees Celsius (°C) relative to the Vostok base Kyr1 = 0.00 °C. 69

Table 8 (Dataset 12)—extended global temperature deviations °C. This table shows the statistics for the following variables: (TDg)—global annual atmospheric temperature deviations from 1880 to 2014, in degrees Celsius (°C) relative to the global base 1951-1980 = 0.00 °C;

(ExtPol)—second order polynomial function of TDg from 1850 to 1879; (ExtLin)—linear function of TDg from 1850 to 1879; and (TDgExt)—backwards extension of TDg comprising the means of ExtPol and ExtLin from 1850 to 1879. 84

Table 9—uninterrupted increasing slopes of the 50-years window-scale temperatures. This table shows the statistics for the following variables: (TDvLin)—linear function of TDvSW; (TDvSW (Dataset 16))—Vostok window-scale atmospheric temperature deviations from calendar year 7962 BCE to 1989; (TDvLinR)—explained variance of TDvLin; (TDgLin)—linear function of TDgSW; (TDgSW (Dataset 15))—global window-scale atmospheric temperature deviations from calendar year 1929 to 2014 (n = 86); (TDgLinR)—explained variance of TDgLin. Window-scale temperature deviations are mean temperature deviations obtained with a window of 50 years sliding in steps of 1 year, which are expressed in degrees Celsius (°C) relative to the Vostok base Kyr1 = 0.00 °C. 99

Table 10 (Dataset 16)—increasing slopes of the Vostok 50-years window-scale temperatures, computed for an observation window of 86 years, sliding across the last 10 Kyrs in steps of 1 year. This table shows the statistics for the following variables: (TDvLin (Dataset 16))—linear function of TDvSW; (TDvSW (Dataset 16))—Vostok window-scale atmospheric temperature deviations from 7877 BCE to 1989; (TDvLinR (Dataset 16))—variance explained by TDvLin. 100

Table 11—duration uninterrupted increasing slopes of the 50-years window-scale temperatures. This table shows the statistics for the following variables: (TDvd)—duration uninterrupted increasing slope of TDvSW; (TDvSW (Dataset 16))—Vostok window-scale atmospheric temperature deviations from calendar year 7962 BCE to 1989; (TDgd)—duration uninterrupted increasing of

TDgSW; (TDgSW (Dataset 15))—global window-scale atmospheric temperature deviations from calendar year 1929 to 2014. Window-scale temperature deviations are mean temperature deviations obtained with a window of 50 years sliding in steps of 1 year, which are expressed in degrees Celsius (°C) relative to the Vostok base Kyr1 = 0.00 °C. 101

Table 12—statistics of CO2hz, TDgz, CO2p, and TDc. This table shows the statistics for the following variables: (CO2hz)—z-score of CO2h; (CO2h (Dataset 24))—Hawaiian atmospheric carbon dioxide concentrations, from 1959 to 2014, in parts per million (ppm); (TDgz)—z-score of TDg; (TDg (Dataset 24))—global annual atmospheric temperature deviations, from 1959 to 2014, in degrees Celsius (°C) relative to the Vostok base Kyr1 = 0.00 °C; (CO2p)—predicts TDc (CO2p = CO2hz – TDgz); (TDc)—change in next year's temperature deviation (TDc = TDgz1 – TDgz); and TDgz1—z-score of next year's temperature deviation. 134

Table 13—decision matrix and Type I and II errors. This table shows the decision matrix concerning the null hypothesis (Ho). A systematic effect is an effect attributed to a specific variable. In contrast, a random effect is an effect attributed to random variation around the mean. A hypothesis is an untested, but testable description of reality. The null hypothesis (Ho) states that the effect is random, rather than systematic. In contrast, the alternative hypothesis (Ha) states that the effect is systematic, rather than random. 220

LIST OF DATASETS

Dataset 1 (VostokDeutTempDepth423Kyrs). This dataset contains the following Vostok variables: (Depth corrected)—ice depth from 3,310 metres to 0 metre in steps of 1 metre; (Ice Age (GT4))—ice age in years from 422,766 years before present (BP) to the present calendar year 1989; (deut)—depth-scale deuterium concentrations in ‰ Standard Mean Ocean Sea Water (SMOW); and (DeltaTS)—depth-scale atmospheric temperature deviations in degrees Celsius (°C) relative to the Vostok base 1850-1989 = 0.00 °C. 37

Dataset 2 (VostokTempDepth423Kyrs). This dataset contains the following Vostok variables: (Depth)—ice depth from 3,310 metres to 0 metre in steps of 1 metre; (Age)—ice age in years from 422,766 years before present (BP) to the present calendar year 1989; and (TDv)—depth-scale atmospheric temperature deviations in degrees Celsius (°C) relative to the Vostok base 1850-1989 = 0.00 °C. This dataset is included in appendix: Dataset 2. 37

Dataset 3 (VostokTempMill423Kyrs). This dataset contains the following Vostok variables: (Age)—ice age in millennia or Kyrs from 423 Kyrs before present (BP) to the present calendar year 1989; (TDv)—millennial-scale atmospheric temperature deviations in degrees Celsius (°C) relative to the Vostok base 1850-1989 = 0.00 °C................................. 41

Dataset 4 (VostokTempMillBase423Kyrs). This dataset contains the following Vostok variables: (Age)—ice age in millennia or Kyrs from 423 Kyrs before present (BP) to the present calendar year 1989; (TDv)—millennial-scale atmospheric temperature deviations in degrees Celsius (°C) relative to the Vostok base Kyr1 = 0.00 °C. 44

Dataset 5 (VostokTempMillBaseDetrend423kyrs). This dataset contains the following Vostok variables: (Age)—ice age in millennia or Kyrs from 423 Kyrs before present (BP) to the present calendar year 1989; (TDv)—detrended millennial-scale atmospheric temperature deviations in degrees Celsius (°C) relative to the Vostok base Kyr1 = 0.00 °C. This dataset is included in appendix: Dataset 5...47

Dataset 6 (VostokTempAnnualRaw10Kyrs). This dataset contains the following Vostok variables: (Age)—in years from 10,000 years before present (BP) to the present calendar year 1989; and (TDv)—Vostok semi-annual atmospheric temperature deviations in degrees Celsius (°C) relative to the Vostok base 1850-1989 = 0.00 °C...............66

Dataset 7 (VostokTempAnnualDetrend10Kyrs). This dataset contains the following Vostok variables: (Age)—in years from 10,000 years before present (BP) to the present calendar year 1989; (TDv)—Vostok semi-annual atmospheric temperature deviations in degrees Celsius relative to the Vostok base 1850-1989 = 0.00 °C; and (TDvd)—detrended semi-annual atmospheric temperature deviations in degrees Celsius (°C) relative to the Vostok base Kyr1 = 0.00 °C. 68

Dataset 8 (VostokWindow10Kyrs). This dataset contains the following Vostok variables: (Age)—in years from 10,000 years before the present (BP) to the present calendar year 1989; (TDv)—semi-annual Vostok atmospheric temperature deviations; and (TDvSW)—Vostok window-scale atmospheric temperature deviations, which are the mean temperature deviations obtained with a sliding window of 50 years. Temperature deviations are in degrees Celsius (°C) relative to Vostok base Kyr1 = 0.00 °C. This dataset is included in appendix: Dataset 8. 73

Dataset 9 (GlobalTempMonthly). This dataset contains the monthly and annual global land-ocean temperature index in 0.01 degrees Celsius (°C) relative to the base period: 1951-1980, during the observation period from calendar year 1880 to 2014...............76

Dataset 10 (GlobalTempAnnual). This dataset contains the following global variables: (Year)—calendar years from 1880 to 2014; (TDg)—global annual atmospheric temperature deviations in degrees Celsius (°C) relative to the global base temperature 1951-1980 = 0.00 °C. 76

Dataset 11 (VostokTempAnnualKyr1). This dataset contains the following Vostok variables: (Age)—in years from 1,000 years before present (BP) to the present calendar year 1989; and (TDv)—Vostok detrended semi-annual atmospheric temperature deviations in degrees Celsius (°C) relative to the Vostok base Kyr1 = 0.00 °C. 79

Dataset 12 (GlobalTempAnnualExtended). This dataset contains the following global variables: (Year)—Calendar year 1850 to 2014; (TDg)—global annual atmospheric temperature deviations from 1880 to 2014, in degrees Celsius (°C) relative to the global base 1951-1980 = 0.00 °C; (ExtPol)—second order polynomial function of TDg from 1850 to 1879; (ExtLin)—linear function of TDg from 1850 to 1879; and (TDgExt)—backwards extension of TDg comprising the means of ExtPol and ExtLin from 1850 to 1879. 82

Dataset 13 (VostokGlobalTempCombined). This dataset contains the following variables: (Year)—calendar years from 989 to 2014; (TDv)—Vostok detrended semi-annual atmospheric temperature deviations from 989 to 1989, in degrees Celsius (°C) relative to the Vostok base Kyr1 = 0.00 °C; (TDvSyn)—TDv from 1850 to 1989; (TDgExt)—extended global annual atmospheric temperature deviations from 1850 to 2014, in degrees Celsius (°C) relative to the global base 1951-1980 = 0.00 °C; and (TDgSyn)—TDgExt from 1850 to 1989. 85

Dataset 14 (VostokGlobalTempSynchronized). This dataset contains the following variables: (Year)—calendar years from 989 to 2014; (TDv)—Vostok semi-annual atmospheric temperature deviations from 989 to 1989;

(TDg)—extended global annual atmospheric temperature deviations from calendar year 1850 to 2014; (Syn)—means of TDv and TDg during synchronization period from calendar year 1850 to 1989. TDg and TDv are expressed in degrees Celsius (°C) relative to the Vostok base Kyr1 = 0.00 °C. 88

Dataset 15 (GlobalWindow°C). This dataset contains the following global variables: (Year)—calendar years from 1880 to 2014; (TDg)—global annual-scale atmospheric temperature deviations from calendar year 1880 to 2014; (TDgSW)—global window-scale atmospheric temperature deviations from calendar year 1929 to 2014, which are the mean temperature deviations obtained with a window of 50 years sliding in steps of 1 year. TDg and TDgSW are in degrees Celsius (°C) relative to the Vostok base Kyr1 = 0.00 °C. This dataset is included in the appendix: Dataset 15. 91

Dataset 16 (VostokGlobalWindow°C). This dataset contains the following variables: (Year)—calendar years from 8011 BCE to 2014; (TDvSW)—Vostok window-scale atmospheric temperature deviations from calendar year 7962 BCE to 1989; (TDgSW)—global window-scale atmospheric temperature deviations from calendar year 1929 to 2014. Window-scale temperature deviations are mean temperature deviations obtained with a window of 50 years sliding in steps of 1 year, which are expressed in degrees Celsius (°C) relative to the Vostok base Kyr1 = 0.00 °C. 95

Dataset 17 (Vostok423Kyrs CO2). This dataset contains the following Vostok variables: (Gas age)—gas age from 414,085 to 2,342 years before calendar year 1989; and (CO2v)—Vostok atmospheric carbon dioxide concentrations, from 414,085 to 2,342 years before calendar year 1989, in parts per million (ppm). The data intervals are irregular. 105

Dataset 18 (VostokTempCO2Synch412Kyrs). This dataset contains the following Vostok variables: (Age)—from 414,085 to 2,342 years before calendar year 1989; (CO2v)—irregular-scale Vostok atmospheric carbon dioxide concentrations in parts per million (ppm); (TDv)—irregular-scale Vostok atmospheric temperature deviations in degrees Celsius (°C) relative to the Vostok base Kyr1 = 0.00 °C. This dataset is included in appendix: Dataset 18. 108

Dataset 19 (TemperaturesHawaiiAbsolute°C). This dataset contains the following Hawaiian variables: (Th)—Hawaiian annual atmospheric temperatures from calendar year 1955 to 2014, in absolute degrees Celsius (°C). 115

Dataset 20 (TemperatureDeviationsHawaii). This dataset contains the following Hawaiian variables: (Year)—calendar years from 1955 to 2014; (TDh)—Hawaiian annual atmospheric temperature deviations in degrees Celsius (°C) relative to the Vostok base Kyr1 = 0.00 °C. 116

Dataset 21 (Hawaii CO2). This dataset contains the following Hawaiian variables: (Year)—calendar years from 1959 to 2014; (CO2h)—Hawaiian annual atmospheric carbon dioxide concentrations in part per million (ppm).122

Dataset 22 (Hawaii temperature and CO2). This dataset contains the following Hawaiian variables: (Year)—calendar years from 1959 to 2014; (TDh)—Hawaiian atmospheric temperature deviations in degrees Celsius (°C) relative to the Vostok base Kyr1 = 0.00 °C; and (CO2h)—Hawaiian atmospheric CO2 concentrations in parts per million (ppm). This dataset is included in appendix: Dataset 22. 124

Dataset 23 (CO2Hawaii1959-2014). This dataset contains the following Hawaiian variables: Year; CO2 Concentration Posted by NOAA; Uncertainty Posted by NOAA; and CO2 Concentration Calculated. 128

Dataset 23 was downloaded for this study on the 2nd of
September 2105 from the website
http://co2now.org/Current-CO2/CO2-Now/noaa-
mauna-loa-co2-data.html "At CO2Now.org, data for
March 1958 - April 1974 was obtained by Charles David
Keeling of the Scripps Institution of Oceanography
(Scripps). Data for CO2 since May 1974 was obtained by
the National Oceanic and Atmospheric Administration
(NOAA). The Scripps Institution of Oceanography also
maintains a CO2 monitoring program at the Mauna Loa
Observatory." Dataset 24 (GlobalTempHawaiiCO2-1959-
2014). This dataset contains the following variables:
(Year)—calendar years from 1959 to 2014; (TDg)—global
annual atmospheric temperature deviations in degrees
Celsius (°C) relative to the Vostok base Kyr1 = 0.00 °C;
(CO2h)—Hawaiian atmospheric carbon dioxide
concentrations in parts per million (ppm). 128

Dataset 25 (VostokTempCO2Synch403Kyrs). This dataset
contains the following Vostok variables: (Age)—from
414,085 to 11,013 years before calendar year 1989;
(CO2v)—irregular-scale Vostok atmospheric carbon
dioxide concentrations in parts per million (ppm); (TDv)—
irregular-scale Vostok atmospheric temperature deviations
in degrees Celsius (°C) relative to the Vostok base Kyr1 =
0.00 °C; (TDvCO2v)—irregular-scale Vostok atmospheric
temperature deviation in degrees Celsius (°C) relative to the
Vostok base Kyr1 = 0.00 °C, as predicted by the Vostok
atmospheric CO2 concentrations substituted in Equation 6.
....139

Dataset 26 (HawaiiCO2PredictGlobal°C). This dataset
contains the following variables: (Year)—calendar years
from 1959 to 2014; (TDg)—global annual atmospheric
temperature deviations in degrees Celsius (°C) relative to
the Vostok base Kyr1 = 0.00 °C; (CO2h)—Hawaiian
annual atmospheric carbon dioxide concentrations in parts
per million (ppm); (TDgCO2h)—global atmospheric

temperature deviations in degrees Celsius (°C) relative to the Vostok base Kyr1 = 0.00 °C, as predicted by the Hawaiian atmospheric CO2 concentrations substituted in Equation 8. 144

Dataset 27 (ThermalGap). This dataset contains the following variables: (Year)—calendar years from 412,096 BCE to 2014; (TDg)—global annual atmospheric temperature deviations in degrees Celsius (°C) relative to the Vostok base Kyr1 = 0.00 °C, and (CO2h)—Hawaiian atmospheric carbon dioxide concentrations in parts per million (ppm), from 1959 to 2014; (CO2v)—irregular-scale Vostok atmospheric carbon dioxide concentrations in parts per million, and (TDv)—irregular-scale Vostok atmospheric temperature deviations in degrees Celsius (°C) relative to The Vostok base Kyr1 = 0.00 °C, from 412,096 to 9,024 BCE. This dataset is included in appendix: Dataset 27 148

Dataset 28 (CO2 forged thermal decrease). This dataset contains the following Vostok variables: (CO2)—atmospheric carbon dioxide concentrations in parts per million (ppm); (TDvLin)—linear function of the Vostok atmospheric temperature deviations, from 414,085 to 11,013 years before calendar year 1989, in degrees Celsius (°C) relative to the Vostok base Kyr1 = 0.00 °C; (TDgLin)—linear function of the global annual atmospheric temperature deviations, from 1959 to 2014, in degrees Celsius (°C) relative to the Vostok base Kyr1 = 0.00 °C; (TDa)—artificial atmospheric temperature deviation, as predicted by Equation 8, in degrees Celsius (°C) relative to the Vostok base Kyr1 = 0.00 °C, at an atmospheric CO2 concentration of 400 parts per million (ppm) including artificial CO2; (TDn)—natural atmospheric temperature deviation in degrees Celsius (°C) relative to the Vostok base Kyr1 = 0.00 °C (TDn = TDa—thermal gap); and (TDdLin)—atmospheric temperature deviations in degrees Celsius (°C) relative to the Vostok base Kyr1 =

0.00 °C as a linear function of CO_2 concentrations between TDa and TDn. 163

LIST OF TESTS

Test 1 (Dataset 5)—randomized correlation TDxSWr and TSISW (RCT). In which: (TDxSW)—Vostok ranked mean millennial temperature deviation obtained with a sliding window of 10 Kyrs; and (TSISW)—Vostok ranked thermal stability index (1/STD) of the millennial temperature deviation obtained with a sliding window of 10 Kyrs; observation period from Kyr423 to Kyr11. At each iteration, the correlation coefficient was computed after TDxSW was randomized relative to TSISW. Using ranked data avoids the statistical problems associated with asymmetric frequency distributions, unequal variances, and outliers. The chosen observation period reduces the confounding effects of human activity on the variables. ... 64

Test 2 (Dataset 18)—randomized correlation TDv and CO2v (RCT). In which: (TDv)—Vostok ranked irregular-scale temperature deviation; (CO2v)—Vostok ranked irregular-scale CO2 concentration; observation period from 414,085 to 2,342 before 1989. At each iteration, the correlation coefficient was computed after CO2v was randomized relative to TDv. Using ranked data avoids the statistical problems associated with asymmetric frequency distributions, unequal variances, and outliers. 112

Test 3—randomized correlation TDh and TDg. In which: (TDh (Dataset 20))—Hawaii ranked annual temperature deviation; (TDg (Dataset 14))—Global ranked annual temperature deviation; observation period from 1955 to 2014. At each iteration, the correlation coefficient was computed after TDh was randomized relative to TDg. Using ranked data avoids the statistical problems associated with asymmetric frequency distributions, unequal variances, and outliers. 118

Test 4 (Dataset 22)—randomized correlation CO2h and TDh. In which: (CO2h)—Hawaii ranked annual CO2 concentration; (TDh)—Hawaii ranked annual temperature deviation; observation period from 1959 to 2014. At each iteration, the correlation coefficient was computed after CO2h was randomized relative to TDh. Using ranked data avoids the statistical problems associated with asymmetric frequency distributions, unequal variances, and outliers. .126

Test 5 (Dataset 24)—randomized correlation TDg and CO2h. In which: (TDg)—Global ranked annual temperature deviation; (CO2h)—Hawaii ranked annual CO2 concentration; observation period from 1959 to 2014. At each iteration, the correlation coefficient was computed after TDg was randomized relative to CO2h. Using ranked data avoids the statistical problems associated with asymmetric frequency distributions, unequal variances, and outliers..131

Test 6 (Dataset 24)—randomized correlation CO2p and TDc. In which: (CO2p)—predictor of annual temperature deviations (CO2p = CO2hz – TDgz); (CO2hz)—Hawaii z-score annual CO2 concentration; (TDgz)—Global z-score annual temperature deviation; (TDc)—change in annual temperature deviations (TDc = TDgz next year – TDgz current year); observation period from 1959 to 2014. At each iteration, the correlation coefficient was computed after CO2p was randomized relative to TDc. Using z-scores avoids the statistical problems associated with different units of measurement. 135

Test 7 (Dataset 24)—randomized probability test evaluated the significance of the percentage correctly predicted TDc by CO2p one year into the future. In which: (CO2p)—predictor of annual temperature deviations (CO2p = CO2hz – TDgz); (CO2hz)—Hawaii z-score annual CO2 concentration; (TDgz)—Global z-score annual temperature deviation; (TDc)—change in annual temperature deviations (TDc = (TDgz at year + 1) – (TDgz at year)); observation period from 1959 to 2014. At each iteration, the percentage correctly predicted TDc by CO2p was computed after randomization. Using z-scores avoids the statistical problems associated with different units of measurement. 136

Test 8 (Dataset 24)—randomized probability test evaluated the significance of the percentage correctly predicted TDc by CO2p ten year into the future. In which: (CO2p)— predictor of annual temperature deviations (CO2p = CO2hz – TDgz); (CO2hz)—Hawaii z-score annual CO2 concentration; (TDgz)—Global z-score annual temperature deviation; (TDc)—change in annual temperature deviations (TDc = (TDgz at year + 10) – (TDgz at year)); observation period from 1959 to 2014. At each iteration, the percentage correctly predicted TDc by CO2p was computed after randomization. Using z-scores avoids the statistical problems associated with different units of measurement. 137

Test 9 (Dataset 25)—randomized correlation TDv and TDvCO2v. In which: (TDv)—Vostok ranked irregular-scale atmospheric temperature deviations in degrees Celsius (°C) relative to the Vostok base Kyr1 = 0.00 °C; (TDvCO2v)—Vostok ranked irregular-scale atmospheric temperature deviation in degrees Celsius (°C) relative to the Vostok base Kyr1 = 0.00 °C, as predicted by the Vostok atmospheric CO2 concentrations substituted in Equation 6; observation period from 414,085 to 11,013 before 1989. At each iteration, the correlation coefficient was computed after TDv was randomized relative to TDvCO2v. Using ranked data avoids the statistical problems associated with asymmetric frequency distributions, unequal variances, and outliers. 142

Test 10 (Dataset 26)—randomized correlation TDg and TDgCO2h. In which: (TDg)—global ranked annual atmospheric temperature deviations in degrees Celsius (°C) relative to the Vostok base Kyr1 = 0.00 °C; (TDgCO2h)— global ranked annual atmospheric temperature deviations in degrees Celsius (°C) relative to the Vostok base Kyr1 = 0.00 °C, as predicted by the Hawaiian atmospheric CO2 concentrations substituted in Equation 8; observation period from 1959 to 2014. At each iteration, the correlation coefficient was computed after TDg was randomized relative to TDgCO2h. Using ranked data avoids the statistical problems associated with asymmetric frequency distributions, unequal variances, and outliers....................146

Test 11 (Dataset 27)—sliding window test (SWT), comprising 56 data pairs shifting in steps of 1 pair, evaluated whether the Vostok natural long-term linear function (TDvLin = 0.09184 °C/ppm) is significantly larger than the global artificial short-term linear function of (TDgLin = 0.00945 °/ppm). In which: (TDvLin)—linear function of TDv as a function of CO2v; (TDv)—irregular-scale Vostok atmospheric temperature deviations in degrees Celsius (°C) relative to the Vostok base Kyr1 = 0.00 °C; (CO2v)—irregular-scale Vostok atmospheric carbon dioxide concentrations in parts per million; and observation period TDv and CO2v from 412,096 to 9,024 BCE; versus (TDgLin)—linear function of TDg as a function of CO2h; (TDg)—global annual atmospheric temperature deviations in degrees Celsius (°C) relative to the Vostok base Kyr1 = 0.00 °C; (CO2h)—Hawaiian atmospheric carbon dioxide concentrations in parts per million (ppm); and observation period TDg and CO2h from 1959 to 2014. 149

Test 12 (Dataset 25)—randomized correlation TDv and CO2v. In which: (TDv)—Vostok ranked irregular-scale atmospheric temperature deviations in degrees Celsius (°C) relative to the Vostok base Kyr1 = 0.00 °C; (CO2v)—Vostok ranked irregular-scale atmospheric carbon dioxide concentrations in parts per million (ppm); observation period from 414,085 to 11,013 years before 1989. At each iteration, the correlation coefficient was computed after CO2v was randomized relative to TDv. Using ranked data avoids the statistical problems associated with asymmetric frequency distributions, unequal variances, and outliers.. 150

Test 13 (Dataset 26)—randomized correlation TDg and CO2h. In which: (TDg)—global ranked annual atmospheric temperature deviations in degrees Celsius (°C) relative to the Vostok base Kyr1 = 0.00 °C; (CO2h)— Hawaiian ranked annual atmospheric carbon dioxide concentrations in parts per million (ppm); observation period from 1959 to 2014. At each iteration, the correlation coefficient was computed after TDg was randomized relative to CO2h. Using ranked data avoids the statistical problems associated with asymmetric frequency distributions, unequal variances, and outliers. 151

LIST OF EQUATIONS

Equation 1 (Dataset 2)—computes the millennial thermal product: $\sum Tp = (TD * DI)$. In which, $(\sum Tp)$—sum thermal products for an observation period of 1,000 years; (TD)—depth-scale temperature deviation in degrees Celsius (°C) relative to the Vostok base Kyr1 = 0.00 °C; and (DI)—depth-scale data interval in years. 41

Equation 2 (Dataset 2)—computes the millennial temperature deviations: $TDw = TP / 1000$. In which (TDw)— millennial weighted mean temperature deviation in degrees Celsius (°C) relative to the Vostok base Kyr1 = 0.00 °C; (Tp)—thermal product Equation 1; and (1,000)—duration of the observation period in years. If a data interval overlapped the boundary between two millennial periods, then TP was allocated proportionally to each millennium. 41

Equation 3 (Dataset 5)—computes the critical duration of the interglacials: $CD = DIx + (STD * 1.645)$. In which, (CD)—critical interglacial duration; (DIx)—mean duration of the previous interglacials in years; (STD) = standard deviation of previous interglacial durations in years; and (1.645)—critical z-score at $\alpha = 0.05$. 53

Equation 4 (Dataset 5)—computes the thermal stability index: $TSI = 1 / STD$. In which, (TSI)—Thermal Stability Index; (STD)—standard deviation of TDv; and (TDv)— detrended millennial-scale atmospheric temperature deviations in degrees Celsius (°C) relative to the Vostok base Kyr1 = 0.00 °C. 61

Equation 5 (Dataset 19)—converts degrees Fahrenheit into degrees Celsius: $°C = (°F - 32) * 5/9$. In which, (°C)— degrees Celsius; and (°F)—degrees Fahrenheit. 115

Equation 6 (Dataset 25)—estimates the long-term natural Vostok temperature for a given Vostok CO2 concentration: $TDvCO2v = (0.0918422886772854 * CO2v) -$

25.1267163418446. In which: (TDvCO2v)—estimated Vostok atmospheric temperature deviation in degrees Celsius (°C) relative to the Vostok base Kyr1 = 0.00 °C; and (CO2v)—observed Vostok atmospheric carbon dioxide concentrations in parts per million (ppm). 141

Equation 7 (Dataset 25)—estimates the long-term natural Vostok CO_2 concentration for a given Vostok temperature: CO2vTDv = (TDv + 25.1267163418446) / 0.0918422886772854. In which: (CO2vTDv)—estimated Vostok atmospheric carbon dioxide concentration in parts per million (ppm); (TDv)—observed Vostok atmospheric temperature deviation in degrees Celsius (°C) relative to the Vostok base Kyr1 = 0.00 °C.. 141

Equation 8 (Dataset 26)—estimates the short-term artificial global temperature for a given Hawaiian CO_2 concentration: TDgCO2h = (0.00944709321227046 * CO2h) - 2.41351639838839. In which: (TDgCO2h)—estimated global atmospheric temperature deviation in degrees Celsius (°C) relative to the Vostok base Kyr1 = 0.00 °C; (CO2h)—observed Hawaiian atmospheric carbon dioxide concentration in parts per million (ppm). 145

Equation 9 (Dataset 28)—estimates the temperature for a given CO_2 concentration within the range from TDa to TDn: TDdCO2d = (0.0459211443386428 * CO2d) - 17.0031368489373. In which: (TDdCO2d)—estimated decreased atmospheric temperature deviation in degrees Celsius (°C) relative to the Vostok base Kyr1 = 0.00 °C; (CO2d)—decreased atmospheric carbon dioxide concentration in parts per million (ppm); (TDa)—artificial atmospheric temperature deviation, as estimated by Equation 8, in degrees Celsius (°C) relative to the Vostok base Kyr1 = 0.00 °C; (TDn)—estimated natural atmospheric temperature deviation in degrees Celsius (°C) relative to the Vostok base Kyr1 = 0.00 °C (TDn = TDa - thermal gap)... 168

Equation 10 (Dataset 28)—estimates the CO2 concentration for a given temperature within the range from TDa to TDn: CO2dTDd = (TDd + 17.0031368489373) / 0.0459211443386427. In which: (CO2dTDd)—estimated decreased atmospheric carbon dioxide concentration in parts per million (ppm); (TDd)—decreased atmospheric temperature deviation in degrees Celsius (°C) relative to the Vostok base Kyr1 = 0.00 °C. (TDa) —artificial atmospheric temperature deviation, as estimated by Equation 8, in degrees Celsius (°C) relative to the Vostok base Kyr1 = 0.00 °C; (TDn)—natural atmospheric temperature deviation in degrees Celsius (°C) relative to the Vostok base Kyr1 = 0.00 °C (TDn = TDa - thermal gap).

....168

STATISTICS

Statistics is a field of study concerned with collecting, organizing, describing, analysing, and evaluating numerical facts called data.

Statistical tests of significance do not make decisions. Instead, they are decision-making tools used to compute the probability of making an incorrect decision. This involves primarily the probability of making Type I and II errors.

Decision	Null hypothesis (Ho)	
	Ho = true (random)	Ho = false (systematic)
Reject Ho & accept Ha	Type I error	Correct rejection
Accept Ho & reject Ha	Correct acceptance	Type II error

Table 13—decision matrix and Type I and II errors. This table shows the decision matrix concerning the null hypothesis (Ho). A systematic effect is an effect attributed to a specific variable. In contrast, a random effect is an effect attributed to random variation around the mean. A hypothesis is an untested, but testable description of reality. The null hypothesis (Ho) states that the effect is random, rather than systematic. In contrast, the alternative hypothesis (Ha) states that the effect is systematic, rather than random.

The truth of the null hypothesis (Ho) is evaluated with a statistical test of significance. The alternative hypothesis (Ha) is accepted if the null hypothesis is rejected by the test of significance. *Accept* and *reject* are two of the 17 nemoniks.

Objective and reactive mindmodes

Statistics is a science because it computes objectively the probability of making a Type I error (p). Objective refers to a description of reality that is independent of what anyone believes. For example, given particular datasets, the magnitudes of statistics such as the mean; standard deviation; maximum; minimum; range; and p-value remain the same, independent of what people believe. Even a computer would obtain the same values.

Objective is the mental nemonik that refers to the objective mindmode. The objective mindmode is a way of rational thinking that deals with the natural order of reality, which can be described by natural laws and facts. Natural laws are objective descriptions of the un-changeable cause-effect relationships of nature. Facts are testable descriptions of reality that are supported adequately by sensory perception and reason. Objectivity is the basis of scientific agreement.

The Greek letter alpha (α) denotes the probability that a decision maker accepts to make an error. For an effect to be significant, the computed probability of making a Type I error (p) should be smaller than the accepted probability to make such an error (α). This decision rule could be written as: $p < \alpha$.

Setting the magnitude of the critical α-value is a personal choice, which traditionally varies widely from a lax criterion of 0.05 to a strict criterion of 0.001. Although the p-value is objectively computed, the α-value is subjectively determined by the amount of risk that the decision maker is prepared to take. Consequently, even if the statistics are computed by artificial intelligence, it is still a human decision maker who determines the critical α-value. By blindly relying on statistics, decision makers abandon their responsibilities and are playing a sophisticated form of Russian roulette. They should at least know the weaknesses of statistics.

Risk assessment is based on a person's believe system, which is created and maintained by the reactive mindmode. The reactive mindmode is a way of affectorial thinking that deals with the disorder of reality by mindsets that generate reactive affecters. Mindsets are internalized sets of rules or algorithms that generate reactive affecters. Reactive affecters are physical and mental responses that deal with the disorder of reality. They include beliefs, common sense, desires, emotions, feelings, habits, heuristics, impulses, intuitions, reactions, reflexes, routines, sensibilities, skills, etc. Reactivity is subjective and, therefore, it is the basis of disagreement.

Parametric tests

Parametric statistics is a branch of statistics concerned with evaluating the probability that an effect is either systematic or random, by comparing the statistics of a sample with the parameters of a normal dataset.[372]   A normal dataset is a standard dataset comprising a symmetric, unimodal, and bell shaped frequency curve. The mean and the median of a normal dataset are the same. The prime weakness of parametric statistics is the assumption that the sample is part of a population with a normal distribution. That assumption is difficult to test and might be incorrect.

Non-parametric tests

Non-parametric statistics is a branch of statistics concerned with evaluating the probability that an effect is either systematic or random without relying on a normal dataset. Therefore, non-parametric statistics avoid the assumption that the sample is part of a population with a normal dataset. Although less sensitive, the results of non-parametric test are often more valid and reliable than the results of parametric test.

223

DATASETS

The datasets and analyses mentioned in this section can be downloaded from: nemonik-thinking.org

DATASET 1

Dataset 1 (VostokDeutTempDepth423Kyrs). This dataset contains the following Vostok variables: (Depth corrected)—ice depth from 3,310 metres to 0 metre in steps of 1 metre; (Ice Age (GT4))—ice age in years from 422,766 years before present (BP) to the present calendar year 1989; (deut)—depth-scale deuterium concentrations in ‰ Standard Mean Ocean Sea Water (SMOW); and (DeltaTS)—depth-scale atmospheric temperature deviations in degrees Celsius (°C) relative to the Vostok base 1850-1989 = 0.00 °C.

Citation: Petit, J.R., et al., 2001, Vostok Ice Core Data for 420,000 Years, IGBP PAGES/World Data Center for Paleoclimatology Data Contribution Series #2001-076. NOAA/NGDC Paleoclimatology Program, Boulder CO, USA. Original reference: Petit J.R., Jouzel J., Raynaud D., Barkov N.I., Barnola J.M., Basile I., Bender M., Chappellaz J., Davis J., Delaygue G., Delmotte M., Kotlyakov V.M., Legrand M., Lipenkov V., Lorius C., Pépin L., Ritz C., Saltzman E., Stievenard M., 1999, Climate and Atmospheric History of the Past 420,000 years from the Vostok Ice Core, Antarctica, Nature, 399, pp. 429-436. Please cite the original reference when using this data.

The dataset was derived from an ice core drilled by a French-Russian team at the Vostok station in Antarctica. Vostok is situated at North-bound latitude -78.47 * South-bound latitude -78.47; West-bound longitude 106.8 * East-bound longitude 106.8.

The base was derived from the surface ice down to 7 metres depth. Each Ice Age (GT4) is a midpoint of a period determined by 1 metre of ice. At seven metres depth, the temporal midpoint of the corresponding metre of ice equals

129 years and at eight meters 149 years BP. In accord, the meter of ice starts at $((129 + 149) / 2) = 139$ years BP. Hence, the base period ranges from 139 years Before-Present (BP) to the present calendar year 1989. Therefore, the base includes the period from calendar year 1850 to 1989 with a mean temperature deviation of $0.00\,°C$.

Quotes from the NOAA website: "http://www.ncdc.noaa.gov/paleo/icecore/antarctica/vostok/vostok_isotope.html. Downloaded Friday, 24-Jun-2011 22:33:43 EDT. Last Updated Wednesday, 20-Aug-2008 11:24:22 EDT by paleo@noaa.gov. Please see the Paleoclimatology Contact Page or the NCDC Contact Page if you have questions or comments." "NAME OF DATASET: Vostok Ice Core Data for 420,000 Years. LAST UPDATE: 11/2001 (Original Receipt by WDC Paleo). CONTRIBUTOR: Jean Robert Petit, LGGE-CNRS. IGBP PAGES/WDCA CONTRIBUTION SERIES NUMBER: 2001-076. SUGGESTED DATA CITATION: Petit, J.R., et al., 2001, Vostok Ice Core Data for 420,000 Years, IGBP PAGES/World Data Center for Paleoclimatology Data Contribution Series #2001-076. NOAA/NGDC Paleoclimatology Program, Boulder CO, USA." "PLEASE CITE ORIGINAL REFERENCE WHEN USING THIS DATA!"

DATASET 2

Dataset 2 (VostokTempDepth423Kyrs). This dataset contains the following Vostok variables: (Depth)—ice depth from 3,310 metres to 0 metre in steps of 1 metre; (Age)—ice age in years from 422,766 years before present (BP) to the present calendar year 1989; and (TDv)—depth-scale atmospheric temperature deviations in degrees Celsius (°C) relative to the Vostok base 1850-1989 = 0.00 °C.[373]

Depth | Age | TDv

Depth	Age	TDv
0	0	0.00
1	17	0.00
2	35	0.00
3	53	0.00
4	72	0.00
5	91	0.00
6	110	0.00
7	129	0.00
8	149	-0.81
9	170	0.02
10	190	0.36
11	211	-0.95
12	234	-1.84
13	258	-1.09
14	281	-0.75
15	304	-0.22
16	327	-0.48
17	351	-0.75
18	375	0.23
19	397	1.33
20	420	0.35
21	444	0.18
22	469	-0.08
23	495	-1.08
24	523	-1.39
25	552	-1.61
26	581	-0.90
27	609	-0.60
28	637	-0.02
29	665	-0.18
30	695	-1.23
31	726	-1.54
32	757	-0.85
33	788	-0.10
34	817	-0.17
35	848	-0.78
36	881	-0.78
37	912	-0.17
38	944	-0.25
39	976	-0.28
40	1,009	-0.81
41	1,042	-0.05
42	1,074	0.27
43	1,107	-0.83
44	1,142	-0.65
45	1,176	-0.27
46	1,211	-0.65
47	1,247	-1.48
48	1,285	-1.64
49	1,321	-0.40
50	1,356	0.33
51	1,390	-0.05
52	1,426	-0.13
53	1,461	-0.03
54	1,497	-0.46
55	1,535	-1.00
56	1,573	-0.96
57	1,612	-1.33
58	1,652	-1.51

59	1,692	-0.88	89	2,934	-1.34
60	1,732	-1.06	90	2,980	-1.41
61	1,772	-1.21	91	3,026	-0.73
62	1,812	-1.24	92	3,070	-0.25
63	1,853	-0.96	93	3,114	-0.53
64	1,893	-0.53	94	3,158	-0.08
65	1,931	0.15	95	3,201	0.15
66	1,970	-0.28	96	3,245	-0.43
67	2,009	-0.45	97	3,289	-0.61
68	2,049	-0.38	98	3,334	-0.30
69	2,089	-0.53	99	3,379	-0.13
70	2,129	-0.61	100	3,422	0.22
71	2,171	-0.95	101	3,466	0.17
72	2,212	-0.73	102	3,511	-0.91
73	2,253	0.12	103	3,558	-0.90
74	2,291	1.16	104	3,603	0.30
75	2,331	-0.98	105	3,646	0.65
76	2,374	-1.44	106	3,689	0.58
77	2,418	-0.73	107	3,732	0.03
78	2,460	-0.63	108	3,778	-0.63
79	2,501	0.30	109	3,824	-0.46
80	2,542	-0.27	110	3,870	0.00
81	2,585	-1.36	111	3,915	-0.23
82	2,628	-0.17	112	3,962	-0.66
83	2,670	-0.02	113	4,009	-0.73
84	2,713	-1.19	114	4,057	-0.75
85	2,760	-1.97	115	4,104	-0.91
86	2,805	-0.61	116	4,153	-1.09
87	2,847	0.58	117	4,202	-1.00
88	2,889	-0.38	118	4,250	-0.53

119	4,295	0.55	149	5,674	-0.06
120	4,339	1.13	150	5,721	-0.15
121	4,381	1.13	151	5,769	-0.25
122	4,423	1.13	152	5,816	-0.22
123	4,466	1.13	153	5,863	0.15
124	4,509	1.13	154	5,909	0.45
125	4,552	0.88	155	5,955	-0.13
126	4,596	0.20	156	6,004	-0.73
127	4,642	-0.33	157	6,052	-0.21
128	4,690	-0.88	158	6,099	0.32
129	4,739	-0.58	159	6,145	-0.17
130	4,786	0.25	160	6,193	-0.33
131	4,831	0.00	161	6,241	0.09
132	4,880	-1.14	162	6,287	0.28
133	4,929	-0.66	163	6,334	-0.55
134	4,977	-0.36	164	6,385	-1.59
135	5,025	-0.66	165	6,436	-1.09
136	5,072	0.13	166	6,486	-0.31
137	5,118	0.18	167	6,534	-0.62
138	5,165	-0.38	168	6,583	-0.65
139	5,212	-0.22	169	6,631	0.37
140	5,259	0.28	170	6,677	0.37
141	5,305	0.10	171	6,724	-0.36
142	5,351	0.08	172	6,773	-0.84
143	5,397	0.50	173	6,823	-1.28
144	5,442	0.50	174	6,874	-1.14
145	5,488	0.20	175	6,924	-0.58
146	5,534	0.02	176	6,973	-0.59
147	5,581	0.08	177	7,023	-1.05
148	5,627	0.03	178	7,074	-1.20

179	7,124	-0.63
180	7,172	0.02
181	7,220	-0.04
182	7,267	-0.11
183	7,315	-0.29
184	7,364	-0.81
185	7,413	-0.58
186	7,462	0.02
187	7,509	-0.03
188	7,555	0.48
189	7,602	-0.02
190	7,649	0.02
191	7,697	-0.04
192	7,745	-0.72
193	7,794	-0.57
194	7,844	-0.90
195	7,894	-0.93
196	7,944	-0.69
197	7,994	-0.54
198	8,043	-0.64
199	8,091	0.30
200	8,135	2.06
201	8,178	0.59
202	8,226	-0.87
203	8,276	-0.78
204	8,325	-0.55
205	8,374	-0.21
206	8,422	-0.26
207	8,471	-0.52
208	8,520	-0.65
209	8,569	-0.68
210	8,619	-0.73
211	8,668	-0.56
212	8,716	0.34
213	8,763	0.13
214	8,811	-0.81
215	8,861	-0.53
216	8,910	-0.54
217	8,960	-0.88
218	9,009	-0.48
219	9,058	-0.41
220	9,107	-0.60
221	9,156	-0.51
222	9,204	-0.13
223	9,252	0.26
224	9,298	0.03
225	9,346	-0.39
226	9,396	-0.82
227	9,445	-0.70
228	9,494	-0.16
229	9,542	-0.16
230	9,589	0.43
231	9,635	0.51
232	9,682	-0.46
233	9,732	-0.80
234	9,782	-0.85
235	9,831	-0.74
236	9,881	-0.36
237	9,929	-0.37
238	9,978	-0.65

239 | 10,027 | -0.28
240 | 10,075 | -0.47
241 | 10,124 | -0.58
242 | 10,172 | 0.31
243 | 10,218 | 0.52
244 | 10,265 | -0.18
245 | 10,315 | -1.30
246 | 10,366 | -0.97
247 | 10,415 | -0.43
248 | 10,465 | -0.76
249 | 10,515 | -0.77
250 | 10,564 | -0.79
251 | 10,614 | -0.90
252 | 10,665 | -0.96
253 | 10,715 | -0.70
254 | 10,764 | -0.26
255 | 10,812 | -0.44
256 | 10,861 | -0.62
257 | 10,910 | -0.15
258 | 10,957 | -0.12
259 | 11,005 | -0.23
260 | 11,053 | 0.02
261 | 11,100 | 0.09
262 | 11,146 | 0.74
263 | 11,191 | 0.81
264 | 11,237 | -0.18
265 | 11,286 | -0.48
266 | 11,334 | -0.27
267 | 11,383 | -0.73
268 | 11,434 | -1.22

269 | 11,485 | -1.29
270 | 11,537 | -1.57
271 | 11,590 | -1.59
272 | 11,642 | -1.61
273 | 11,695 | -1.83
274 | 11,749 | -2.20
275 | 11,805 | -2.37
276 | 11,861 | -2.85
277 | 11,918 | -2.55
278 | 11,973 | -2.19
279 | 12,029 | -2.84
280 | 12,087 | -3.09
281 | 12,144 | -2.77
282 | 12,202 | -2.90
283 | 12,261 | -3.88
284 | 12,323 | -4.18
285 | 12,385 | -3.96
286 | 12,446 | -3.61
287 | 12,507 | -3.76
288 | 12,569 | -4.25
289 | 12,632 | -4.25
290 | 12,694 | -3.91
291 | 12,755 | -3.57
292 | 12,815 | -3.24
293 | 12,874 | -3.40
294 | 12,934 | -3.50
295 | 12,994 | -3.67
296 | 13,055 | -3.85
297 | 13,116 | -3.75
298 | 13,177 | -3.72

299 | 13,237 | -3.45
300 | 13,296 | -3.17
301 | 13,355 | -3.17
302 | 13,414 | -3.67
303 | 13,476 | -4.28
304 | 13,539 | -4.02
305 | 13,600 | -3.57
306 | 13,659 | -3.27
307 | 13,718 | -2.91
308 | 13,774 | -2.19
309 | 13,828 | -2.04
310 | 13,883 | -2.24
311 | 13,938 | -2.06
312 | 13,992 | -2.32
313 | 14,048 | -2.57
314 | 14,105 | -2.83
315 | 14,163 | -3.08
316 | 14,221 | -3.34
317 | 14,281 | -3.60
318 | 14,342 | -3.85
319 | 14,404 | -4.11
320 | 14,466 | -3.64
321 | 14,526 | -3.82
322 | 14,589 | -4.34
323 | 14,651 | -3.98
324 | 14,713 | -3.92
325 | 14,775 | -4.14
326 | 14,839 | -4.52
327 | 14,904 | -4.68
328 | 14,968 | -4.44

329 | 15,032 | -4.66
330 | 15,099 | -5.32
331 | 15,167 | -5.40
332 | 15,234 | -5.00
333 | 15,300 | -4.81
334 | 15,366 | -4.74
335 | 15,432 | -5.15
336 | 15,501 | -5.71
337 | 15,570 | -5.60
338 | 15,639 | -5.44
339 | 15,708 | -5.57
340 | 15,777 | -5.29
341 | 15,845 | -5.35
342 | 15,915 | -6.19
343 | 15,987 | -5.99
344 | 16,057 | -5.75
345 | 16,128 | -6.08
346 | 16,201 | -6.34
347 | 16,275 | -6.74
348 | 16,350 | -6.88
349 | 16,426 | -6.88
350 | 16,502 | -6.74
351 | 16,577 | -6.77
352 | 16,653 | -6.97
353 | 16,729 | -7.16
354 | 16,808 | -7.49
355 | 16,889 | -8.33
356 | 16,974 | -8.51
357 | 17,058 | -8.04
358 | 17,139 | -7.60

359	17,219	-7.53
360	17,298	-7.45
361	17,379	-8.09
362	17,462	-8.27
363	17,544	-7.68
364	17,625	-7.78
365	17,706	-7.84
366	17,787	-7.82
367	17,868	-7.72
368	17,949	-7.75
369	18,031	-8.19
370	18,116	-8.63
371	18,201	-8.16
372	18,283	-7.89
373	18,365	-7.74
374	18,446	-7.97
375	18,530	-8.50
376	18,615	-8.43
377	18,701	-8.73
378	18,787	-8.46
379	18,870	-7.58
380	18,950	-7.63
381	19,032	-8.11
382	19,116	-8.26
383	19,199	-8.03
384	19,282	-7.81
385	19,362	-7.54
386	19,443	-7.70
387	19,525	-8.15
388	19,610	-8.45
389	19,696	-8.71
390	19,782	-8.52
391	19,868	-8.61
392	19,953	-7.99
393	20,035	-7.62
394	20,116	-7.84
395	20,197	-7.52
396	20,278	-7.82
397	20,361	-8.11
398	20,444	-8.08
399	20,528	-8.03
400	20,611	-8.04
401	20,694	-7.96
402	20,777	-7.87
403	20,859	-7.94
404	20,943	-8.12
405	21,026	-8.10
406	21,110	-7.93
407	21,192	-7.91
408	21,275	-7.91
409	21,358	-8.00
410	21,442	-8.09
411	21,525	-7.73
412	21,605	-7.36
413	21,686	-7.66
414	21,769	-8.24
415	21,854	-8.33
416	21,939	-8.12
417	22,023	-8.15
418	22,108	-8.41

419	22,196	-8.82	449	24,781	-7.25
420	22,284	-8.82	450	24,860	-7.00
421	22,371	-8.45	451	24,941	-7.57
422	22,457	-8.30	452	25,024	-8.14
423	22,543	-8.37	453	25,109	-7.98
424	22,629	-8.45	454	25,193	-7.83
425	22,716	-8.52	455	25,277	-7.68
426	22,803	-8.59	456	25,360	-7.89
427	22,888	-7.97	457	25,445	-8.27
428	22,972	-7.92	458	25,531	-8.13
429	23,057	-8.50	459	25,615	-7.70
430	23,145	-8.80	460	25,697	-7.05
431	23,234	-8.89	461	25,776	-6.98
432	23,324	-9.01	462	25,855	-7.26
433	23,412	-8.54	463	25,936	-7.29
434	23,497	-7.76	464	26,017	-7.47
435	23,581	-8.20	465	26,099	-7.48
436	23,668	-8.63	466	26,180	-7.16
437	23,755	-8.63	467	26,261	-7.24
438	23,843	-8.63	468	26,342	-7.55
439	23,931	-8.44	469	26,425	-7.82
440	24,017	-8.26	470	26,510	-8.05
441	24,102	-8.03	471	26,595	-8.09
442	24,186	-7.87	472	26,681	-8.09
443	24,272	-8.68	473	26,766	-7.74
444	24,363	-9.39	474	26,849	-7.56
445	24,453	-8.46	475	26,933	-7.90
446	24,537	-7.53	476	27,019	-8.64
447	24,619	-7.51	477	27,110	-9.02
448	24,700	-7.50	478	27,202	-9.02

479 | 27,293 | -8.82
480 | 27,382 | -8.62
481 | 27,470 | -8.14
482 | 27,555 | -7.84
483 | 27,641 | -8.28
484 | 27,730 | -8.66
485 | 27,819 | -8.52
486 | 27,907 | -8.27
487 | 27,993 | -8.09
488 | 28,079 | -7.90
489 | 28,163 | -7.69
490 | 28,247 | -7.49
491 | 28,331 | -8.16
492 | 28,420 | -8.82
493 | 28,510 | -8.47
494 | 28,597 | -8.12
495 | 28,683 | -7.77
496 | 28,766 | -7.42
497 | 28,848 | -7.07
498 | 28,927 | -6.71
499 | 29,006 | -6.71
500 | 29,085 | -6.98
501 | 29,166 | -7.25
502 | 29,249 | -7.52
503 | 29,332 | -7.79
504 | 29,418 | -8.05
505 | 29,505 | -8.35
506 | 29,594 | -8.37
507 | 29,682 | -8.33
508 | 29,771 | -8.58
509 | 29,861 | -8.43
510 | 29,949 | -8.30
511 | 30,036 | -7.95
512 | 30,120 | -7.14
513 | 30,201 | -7.02
514 | 30,283 | -7.62
515 | 30,368 | -7.83
516 | 30,453 | -7.75
517 | 30,538 | -7.64
518 | 30,622 | -7.52
519 | 30,705 | -7.41
520 | 30,788 | -7.30
521 | 30,871 | -7.63
522 | 30,957 | -7.95
523 | 31,043 | -8.00
524 | 31,130 | -8.05
525 | 31,216 | -7.85
526 | 31,302 | -7.65
527 | 31,387 | -7.77
528 | 31,473 | -8.03
529 | 31,560 | -8.20
530 | 31,648 | -7.93
531 | 31,733 | -7.40
532 | 31,815 | -7.14
533 | 31,896 | -6.77
534 | 31,976 | -6.74
535 | 32,056 | -6.79
536 | 32,137 | -7.09
537 | 32,219 | -7.27
538 | 32,300 | -6.61

539 | 32,380 | -7.07
540 | 32,463 | -7.49
541 | 32,546 | -6.99
542 | 32,626 | -6.62
543 | 32,705 | -6.71
544 | 32,787 | -7.47
545 | 32,871 | -7.47
546 | 32,955 | -7.29
547 | 33,039 | -7.59
548 | 33,124 | -7.60
549 | 33,209 | -7.60
550 | 33,293 | -7.34
551 | 33,376 | -7.07
552 | 33,459 | -7.44
553 | 33,544 | -7.54
554 | 33,628 | -7.31
555 | 33,711 | -7.12
556 | 33,793 | -7.17
557 | 33,877 | -7.36
558 | 33,960 | -7.17
559 | 34,041 | -6.53
560 | 34,121 | -6.88
561 | 34,202 | -6.76
562 | 34,281 | -6.25
563 | 34,360 | -6.59
564 | 34,438 | -6.08
565 | 34,515 | -6.01
566 | 34,592 | -6.15
567 | 34,669 | -5.87
568 | 34,744 | -5.70

569 | 34,818 | -5.14
570 | 34,889 | -4.82
571 | 34,960 | -5.05
572 | 35,033 | -5.52
573 | 35,108 | -5.98
574 | 35,185 | -5.90
575 | 35,261 | -5.82
576 | 35,336 | -5.39
577 | 35,409 | -5.24
578 | 35,482 | -5.37
579 | 35,556 | -5.55
580 | 35,631 | -5.65
581 | 35,705 | -5.41
582 | 35,780 | -5.62
583 | 35,855 | -5.72
584 | 35,930 | -5.77
585 | 36,007 | -6.22
586 | 36,086 | -6.40
587 | 36,166 | -6.70
588 | 36,248 | -7.03
589 | 36,331 | -7.22
590 | 36,416 | -7.70
591 | 36,502 | -7.30
592 | 36,585 | -6.67
593 | 36,666 | -6.97
594 | 36,748 | -6.65
595 | 36,827 | -6.02
596 | 36,905 | -6.32
597 | 36,985 | -6.62
598 | 37,066 | -6.76

599 | 37,148 | -6.90
600 | 37,230 | -6.90
601 | 37,313 | -6.97
602 | 37,396 | -7.20
603 | 37,479 | -6.80
604 | 37,558 | -5.97
605 | 37,637 | -6.38
606 | 37,718 | -6.88
607 | 37,799 | -6.60
608 | 37,880 | -6.45
609 | 37,959 | -6.30
610 | 38,039 | -6.43
611 | 38,120 | -6.66
612 | 38,201 | -6.80
613 | 38,284 | -6.96
614 | 38,367 | -7.09
615 | 38,451 | -7.18
616 | 38,535 | -7.18
617 | 38,619 | -6.94
618 | 38,702 | -6.73
619 | 38,783 | -6.69
620 | 38,865 | -6.64
621 | 38,946 | -6.60
622 | 39,027 | -6.56
623 | 39,107 | -6.21
624 | 39,185 | -5.93
625 | 39,263 | -5.93
626 | 39,341 | -6.16
627 | 39,420 | -6.24
628 | 39,499 | -6.24

629 | 39,579 | -6.13
630 | 39,657 | -6.13
631 | 39,735 | -5.91
632 | 39,812 | -5.48
633 | 39,889 | -6.24
634 | 39,971 | -6.96
635 | 40,054 | -6.91
636 | 40,138 | -7.21
637 | 40,223 | -7.07
638 | 40,305 | -6.31
639 | 40,386 | -6.51
640 | 40,468 | -6.91
641 | 40,551 | -6.71
642 | 40,632 | -6.56
643 | 40,714 | -6.67
644 | 40,796 | -6.46
645 | 40,875 | -5.74
646 | 40,952 | -5.74
647 | 41,031 | -6.37
648 | 41,112 | -6.59
649 | 41,191 | -5.89
650 | 41,268 | -5.19
651 | 41,342 | -5.11
652 | 41,415 | -5.02
653 | 41,489 | -5.09
654 | 41,564 | -5.48
655 | 41,638 | -4.80
656 | 41,710 | -4.53
657 | 41,782 | -5.08
658 | 41,855 | -4.81

659 | 41,927 | -4.68
660 | 42,001 | -5.27
661 | 42,077 | -5.69
662 | 42,152 | -4.99
663 | 42,225 | -4.65
664 | 42,298 | -5.30
665 | 42,374 | -5.70
666 | 42,451 | -5.40
667 | 42,527 | -5.49
668 | 42,603 | -5.43
669 | 42,679 | -5.69
670 | 42,758 | -6.13
671 | 42,837 | -5.95
672 | 42,916 | -5.88
673 | 42,994 | -5.93
674 | 43,073 | -5.80
675 | 43,150 | -5.53
676 | 43,226 | -5.50
677 | 43,304 | -6.04
678 | 43,385 | -6.45
679 | 43,464 | -5.42
680 | 43,540 | -5.47
681 | 43,619 | -6.68
682 | 43,702 | -6.60
683 | 43,785 | -6.48
684 | 43,868 | -6.94
685 | 43,952 | -6.88
686 | 44,036 | -6.72
687 | 44,120 | -6.99
688 | 44,206 | -7.27

689 | 44,292 | -6.94
690 | 44,377 | -6.97
691 | 44,463 | -7.51
692 | 44,551 | -7.13
693 | 44,634 | -6.15
694 | 44,715 | -6.55
695 | 44,800 | -7.44
696 | 44,887 | -7.04
697 | 44,972 | -6.68
698 | 45,055 | -6.72
699 | 45,139 | -6.69
700 | 45,225 | -7.40
701 | 45,315 | -8.16
702 | 45,407 | -7.91
703 | 45,497 | -7.64
704 | 45,587 | -7.94
705 | 45,677 | -7.38
706 | 45,762 | -6.55
707 | 45,846 | -6.93
708 | 45,932 | -7.01
709 | 46,018 | -7.02
710 | 46,104 | -7.39
711 | 46,193 | -7.38
712 | 46,281 | -7.47
713 | 46,370 | -7.51
714 | 46,457 | -7.15
715 | 46,543 | -6.78
716 | 46,628 | -6.90
717 | 46,713 | -6.91
718 | 46,797 | -6.48

719 | 46,881 | -6.71
720 | 46,966 | -6.92
721 | 47,050 | -6.67
722 | 47,133 | -6.29
723 | 47,214 | -5.92
724 | 47,295 | -6.17
725 | 47,378 | -6.73
726 | 47,462 | -6.55
727 | 47,546 | -6.78
728 | 47,631 | -6.99
729 | 47,716 | -6.36
730 | 47,798 | -6.28
731 | 47,881 | -6.74
732 | 47,966 | -6.85
733 | 48,050 | -6.46
734 | 48,134 | -6.65
735 | 48,218 | -6.49
736 | 48,300 | -6.02
737 | 48,381 | -6.05
738 | 48,461 | -5.80
739 | 48,541 | -5.83
740 | 48,622 | -6.26
741 | 48,704 | -6.36
742 | 48,787 | -6.48
743 | 48,870 | -6.39
744 | 48,953 | -6.24
745 | 49,034 | -5.76
746 | 49,113 | -5.77
747 | 49,193 | -5.89
748 | 49,272 | -5.51

749 | 49,350 | -5.51
750 | 49,430 | -5.89
751 | 49,510 | -5.97
752 | 49,591 | -6.07
753 | 49,672 | -5.78
754 | 49,750 | -5.30
755 | 49,828 | -5.58
756 | 49,907 | -5.60
757 | 49,985 | -5.18
758 | 50,061 | -5.03
759 | 50,138 | -5.45
760 | 50,217 | -5.70
761 | 50,295 | -5.37
762 | 50,374 | -5.54
763 | 50,452 | -5.54
764 | 50,531 | -5.65
765 | 50,610 | -5.57
766 | 50,688 | -5.19
767 | 50,766 | -5.72
768 | 50,847 | -6.07
769 | 50,928 | -5.89
770 | 51,009 | -5.74
771 | 51,086 | -4.91
772 | 51,159 | -3.77
773 | 51,230 | -4.34
774 | 51,306 | -5.38
775 | 51,384 | -5.60
776 | 51,465 | -6.21
777 | 51,549 | -6.60
778 | 51,633 | -6.43

779 | 51,715 | -5.47
780 | 51,794 | -5.59
781 | 51,874 | -6.02
782 | 51,955 | -5.56
783 | 52,034 | -5.56
784 | 52,114 | -5.81
785 | 52,195 | -5.84
786 | 52,275 | -5.55
787 | 52,352 | -5.03
788 | 52,430 | -5.60
789 | 52,511 | -6.13
790 | 52,594 | -6.23
791 | 52,679 | -6.71
792 | 52,763 | -6.15
793 | 52,846 | -6.10
794 | 52,931 | -6.93
795 | 53,019 | -7.18
796 | 53,107 | -7.07
797 | 53,193 | -6.36
798 | 53,275 | -5.65
799 | 53,355 | -5.55
800 | 53,436 | -6.05
801 | 53,520 | -6.65
802 | 53,606 | -6.65
803 | 53,692 | -6.55
804 | 53,775 | -6.04
805 | 53,858 | -6.22
806 | 53,942 | -6.49
807 | 54,026 | -6.14
808 | 54,111 | -6.60

809 | 54,198 | -7.06
810 | 54,286 | -6.96
811 | 54,372 | -6.40
812 | 54,457 | -6.53
813 | 54,544 | -6.80
814 | 54,629 | -6.17
815 | 54,713 | -6.35
816 | 54,797 | -6.29
817 | 54,881 | -6.36
818 | 54,967 | -6.71
819 | 55,052 | -6.31
820 | 55,136 | -6.16
821 | 55,218 | -5.58
822 | 55,297 | -5.34
823 | 55,377 | -5.65
824 | 55,457 | -5.42
825 | 55,536 | -5.21
826 | 55,615 | -5.43
827 | 55,694 | -5.54
828 | 55,775 | -5.81
829 | 55,855 | -5.23
830 | 55,933 | -4.80
831 | 56,009 | -5.09
832 | 56,087 | -4.95
833 | 56,163 | -4.71
834 | 56,239 | -4.92
835 | 56,317 | -5.11
836 | 56,394 | -4.76
837 | 56,471 | -5.09
838 | 56,548 | -5.10

839 | 56,624 | -4.35
840 | 56,698 | -4.42
841 | 56,774 | -4.88
842 | 56,850 | -4.47
843 | 56,922 | -3.71
844 | 56,993 | -3.68
845 | 57,065 | -4.18
846 | 57,139 | -4.54
847 | 57,215 | -4.73
848 | 57,289 | -4.01
849 | 57,362 | -4.05
850 | 57,436 | -4.85
851 | 57,513 | -4.85
852 | 57,590 | -4.86
853 | 57,669 | -5.78
854 | 57,750 | -5.51
855 | 57,828 | -4.65
856 | 57,905 | -4.98
857 | 57,981 | -4.65
858 | 58,057 | -4.62
859 | 58,133 | -4.90
860 | 58,211 | -4.96
861 | 58,286 | -4.27
862 | 58,360 | -4.22
863 | 58,436 | -4.93
864 | 58,514 | -5.21
865 | 58,594 | -5.66
866 | 58,676 | -5.86
867 | 58,759 | -6.06
868 | 58,842 | -5.93

869 | 58,925 | -5.85
870 | 59,008 | -6.04
871 | 59,093 | -6.41
872 | 59,178 | -6.31
873 | 59,262 | -5.86
874 | 59,345 | -5.85
875 | 59,428 | -6.03
876 | 59,511 | -5.86
877 | 59,593 | -5.53
878 | 59,674 | -5.53
879 | 59,755 | -5.53
880 | 59,836 | -5.81
881 | 59,921 | -6.49
882 | 60,007 | -6.53
883 | 60,093 | -6.15
884 | 60,177 | -6.11
885 | 60,262 | -6.20
886 | 60,348 | -6.45
887 | 60,433 | -6.30
888 | 60,519 | -6.28
889 | 60,605 | -6.53
890 | 60,692 | -6.58
891 | 60,779 | -6.61
892 | 60,869 | -7.31
893 | 60,961 | -7.48
894 | 61,052 | -6.98
895 | 61,141 | -6.81
896 | 61,229 | -6.53
897 | 61,317 | -6.88
898 | 61,406 | -6.86

899 | 61,496 | -7.00
900 | 61,586 | -7.26
901 | 61,677 | -6.86
902 | 61,765 | -6.50
903 | 61,852 | -6.60
904 | 61,941 | -7.00
905 | 62,031 | -7.10
906 | 62,121 | -7.05
907 | 62,213 | -7.44
908 | 62,307 | -7.72
909 | 62,401 | -7.64
910 | 62,497 | -8.09
911 | 62,594 | -8.02
912 | 62,689 | -7.67
913 | 62,783 | -7.54
914 | 62,877 | -7.51
915 | 62,970 | -7.51
916 | 63,064 | -7.51
917 | 63,157 | -7.51
918 | 63,249 | -7.11
919 | 63,338 | -6.43
920 | 63,424 | -6.26
921 | 63,512 | -6.84
922 | 63,600 | -6.66
923 | 63,688 | -6.41
924 | 63,776 | -6.83
925 | 63,865 | -6.82
926 | 63,954 | -6.79
927 | 64,043 | -6.94
928 | 64,134 | -7.27

929 | 64,228 | -7.70
930 | 64,322 | -7.55
931 | 64,416 | -7.40
932 | 64,510 | -7.91
933 | 64,605 | -7.43
934 | 64,698 | -7.49
935 | 64,793 | -7.97
936 | 64,890 | -7.92
937 | 64,987 | -8.17
938 | 65,085 | -7.98
939 | 65,182 | -7.95
940 | 65,278 | -7.76
941 | 65,371 | -7.25
942 | 65,464 | -7.49
943 | 65,559 | -7.71
944 | 65,655 | -8.15
945 | 65,756 | -8.85
946 | 65,855 | -7.75
947 | 65,949 | -7.40
948 | 66,045 | -8.21
949 | 66,144 | -8.31
950 | 66,243 | -8.06
951 | 66,338 | -7.33
952 | 66,429 | -6.88
953 | 66,520 | -7.07
954 | 66,613 | -7.73
955 | 66,708 | -7.76
956 | 66,803 | -7.54
957 | 66,898 | -7.73
958 | 66,992 | -7.44

959 | 67,085 | -7.22
960 | 67,178 | -7.47
961 | 67,272 | -7.63
962 | 67,367 | -7.52
963 | 67,462 | -7.80
964 | 67,557 | -7.71
965 | 67,653 | -7.86
966 | 67,750 | -8.15
967 | 67,848 | -8.04
968 | 67,945 | -7.89
969 | 68,041 | -7.55
970 | 68,136 | -7.80
971 | 68,233 | -8.27
972 | 68,330 | -7.65
973 | 68,424 | -7.45
974 | 68,520 | -8.01
975 | 68,617 | -7.94
976 | 68,715 | -8.20
977 | 68,811 | -7.59
978 | 68,902 | -6.36
979 | 68,987 | -5.81
980 | 69,071 | -6.12
981 | 69,155 | -5.62
982 | 69,238 | -5.75
983 | 69,324 | -6.69
984 | 69,413 | -6.92
985 | 69,506 | -7.54
986 | 69,599 | -7.33
987 | 69,691 | -6.97
988 | 69,780 | -6.55

989 | 69,869 | -6.75
990 | 69,959 | -7.31
991 | 70,054 | -7.84
992 | 70,149 | -7.72
993 | 70,243 | -7.05
994 | 70,334 | -7.08
995 | 70,425 | -7.26
996 | 70,515 | -6.52
997 | 70,600 | -5.67
998 | 70,683 | -5.67
999 | 70,766 | -5.67
1000 | 70,848 | -5.66
1001 | 70,931 | -5.66
1002 | 71,013 | -5.70
1003 | 71,096 | -5.75
1004 | 71,180 | -5.93
1005 | 71,264 | -5.99
1006 | 71,347 | -5.44
1007 | 71,427 | -5.15
1008 | 71,506 | -4.60
1009 | 71,584 | -5.30
1010 | 71,664 | -5.20
1011 | 71,743 | -4.81
1012 | 71,822 | -5.36
1013 | 71,905 | -6.13
1014 | 71,989 | -5.55
1015 | 72,070 | -5.38
1016 | 72,152 | -5.69
1017 | 72,235 | -5.81
1018 | 72,318 | -5.82

1019	72,403	-6.28	1049	74,985	-5.61
1020	72,489	-6.53	1050	75,066	-5.41
1021	72,578	-6.89	1051	75,147	-5.72
1022	72,666	-6.39	1052	75,229	-5.42
1023	72,753	-6.64	1053	75,308	-5.12
1024	72,840	-5.97	1054	75,388	-5.32
1025	72,922	-5.29	1055	75,468	-5.41
1026	73,003	-5.77	1056	75,547	-5.13
1027	73,087	-6.00	1057	75,625	-4.66
1028	73,172	-6.31	1058	75,701	-4.53
1029	73,258	-6.44	1059	75,776	-4.39
1030	73,347	-7.10	1060	75,850	-4.35
1031	73,440	-7.81	1061	75,928	-5.41
1032	73,533	-6.96	1062	76,007	-5.00
1033	73,620	-6.10	1063	76,084	-4.83
1034	73,706	-6.61	1064	76,162	-5.23
1035	73,795	-7.12	1065	76,239	-4.37
1036	73,885	-6.85	1066	76,313	-4.05
1037	73,973	-6.45	1067	76,388	-4.80
1038	74,060	-6.56	1068	76,465	-5.08
1039	74,150	-7.17	1069	76,543	-5.16
1040	74,238	-6.31	1070	76,622	-5.22
1041	74,322	-5.91	1071	76,700	-4.92
1042	74,405	-5.62	1072	76,777	-4.67
1043	74,484	-4.74	1073	76,854	-5.09
1044	74,565	-6.03	1074	76,933	-5.50
1045	74,651	-6.92	1075	77,014	-5.47
1046	74,738	-5.87	1076	77,094	-5.42
1047	74,821	-5.85	1077	77,175	-5.90
1048	74,903	-5.68	1078	77,259	-6.33

1079	77,345	-6.71	1109	79,791	-5.28
1080	77,433	-7.14	1110	79,870	-5.41
1081	77,521	-6.59	1111	79,947	-4.95
1082	77,606	-5.81	1112	80,020	-3.66
1083	77,687	-5.76	1113	80,093	-4.87
1084	77,769	-5.81	1114	80,173	-6.38
1085	77,853	-6.45	1115	80,257	-6.11
1086	77,938	-6.52	1116	80,341	-6.39
1087	78,025	-6.62	1117	80,422	-5.41
1088	78,113	-7.28	1118	80,498	-4.45
1089	78,200	-6.06	1119	80,572	-4.25
1090	78,277	-3.82	1120	80,645	-4.40
1091	78,353	-5.71	1121	80,719	-4.75
1092	78,437	-6.79	1122	80,794	-4.75
1093	78,522	-6.12	1123	80,869	-4.50
1094	78,606	-6.32	1124	80,943	-4.67
1095	78,690	-6.06	1125	81,019	-4.93
1096	78,772	-5.78	1126	81,096	-5.20
1097	78,854	-5.87	1127	81,170	-4.04
1098	78,937	-6.24	1128	81,240	-3.29
1099	79,020	-5.91	1129	81,309	-3.87
1100	79,098	-4.80	1130	81,381	-4.45
1101	79,174	-4.80	1131	81,453	-3.89
1102	79,251	-5.00	1132	81,523	-3.33
1103	79,327	-4.76	1133	81,592	-3.79
1104	79,402	-4.33	1134	81,662	-3.79
1105	79,478	-5.43	1135	81,732	-3.81
1106	79,557	-5.50	1136	81,802	-3.74
1107	79,636	-5.23	1137	81,872	-3.73
1108	79,714	-4.93	1138	81,942	-3.83

1139 | 82,012 | -3.48
1140 | 82,080 | -3.56
1141 | 82,148 | -3.01
1142 | 82,213 | -2.54
1143 | 82,280 | -3.65
1144 | 82,352 | -4.85
1145 | 82,427 | -4.59
1146 | 82,500 | -4.31
1147 | 82,573 | -4.53
1148 | 82,647 | -4.75
1149 | 82,721 | -4.27
1150 | 82,792 | -3.78
1151 | 82,860 | -3.10
1152 | 82,925 | -2.42
1153 | 82,990 | -2.98
1154 | 83,055 | -2.58
1155 | 83,121 | -3.46
1156 | 83,190 | -3.92
1157 | 83,261 | -3.94
1158 | 83,331 | -3.94
1159 | 83,401 | -3.70
1160 | 83,470 | -3.62
1161 | 83,538 | -3.27
1162 | 83,608 | -4.39
1163 | 83,679 | -3.71
1164 | 83,747 | -3.23
1165 | 83,814 | -3.20
1166 | 83,881 | -3.54
1167 | 83,952 | -4.32
1168 | 84,025 | -4.64

1169 | 84,097 | -4.27
1170 | 84,168 | -3.91
1171 | 84,238 | -3.91
1172 | 84,308 | -3.92
1173 | 84,378 | -4.01
1174 | 84,448 | -3.76
1175 | 84,515 | -2.53
1176 | 84,576 | -1.64
1177 | 84,638 | -2.65
1178 | 84,703 | -3.03
1179 | 84,770 | -3.43
1180 | 84,838 | -3.49
1181 | 84,905 | -3.29
1182 | 84,971 | -2.92
1183 | 85,036 | -2.64
1184 | 85,100 | -2.48
1185 | 85,164 | -2.92
1186 | 85,229 | -3.12
1187 | 85,296 | -3.23
1188 | 85,361 | -2.69
1189 | 85,425 | -2.64
1190 | 85,489 | -2.47
1191 | 85,551 | -2.34
1192 | 85,615 | -2.71
1193 | 85,680 | -3.00
1194 | 85,746 | -3.53
1195 | 85,814 | -3.42
1196 | 85,881 | -3.53
1197 | 85,949 | -3.51
1198 | 86,016 | -3.38

1199 | 86,083 | -3.38
1200 | 86,150 | -3.38
1201 | 86,218 | -3.63
1202 | 86,286 | -3.76
1203 | 86,355 | -3.64
1204 | 86,423 | -3.78
1205 | 86,492 | -3.91
1206 | 86,562 | -4.34
1207 | 86,635 | -4.87
1208 | 86,707 | -3.99
1209 | 86,777 | -4.35
1210 | 86,848 | -4.14
1211 | 86,919 | -4.12
1212 | 86,989 | -4.17
1213 | 87,060 | -4.40
1214 | 87,132 | -4.49
1215 | 87,203 | -4.39
1216 | 87,274 | -3.86
1217 | 87,342 | -3.46
1218 | 87,410 | -4.15
1219 | 87,482 | -4.83
1220 | 87,554 | -4.35
1221 | 87,624 | -3.80
1222 | 87,695 | -4.55
1223 | 87,766 | -4.28
1224 | 87,837 | -4.13
1225 | 87,907 | -4.21
1226 | 87,978 | -4.43
1227 | 88,049 | -4.43
1228 | 88,122 | -4.79

1229 | 88,196 | -5.13
1230 | 88,271 | -5.12
1231 | 88,345 | -5.12
1232 | 88,420 | -5.11
1233 | 88,495 | -5.10
1234 | 88,570 | -5.07
1235 | 88,644 | -5.03
1236 | 88,719 | -5.07
1237 | 88,793 | -4.93
1238 | 88,866 | -4.86
1239 | 88,938 | -4.09
1240 | 89,008 | -4.10
1241 | 89,077 | -4.11
1242 | 89,148 | -4.65
1243 | 89,221 | -4.91
1244 | 89,296 | -5.34
1245 | 89,373 | -5.76
1246 | 89,450 | -5.23
1247 | 89,523 | -4.69
1248 | 89,597 | -5.10
1249 | 89,672 | -5.43
1250 | 89,748 | -5.42
1251 | 89,825 | -5.41
1252 | 89,900 | -5.06
1253 | 89,973 | -4.69
1254 | 90,048 | -5.61
1255 | 90,128 | -6.68
1256 | 90,209 | -5.94
1257 | 90,286 | -5.34
1258 | 90,362 | -5.58

1259	90,440	-5.73	1289	92,617	-5.45
1260	90,517	-5.77	1290	92,690	-4.58
1261	90,594	-5.18	1291	92,759	-3.32
1262	90,668	-5.11	1292	92,827	-4.28
1263	90,740	-4.16	1293	92,897	-4.46
1264	90,812	-5.00	1294	92,967	-4.06
1265	90,885	-4.79	1295	93,037	-4.14
1266	90,959	-5.02	1296	93,107	-4.62
1267	91,033	-4.99	1297	93,180	-5.15
1268	91,107	-5.19	1298	93,254	-4.70
1269	91,182	-5.21	1299	93,325	-4.48
1270	91,258	-5.72	1300	93,397	-4.69
1271	91,336	-5.81	1301	93,468	-4.44
1272	91,411	-4.73	1302	93,540	-4.92
1273	91,480	-3.53	1303	93,615	-5.61
1274	91,547	-3.61	1304	93,690	-4.68
1275	91,615	-4.05	1305	93,761	-4.27
1276	91,685	-4.12	1306	93,829	-3.45
1277	91,753	-3.85	1307	93,895	-3.27
1278	91,823	-4.61	1308	93,961	-3.45
1279	91,896	-4.85	1309	94,026	-3.25
1280	91,968	-4.59	1310	94,093	-3.88
1281	92,040	-4.83	1311	94,164	-4.97
1282	92,114	-5.10	1312	94,238	-4.97
1283	92,187	-4.71	1313	94,311	-4.84
1284	92,259	-4.54	1314	94,382	-4.12
1285	92,329	-3.99	1315	94,450	-3.66
1286	92,398	-4.12	1316	94,517	-3.51
1287	92,470	-4.94	1317	94,585	-4.15
1288	92,542	-4.77	1318	94,655	-4.35

1319 | 94,725 | -4.45
1320 | 94,795 | -4.05
1321 | 94,864 | -3.87
1322 | 94,932 | -3.80
1323 | 95,000 | -3.96
1324 | 95,070 | -4.39
1325 | 95,141 | -4.77
1326 | 95,214 | -4.90
1327 | 95,284 | -3.64
1328 | 95,351 | -3.39
1329 | 95,418 | -3.89
1330 | 95,487 | -4.00
1331 | 95,556 | -4.00
1332 | 95,624 | -4.00
1333 | 95,693 | -3.85
1334 | 95,761 | -3.70
1335 | 95,828 | -3.70
1336 | 95,895 | -3.70
1337 | 95,964 | -4.00
1338 | 96,033 | -4.30
1339 | 96,103 | -3.95
1340 | 96,170 | -3.50
1341 | 96,237 | -3.50
1342 | 96,304 | -3.80
1343 | 96,372 | -3.87
1344 | 96,441 | -4.23
1345 | 96,511 | -4.35
1346 | 96,582 | -4.28
1347 | 96,652 | -4.35
1348 | 96,722 | -3.88

1349 | 96,791 | -4.20
1350 | 96,861 | -4.51
1351 | 96,933 | -4.51
1352 | 97,004 | -4.51
1353 | 97,075 | -4.42
1354 | 97,148 | -5.24
1355 | 97,222 | -4.90
1356 | 97,294 | -4.35
1357 | 97,365 | -4.57
1358 | 97,439 | -5.46
1359 | 97,514 | -4.98
1360 | 97,588 | -4.88
1361 | 97,659 | -4.12
1362 | 97,727 | -3.36
1363 | 97,794 | -4.10
1364 | 97,864 | -4.24
1365 | 97,936 | -4.82
1366 | 98,010 | -5.38
1367 | 98,085 | -4.87
1368 | 98,157 | -4.38
1369 | 98,228 | -4.51
1370 | 98,300 | -4.84
1371 | 98,371 | -3.93
1372 | 98,439 | -3.50
1373 | 98,505 | -3.51
1374 | 98,572 | -3.51
1375 | 98,640 | -4.12
1376 | 98,712 | -4.75
1377 | 98,783 | -4.39
1378 | 98,855 | -4.56

1379 | 98,925 | -4.00
1380 | 98,994 | -3.80
1381 | 99,063 | -4.17
1382 | 99,134 | -4.58
1383 | 99,205 | -4.35
1384 | 99,275 | -4.12
1385 | 99,345 | -4.07
1386 | 99,414 | -4.03
1387 | 99,484 | -4.01
1388 | 99,553 | -3.95
1389 | 99,622 | -3.92
1390 | 99,691 | -4.08
1391 | 99,760 | -3.89
1392 | 99,828 | -3.49
1393 | 99,894 | -3.49
1394 | 99,961 | -3.49
1395 | 100,028 | -3.45
1396 | 100,095 | -3.65
1397 | 100,163 | -3.83
1398 | 100,232 | -4.08
1399 | 100,301 | -3.59
1400 | 100,367 | -3.11
1401 | 100,432 | -3.11
1402 | 100,498 | -3.39
1403 | 100,565 | -3.68
1404 | 100,633 | -3.68
1405 | 100,700 | -3.61
1406 | 100,768 | -3.53
1407 | 100,835 | -3.42
1408 | 100,902 | -3.64

1409 | 100,969 | -3.52
1410 | 101,036 | -3.41
1411 | 101,102 | -3.29
1412 | 101,168 | -3.18
1413 | 101,234 | -3.18
1414 | 101,299 | -2.99
1415 | 101,363 | -2.77
1416 | 101,427 | -2.67
1417 | 101,490 | -2.58
1418 | 101,554 | -2.96
1419 | 101,618 | -2.81
1420 | 101,681 | -2.27
1421 | 101,743 | -2.37
1422 | 101,806 | -2.49
1423 | 101,868 | -2.36
1424 | 101,930 | -2.23
1425 | 101,991 | -2.10
1426 | 102,052 | -1.97
1427 | 102,114 | -2.42
1428 | 102,177 | -2.86
1429 | 102,241 | -2.63
1430 | 102,304 | -2.42
1431 | 102,366 | -2.26
1432 | 102,427 | -2.11
1433 | 102,489 | -2.48
1434 | 102,553 | -2.85
1435 | 102,618 | -3.08
1436 | 102,684 | -3.32
1437 | 102,749 | -2.82
1438 | 102,812 | -2.33

1439 | 102,876 | -2.93
1440 | 102,942 | -3.55
1441 | 103,008 | -3.09
1442 | 103,073 | -2.61
1443 | 103,137 | -3.19
1444 | 103,205 | -3.78
1445 | 103,273 | -3.67
1446 | 103,341 | -3.55
1447 | 103,410 | -4.04
1448 | 103,481 | -4.52
1449 | 103,553 | -4.46
1450 | 103,625 | -4.41
1451 | 103,697 | -4.41
1452 | 103,769 | -4.50
1453 | 103,841 | -4.45
1454 | 103,912 | -4.30
1455 | 103,984 | -4.24
1456 | 104,054 | -4.16
1457 | 104,125 | -4.11
1458 | 104,195 | -4.05
1459 | 104,266 | -4.22
1460 | 104,337 | -4.40
1461 | 104,409 | -4.37
1462 | 104,480 | -4.32
1463 | 104,552 | -4.33
1464 | 104,623 | -4.33
1465 | 104,695 | -4.25
1466 | 104,766 | -4.17
1467 | 104,837 | -4.17
1468 | 104,908 | -4.52

1469 | 104,982 | -4.85
1470 | 105,055 | -4.66
1471 | 105,128 | -4.48
1472 | 105,200 | -4.28
1473 | 105,271 | -4.10
1474 | 105,342 | -4.20
1475 | 105,413 | -4.29
1476 | 105,485 | -4.51
1477 | 105,559 | -4.71
1478 | 105,631 | -4.33
1479 | 105,702 | -3.93
1480 | 105,772 | -3.93
1481 | 105,842 | -3.92
1482 | 105,911 | -3.82
1483 | 105,980 | -3.71
1484 | 106,048 | -3.21
1485 | 106,114 | -2.71
1486 | 106,179 | -3.37
1487 | 106,248 | -4.02
1488 | 106,319 | -4.25
1489 | 106,391 | -4.48
1490 | 106,463 | -4.26
1491 | 106,535 | -4.04
1492 | 106,606 | -4.32
1493 | 106,678 | -4.60
1494 | 106,752 | -4.84
1495 | 106,828 | -5.07
1496 | 106,902 | -4.73
1497 | 106,976 | -4.40
1498 | 107,048 | -4.40

1499	107,121	-4.40		1529	109,625	-6.09
1500	107,194	-4.88		1530	109,707	-6.20
1501	107,270	-5.36		1531	109,790	-6.30
1502	107,349	-5.92		1532	109,872	-5.88
1503	107,431	-6.48		1533	109,952	-5.47
1504	107,515	-6.48		1534	110,031	-5.53
1505	107,599	-6.48		1535	110,110	-5.58
1506	107,683	-6.48		1536	110,189	-5.58
1507	107,767	-6.67		1537	110,269	-5.58
1508	107,852	-6.85		1538	110,348	-5.58
1509	107,938	-6.90		1539	110,427	-5.57
1510	108,025	-6.95		1540	110,507	-5.57
1511	108,111	-6.95		1541	110,586	-5.57
1512	108,198	-6.83		1542	110,666	-5.57
1513	108,283	-6.71		1543	110,746	-6.00
1514	108,369	-6.71		1544	110,829	-6.42
1515	108,454	-6.64		1545	110,913	-6.25
1516	108,538	-6.57		1546	110,996	-6.08
1517	108,622	-6.43		1547	111,078	-5.88
1518	108,706	-6.28		1548	111,159	-5.68
1519	108,789	-6.28		1549	111,239	-5.68
1520	108,872	-6.28		1550	111,319	-5.68
1521	108,954	-6.27		1551	111,399	-5.68
1522	109,037	-6.27		1552	111,480	-5.68
1523	109,121	-6.53		1553	111,560	-5.68
1524	109,206	-6.81		1554	111,640	-5.68
1525	109,292	-6.81		1555	111,721	-6.02
1526	109,378	-6.44		1556	111,805	-6.37
1527	109,461	-6.09		1557	111,889	-6.37
1528	109,543	-6.09		1558	111,973	-6.16

1559	112,055	-5.94
1560	112,137	-5.94
1561	112,219	-5.94
1562	112,301	-5.94
1563	112,383	-5.94
1564	112,465	-5.93
1565	112,547	-5.82
1566	112,628	-5.70
1567	112,708	-5.67
1568	112,790	-5.86
1569	112,872	-6.06
1570	112,954	-5.93
1571	113,036	-5.80
1572	113,116	-5.40
1573	113,195	-4.98
1574	113,272	-4.98
1575	113,348	-4.86
1576	113,425	-4.85
1577	113,501	-4.89
1578	113,578	-4.94
1579	113,656	-5.27
1580	113,736	-5.62
1581	113,815	-5.10
1582	113,891	-4.57
1583	113,966	-4.57
1584	114,041	-4.45
1585	114,115	-4.33
1586	114,189	-4.36
1587	114,264	-4.39
1588	114,338	-4.31
1589	114,411	-4.24
1590	114,486	-4.59
1591	114,562	-4.95
1592	114,639	-4.78
1593	114,715	-4.60
1594	114,790	-4.33
1595	114,863	-4.06
1596	114,935	-3.75
1597	115,006	-3.45
1598	115,077	-3.89
1599	115,150	-4.32
1600	115,222	-3.41
1601	115,289	-2.48
1602	115,355	-2.47
1603	115,420	-2.47
1604	115,486	-2.47
1605	115,551	-2.47
1606	115,617	-2.47
1607	115,683	-2.56
1608	115,749	-2.64
1609	115,816	-3.11
1610	115,886	-3.57
1611	115,956	-3.34
1612	116,025	-3.08
1613	116,093	-2.97
1614	116,161	-2.86
1615	116,228	-2.88
1616	116,296	-2.83
1617	116,363	-2.76
1618	116,430	-2.71

1619	116,497	-2.67
1620	116,563	-2.67
1621	116,630	-2.67
1622	116,697	-2.53
1623	116,762	-2.41
1624	116,827	-2.11
1625	116,891	-1.83
1626	116,954	-1.62
1627	117,016	-1.44
1628	117,077	-1.44
1629	117,139	-1.63
1630	117,202	-1.83
1631	117,264	-1.59
1632	117,326	-1.36
1633	117,388	-1.67
1634	117,451	-2.00
1635	117,514	-1.50
1636	117,575	-0.98
1637	117,635	-1.26
1638	117,697	-1.54
1639	117,759	-1.67
1640	117,822	-1.85
1641	117,886	-1.88
1642	117,949	-1.78
1643	118,012	-1.66
1644	118,074	-1.31
1645	118,135	-0.94
1646	118,194	-0.93
1647	118,254	-0.93
1648	118,314	-1.19
1649	118,376	-1.46
1650	118,438	-1.45
1651	118,499	-1.12
1652	118,559	-0.80
1653	118,618	-0.57
1654	118,676	-0.31
1655	118,735	-1.04
1656	118,796	-1.77
1657	118,859	-1.47
1658	118,921	-1.16
1659	118,981	-1.16
1660	119,042	-0.84
1661	119,101	-0.51
1662	119,160	-0.95
1663	119,221	-1.40
1664	119,283	-1.40
1665	119,345	-1.33
1666	119,406	-1.27
1667	119,467	-1.07
1668	119,528	-1.09
1669	119,589	-1.31
1670	119,650	-1.00
1671	119,710	-0.68
1672	119,770	-1.07
1673	119,831	-1.47
1674	119,894	-1.47
1675	119,955	-1.16
1676	120,016	-0.86
1677	120,076	-1.08
1678	120,138	-1.29

1679	120,199	-1.09	1709	121,951	0.59
1680	120,259	-0.87	1710	122,006	0.11
1681	120,319	-0.86	1711	122,064	-0.37
1682	120,379	-0.86	1712	122,122	-0.01
1683	120,439	-0.63	1713	122,178	0.32
1684	120,498	-0.41	1714	122,234	0.41
1685	120,555	-0.14	1715	122,290	0.50
1686	120,612	0.13	1716	122,345	0.32
1687	120,669	0.02	1717	122,402	0.13
1688	120,726	-0.10	1718	122,458	0.13
1689	120,783	-0.24	1719	122,515	0.27
1690	120,841	-0.49	1720	122,571	0.39
1691	120,900	-0.59	1721	122,627	0.18
1692	120,959	-0.58	1722	122,684	-0.05
1693	121,018	-0.75	1723	122,742	-0.39
1694	121,079	-0.93	1724	122,801	-0.75
1695	121,139	-0.92	1725	122,861	-0.30
1696	121,199	-0.47	1726	122,919	-0.09
1697	121,256	0.00	1727	122,977	-0.32
1698	121,314	-0.32	1728	123,036	-0.41
1699	121,373	-0.65	1729	123,095	-0.51
1700	121,432	-0.64	1730	123,154	-0.24
1701	121,492	-0.62	1731	123,211	0.05
1702	121,551	-0.60	1732	123,268	0.54
1703	121,610	-0.60	1733	123,322	1.02
1704	121,669	-0.32	1734	123,377	0.63
1705	121,727	-0.03	1735	123,432	0.22
1706	121,784	-0.03	1736	123,489	0.27
1707	121,840	0.28	1737	123,545	0.34
1708	121,896	0.58	1738	123,602	0.25

1739	123,658	0.16	1769	125,325	0.92
1740	123,715	0.44	1770	125,380	0.61
1741	123,770	0.71	1771	125,435	0.82
1742	123,825	0.72	1772	125,490	1.02
1743	123,880	0.62	1773	125,544	1.07
1744	123,936	0.51	1774	125,598	1.11
1745	123,991	0.40	1775	125,653	0.76
1746	124,046	1.07	1776	125,708	0.43
1747	124,100	1.24	1777	125,765	0.45
1748	124,154	0.59	1778	125,821	0.53
1749	124,209	0.60	1779	125,877	0.61
1750	124,265	0.60	1780	125,933	0.55
1751	124,320	0.50	1781	125,989	0.47
1752	124,376	0.40	1782	126,046	0.38
1753	124,432	0.50	1783	126,103	0.28
1754	124,488	0.58	1784	126,160	0.17
1755	124,544	0.62	1785	126,217	0.53
1756	124,599	0.65	1786	126,272	1.03
1757	124,654	0.62	1787	126,327	1.00
1758	124,710	0.60	1788	126,381	0.97
1759	124,765	0.60	1789	126,436	0.93
1760	124,821	0.65	1790	126,491	0.90
1761	124,877	0.27	1791	126,545	1.26
1762	124,935	-0.14	1792	126,598	1.61
1763	124,993	0.04	1793	126,650	1.93
1764	125,050	0.24	1794	126,700	2.50
1765	125,107	0.47	1795	126,749	2.74
1766	125,162	0.69	1796	126,799	2.07
1767	125,217	0.95	1797	126,851	1.41
1768	125,271	1.22	1798	126,904	1.73

Year	Value	Rate
1799	126,955	2.07
1800	127,006	2.06
1801	127,057	2.05
1802	127,108	2.04
1803	127,159	2.13
1804	127,210	2.22
1805	127,260	2.33
1806	127,310	2.44
1807	127,360	2.44
1808	127,410	2.46
1809	127,460	2.37
1810	127,510	2.29
1811	127,560	2.40
1812	127,610	2.52
1813	127,660	2.37
1814	127,710	2.22
1815	127,760	2.43
1816	127,810	2.58
1817	127,860	2.49
1818	127,910	2.16
1819	127,961	2.06
1820	128,012	2.26
1821	128,062	2.49
1822	128,112	2.72
1823	128,161	2.70
1824	128,210	2.68
1825	128,259	2.50
1826	128,309	2.78
1827	128,357	3.23
1828	128,405	3.16
1829	128,453	3.08
1830	128,501	3.06
1831	128,549	2.71
1832	128,599	2.37
1833	128,650	2.11
1834	128,702	1.87
1835	128,753	2.16
1836	128,804	2.45
1837	128,854	2.20
1838	128,906	1.96
1839	128,958	1.70
1840	129,011	1.44
1841	129,065	1.42
1842	129,119	1.39
1843	129,172	1.83
1844	129,224	2.25
1845	129,274	2.52
1846	129,324	2.78
1847	129,374	2.20
1848	129,428	0.67
1849	129,486	-0.31
1850	129,545	0.56
1851	129,600	1.42
1852	129,653	1.78
1853	129,705	2.16
1854	129,757	1.95
1855	129,809	1.73
1856	129,863	1.60
1857	129,916	1.31
1858	129,971	1.14

1859 | 130,026 | 1.47
1860 | 130,079 | 1.81
1861 | 130,131 | 1.79
1862 | 130,185 | 1.24
1863 | 130,241 | 0.68
1864 | 130,297 | 0.88
1865 | 130,353 | 0.84
1866 | 130,410 | 0.59
1867 | 130,467 | 0.57
1868 | 130,525 | 0.53
1869 | 130,582 | 0.40
1870 | 130,641 | 0.24
1871 | 130,699 | 0.14
1872 | 130,759 | -0.13
1873 | 130,819 | -0.31
1874 | 130,880 | -0.43
1875 | 130,942 | -0.54
1876 | 131,004 | -0.56
1877 | 131,066 | -0.58
1878 | 131,128 | -0.79
1879 | 131,190 | -0.42
1880 | 131,250 | 0.16
1881 | 131,311 | -0.49
1882 | 131,374 | -1.14
1883 | 131,439 | -1.54
1884 | 131,506 | -1.94
1885 | 131,574 | -1.73
1886 | 131,641 | -1.51
1887 | 131,707 | -1.48
1888 | 131,773 | -1.45

1889 | 131,840 | -1.87
1890 | 131,908 | -2.28
1891 | 131,978 | -2.23
1892 | 132,048 | -2.48
1893 | 132,120 | -2.78
1894 | 132,192 | -2.78
1895 | 132,264 | -2.60
1896 | 132,335 | -2.42
1897 | 132,406 | -2.44
1898 | 132,477 | -2.52
1899 | 132,548 | -2.57
1900 | 132,619 | -2.58
1901 | 132,691 | -2.58
1902 | 132,763 | -2.76
1903 | 132,836 | -3.16
1904 | 132,911 | -3.36
1905 | 132,987 | -3.61
1906 | 133,065 | -3.84
1907 | 133,142 | -3.69
1908 | 133,219 | -3.54
1909 | 133,296 | -3.73
1910 | 133,374 | -3.93
1911 | 133,452 | -3.58
1912 | 133,528 | -3.23
1913 | 133,603 | -3.23
1914 | 133,679 | -3.81
1915 | 133,759 | -4.41
1916 | 133,841 | -4.46
1917 | 133,923 | -4.47
1918 | 134,005 | -4.89

1919	134,091	-5.28	1949	137,026	-7.50
1920	134,178	-5.40	1950	137,130	-7.82
1921	134,266	-5.74	1951	137,235	-7.93
1922	134,356	-5.93	1952	137,341	-7.98
1923	134,447	-6.09	1953	137,446	-7.72
1924	134,538	-5.76	1954	137,549	-7.45
1925	134,628	-5.74	1955	137,651	-7.47
1926	134,719	-6.39	1956	137,754	-7.55
1927	134,815	-6.97	1957	137,858	-7.88
1928	134,912	-6.90	1958	137,966	-8.51
1929	135,009	-6.88	1959	138,078	-8.89
1930	135,107	-7.18	1960	138,193	-9.24
1931	135,207	-7.26	1961	138,308	-8.94
1932	135,308	-7.48	1962	138,420	-8.59
1933	135,408	-7.30	1963	138,532	-8.72
1934	135,507	-6.83	1964	138,644	-8.67
1935	135,604	-6.91	1965	138,756	-8.67
1936	135,702	-7.13	1966	138,868	-8.59
1937	135,802	-7.33	1967	138,978	-8.30
1938	135,903	-7.31	1968	139,087	-8.30
1939	136,003	-7.14	1969	139,197	-8.53
1940	136,103	-7.51	1970	139,308	-8.40
1941	136,206	-7.81	1971	139,418	-8.40
1942	136,309	-7.44	1972	139,530	-8.70
1943	136,411	-7.39	1973	139,643	-8.86
1944	136,512	-7.54	1974	139,756	-8.66
1945	136,614	-7.29	1975	139,868	-8.43
1946	136,715	-7.51	1976	139,979	-8.64
1947	136,819	-7.94	1977	140,093	-8.99
1948	136,923	-7.62	1978	140,207	-8.82

1979 | 140,319 | -8.34
1980 | 140,430 | -8.44
1981 | 140,542 | -8.67
1982 | 140,655 | -8.60
1983 | 140,766 | -8.38
1984 | 140,876 | -8.11
1985 | 140,986 | -8.59
1986 | 141,100 | -8.75
1987 | 141,212 | -8.38
1988 | 141,323 | -8.31
1989 | 141,434 | -8.54
1990 | 141,547 | -8.70
1991 | 141,661 | -8.58
1992 | 141,773 | -8.51
1993 | 141,885 | -8.39
1994 | 141,997 | -8.54
1995 | 142,108 | -8.21
1996 | 142,218 | -8.12
1997 | 142,329 | -8.36
1998 | 142,440 | -8.25
1999 | 142,551 | -8.49
2000 | 142,665 | -8.73
2001 | 142,779 | -8.68
2002 | 142,893 | -8.58
2003 | 143,006 | -8.41
2004 | 143,117 | -8.09
2005 | 143,227 | -8.17
2006 | 143,338 | -8.31
2007 | 143,450 | -8.34
2008 | 143,562 | -8.34

2009 | 143,675 | -8.65
2010 | 143,790 | -8.87
2011 | 143,905 | -8.36
2012 | 144,016 | -7.95
2013 | 144,126 | -8.18
2014 | 144,237 | -8.13
2015 | 144,346 | -7.71
2016 | 144,453 | -7.79
2017 | 144,562 | -7.85
2018 | 144,671 | -7.93
2019 | 144,782 | -8.21
2020 | 144,894 | -8.32
2021 | 145,006 | -8.19
2022 | 145,116 | -7.90
2023 | 145,226 | -8.00
2024 | 145,337 | -8.13
2025 | 145,449 | -8.31
2026 | 145,563 | -8.62
2027 | 145,678 | -8.49
2028 | 145,792 | -8.34
2029 | 145,905 | -8.45
2030 | 146,019 | -8.31
2031 | 146,131 | -8.00
2032 | 146,242 | -7.99
2033 | 146,352 | -7.86
2034 | 146,461 | -7.71
2035 | 146,570 | -7.82
2036 | 146,681 | -7.98
2037 | 146,792 | -8.02
2038 | 146,903 | -8.00

2039	147,014	-7.88
2040	147,124	-7.95
2041	147,236	-8.02
2042	147,347	-8.02
2043	147,459	-8.15
2044	147,572	-8.05
2045	147,684	-7.97
2046	147,796	-8.25
2047	147,910	-8.32
2048	148,025	-8.47
2049	148,141	-8.45
2050	148,253	-7.67
2051	148,362	-7.52
2052	148,471	-7.70
2053	148,581	-7.64
2054	148,691	-7.84
2055	148,802	-7.84
2056	148,913	-7.72
2057	149,022	-7.62
2058	149,132	-7.54
2059	149,241	-7.67
2060	149,351	-7.52
2061	149,460	-7.46
2062	149,569	-7.66
2063	149,679	-7.64
2064	149,789	-7.61
2065	149,899	-7.57
2066	150,008	-7.34
2067	150,116	-7.33
2068	150,224	-7.54
2069	150,333	-7.34
2070	150,440	-6.93
2071	150,545	-6.96
2072	150,653	-7.43
2073	150,762	-7.41
2074	150,868	-6.75
2075	150,972	-6.61
2076	151,076	-6.86
2077	151,181	-6.91
2078	151,287	-6.91
2079	151,393	-7.22
2080	151,502	-7.42
2081	151,611	-7.38
2082	151,721	-7.40
2083	151,832	-7.82
2084	151,945	-7.89
2085	152,058	-7.89
2086	152,171	-7.69
2087	152,283	-7.77
2088	152,398	-8.15
2089	152,513	-8.08
2090	152,628	-7.86
2091	152,740	-7.64
2092	152,852	-7.49
2093	152,963	-7.67
2094	153,076	-7.64
2095	153,188	-7.78
2096	153,301	-7.68
2097	153,412	-7.23
2098	153,523	-7.61

2099	153,637	-8.06
2100	153,753	-8.10
2101	153,868	-7.74
2102	153,981	-7.62
2103	154,094	-7.72
2104	154,207	-7.64
2105	154,322	-8.14
2106	154,438	-7.86
2107	154,551	-7.37
2108	154,664	-7.85
2109	154,783	-9.00
2110	154,907	-8.80
2111	155,029	-8.53
2112	155,151	-8.84
2113	155,271	-8.06
2114	155,388	-8.08
2115	155,506	-8.21
2116	155,625	-8.40
2117	155,743	-7.89
2118	155,861	-8.37
2119	155,983	-8.87
2120	156,105	-8.28
2121	156,225	-8.50
2122	156,349	-9.08
2123	156,473	-8.54
2124	156,593	-8.17
2125	156,714	-8.52
2126	156,835	-8.49
2127	156,957	-8.55
2128	157,080	-8.58
2129	157,202	-8.50
2130	157,324	-8.59
2131	157,444	-7.84
2132	157,560	-7.69
2133	157,677	-7.91
2134	157,793	-7.67
2135	157,909	-7.75
2136	158,026	-7.93
2137	158,140	-6.97
2138	158,250	-6.93
2139	158,359	-6.63
2140	158,469	-7.13
2141	158,584	-8.05
2142	158,702	-7.62
2143	158,815	-7.07
2144	158,925	-6.62
2145	159,033	-6.72
2146	159,139	-5.89
2147	159,240	-5.27
2148	159,340	-5.63
2149	159,444	-6.58
2150	159,548	-5.36
2151	159,651	-6.27
2152	159,757	-6.37
2153	159,865	-6.54
2154	159,971	-5.94
2155	160,076	-6.25
2156	160,184	-6.76
2157	160,293	-6.67
2158	160,403	-6.78

2159 | 160,514 | -6.79
2160 | 160,626 | -7.40
2161 | 160,740 | -6.85
2162 | 160,851 | -7.07
2163 | 160,963 | -6.79
2164 | 161,077 | -7.48
2165 | 161,192 | -7.34
2166 | 161,308 | -7.52
2167 | 161,427 | -7.97
2168 | 161,545 | -7.41
2169 | 161,661 | -7.38
2170 | 161,776 | -7.03
2171 | 161,890 | -7.25
2172 | 162,005 | -7.18
2173 | 162,117 | -6.60
2174 | 162,227 | -6.60
2175 | 162,336 | -6.22
2176 | 162,440 | -5.46
2177 | 162,543 | -5.59
2178 | 162,649 | -6.60
2179 | 162,759 | -6.42
2180 | 162,867 | -6.14
2181 | 162,976 | -6.52
2182 | 163,089 | -7.25
2183 | 163,204 | -7.03
2184 | 163,318 | -7.05
2185 | 163,432 | -7.00
2186 | 163,546 | -6.81
2187 | 163,660 | -7.23
2188 | 163,777 | -7.39

2189 | 163,896 | -7.66
2190 | 164,016 | -7.76
2191 | 164,136 | -7.58
2192 | 164,258 | -8.26
2193 | 164,385 | -8.46
2194 | 164,512 | -8.57
2195 | 164,642 | -8.77
2196 | 164,769 | -8.11
2197 | 164,894 | -8.29
2198 | 165,021 | -8.44
2199 | 165,148 | -8.30
2200 | 165,274 | -8.31
2201 | 165,399 | -7.82
2202 | 165,522 | -7.98
2203 | 165,646 | -8.17
2204 | 165,774 | -8.57
2205 | 165,903 | -8.53
2206 | 166,029 | -7.97
2207 | 166,155 | -8.44
2208 | 166,284 | -8.39
2209 | 166,411 | -8.12
2210 | 166,537 | -8.34
2211 | 166,662 | -7.65
2212 | 166,785 | -7.90
2213 | 166,909 | -7.82
2214 | 167,035 | -8.45
2215 | 167,164 | -8.25
2216 | 167,290 | -7.99
2217 | 167,414 | -7.76
2218 | 167,539 | -7.96

2219	167,664	-7.96		2249	171,165	-7.97
2220	167,790	-8.05		2250	171,289	-7.00
2221	167,915	-7.89		2251	171,405	-6.16
2222	168,039	-7.56		2252	171,519	-6.27
2223	168,159	-7.13		2253	171,638	-7.41
2224	168,276	-6.65		2254	171,760	-6.88
2225	168,391	-6.62		2255	171,880	-7.08
2226	168,506	-6.90		2256	172,004	-7.65
2227	168,623	-6.71		2257	172,132	-8.05
2228	168,739	-6.94		2258	172,262	-8.15
2229	168,857	-7.05		2259	172,393	-8.14
2230	168,974	-6.59		2260	172,522	-7.53
2231	169,088	-6.33		2261	172,647	-7.56
2232	169,201	-6.33		2262	172,774	-7.64
2233	169,314	-6.29		2263	172,900	-7.51
2234	169,425	-5.81		2264	173,027	-7.72
2235	169,531	-5.08		2265	173,153	-7.31
2236	169,638	-5.83		2266	173,275	-6.89
2237	169,750	-6.47		2267	173,397	-7.11
2238	169,866	-6.87		2268	173,519	-6.96
2239	169,983	-6.61		2269	173,642	-7.13
2240	170,101	-7.01		2270	173,761	-6.30
2241	170,221	-7.30		2271	173,875	-5.74
2242	170,340	-6.71		2272	173,988	-5.91
2243	170,455	-6.23		2273	174,105	-6.86
2244	170,568	-6.05		2274	174,226	-6.74
2245	170,683	-6.85		2275	174,344	-6.18
2246	170,803	-7.25		2276	174,459	-6.02
2247	170,922	-6.61		2277	174,573	-5.95
2248	171,040	-7.01		2278	174,688	-6.09

2279	174,804	-6.27	2309	178,554	-8.15
2280	174,920	-5.92	2310	178,691	-8.28
2281	175,034	-6.02	2311	178,830	-8.35
2282	175,150	-6.14	2312	178,965	-7.50
2283	175,266	-6.16	2313	179,093	-6.77
2284	175,386	-6.84	2314	179,217	-6.65
2285	175,509	-7.25	2315	179,342	-7.00
2286	175,630	-6.09	2316	179,468	-6.83
2287	175,746	-6.04	2317	179,590	-6.08
2288	175,866	-7.00	2318	179,706	-5.30
2289	175,991	-7.22	2319	179,820	-5.49
2290	176,117	-7.23	2320	179,939	-6.39
2291	176,244	-7.56	2321	180,060	-6.34
2292	176,373	-7.33	2322	180,182	-6.35
2293	176,501	-7.50	2323	180,304	-6.28
2294	176,630	-7.58	2324	180,426	-6.44
2295	176,759	-7.54	2325	180,546	-5.71
2296	176,887	-7.19	2326	180,663	-5.64
2297	177,015	-7.49	2327	180,780	-5.64
2298	177,142	-7.08	2328	180,895	-5.32
2299	177,268	-7.06	2329	181,011	-5.73
2300	177,393	-6.88	2330	181,132	-6.64
2301	177,517	-6.92	2331	181,259	-6.92
2302	177,644	-7.40	2332	181,382	-5.80
2303	177,773	-7.44	2333	181,502	-6.20
2304	177,900	-6.94	2334	181,626	-6.77
2305	178,026	-7.07	2335	181,753	-6.71
2306	178,153	-7.25	2336	181,881	-6.95
2307	178,283	-7.64	2337	182,011	-7.30
2308	178,417	-8.19	2338	182,146	-7.87

2339 | 182,284 | -8.00
2340 | 182,421 | -7.52
2341 | 182,556 | -7.66
2342 | 182,694 | -8.08
2343 | 182,830 | -7.29
2344 | 182,965 | -7.70
2345 | 183,102 | -7.77
2346 | 183,243 | -8.58
2347 | 183,386 | -8.11
2348 | 183,525 | -7.64
2349 | 183,664 | -8.17
2350 | 183,807 | -8.20
2351 | 183,948 | -8.08
2352 | 184,088 | -7.79
2353 | 184,229 | -8.10
2354 | 184,367 | -7.37
2355 | 184,501 | -7.10
2356 | 184,641 | -8.72
2357 | 184,787 | -8.21
2358 | 184,928 | -7.75
2359 | 185,069 | -8.10
2360 | 185,212 | -8.23
2361 | 185,354 | -7.75
2362 | 185,494 | -7.94
2363 | 185,640 | -8.72
2364 | 185,788 | -8.49
2365 | 185,931 | -7.64
2366 | 186,071 | -7.90
2367 | 186,213 | -7.98
2368 | 186,356 | -7.95

2369 | 186,499 | -8.14
2370 | 186,642 | -7.68
2371 | 186,784 | -8.09
2372 | 186,927 | -7.94
2373 | 187,067 | -7.36
2374 | 187,204 | -7.27
2375 | 187,343 | -7.75
2376 | 187,486 | -8.14
2377 | 187,630 | -7.78
2378 | 187,770 | -7.42
2379 | 187,907 | -7.01
2380 | 188,042 | -7.03
2381 | 188,181 | -7.87
2382 | 188,324 | -7.84
2383 | 188,467 | -7.63
2384 | 188,609 | -7.92
2385 | 188,752 | -7.51
2386 | 188,892 | -7.46
2387 | 189,030 | -7.10
2388 | 189,165 | -6.64
2389 | 189,297 | -6.48
2390 | 189,431 | -7.08
2391 | 189,568 | -7.02
2392 | 189,702 | -6.63
2393 | 189,832 | -6.02
2394 | 189,961 | -6.27
2395 | 190,092 | -6.49
2396 | 190,222 | -6.00
2397 | 190,348 | -5.61
2398 | 190,469 | -4.84

2399	190,587	-4.79
2400	190,705	-4.72
2401	190,822	-4.51
2402	190,934	-3.60
2403	191,043	-3.68
2404	191,156	-4.63
2405	191,276	-5.18
2406	191,398	-5.35
2407	191,522	-5.51
2408	191,647	-5.55
2409	191,771	-5.36
2410	191,895	-5.46
2411	192,018	-4.97
2412	192,138	-4.83
2413	192,258	-5.02
2414	192,382	-5.51
2415	192,509	-5.85
2416	192,638	-6.01
2417	192,768	-5.97
2418	192,895	-5.35
2419	193,022	-5.84
2420	193,152	-6.21
2421	193,285	-6.28
2422	193,421	-6.78
2423	193,556	-6.25
2424	193,687	-5.68
2425	193,816	-5.89
2426	193,946	-5.88
2427	194,075	-5.63
2428	194,201	-5.19
2429	194,324	-4.80
2430	194,444	-4.53
2431	194,565	-4.96
2432	194,687	-4.79
2433	194,806	-4.23
2434	194,926	-5.07
2435	195,051	-5.22
2436	195,175	-4.90
2437	195,298	-4.81
2438	195,420	-4.92
2439	195,546	-5.48
2440	195,676	-5.88
2441	195,808	-5.95
2442	195,940	-5.78
2443	196,071	-5.73
2444	196,202	-5.59
2445	196,332	-5.72
2446	196,463	-5.72
2447	196,593	-5.45
2448	196,722	-5.37
2449	196,846	-4.49
2450	196,965	-4.04
2451	197,086	-4.86
2452	197,211	-5.17
2453	197,338	-4.97
2454	197,462	-4.79
2455	197,587	-4.87
2456	197,711	-4.83
2457	197,833	-4.30
2458	197,954	-4.48

2459 | 198,075 | -4.07
2460 | 198,192 | -3.85
2461 | 198,309 | -3.80
2462 | 198,426 | -3.81
2463 | 198,542 | -3.68
2464 | 198,656 | -3.20
2465 | 198,768 | -3.29
2466 | 198,881 | -3.08
2467 | 198,992 | -2.96
2468 | 199,104 | -3.37
2469 | 199,220 | -3.84
2470 | 199,337 | -3.62
2471 | 199,452 | -3.40
2472 | 199,567 | -3.63
2473 | 199,682 | -3.17
2474 | 199,793 | -2.80
2475 | 199,904 | -2.95
2476 | 200,015 | -2.68
2477 | 200,124 | -2.78
2478 | 200,236 | -3.07
2479 | 200,345 | -2.23
2480 | 200,451 | -2.02
2481 | 200,558 | -2.42
2482 | 200,666 | -2.48
2483 | 200,776 | -2.82
2484 | 200,887 | -2.84
2485 | 200,998 | -2.78
2486 | 201,109 | -2.59
2487 | 201,216 | -1.90
2488 | 201,319 | -1.49

2489 | 201,423 | -1.96
2490 | 201,529 | -2.11
2491 | 201,635 | -1.99
2492 | 201,742 | -2.26
2493 | 201,849 | -2.08
2494 | 201,957 | -2.34
2495 | 202,065 | -2.02
2496 | 202,169 | -1.53
2497 | 202,275 | -2.40
2498 | 202,386 | -2.59
2499 | 202,496 | -2.51
2500 | 202,607 | -2.54
2501 | 202,717 | -2.35
2502 | 202,826 | -2.29
2503 | 202,936 | -2.49
2504 | 203,046 | -2.38
2505 | 203,155 | -1.98
2506 | 203,262 | -1.90
2507 | 203,370 | -2.34
2508 | 203,481 | -2.56
2509 | 203,592 | -2.43
2510 | 203,704 | -2.73
2511 | 203,818 | -2.92
2512 | 203,933 | -3.03
2513 | 204,049 | -3.10
2514 | 204,164 | -2.71
2515 | 204,278 | -2.84
2516 | 204,394 | -3.25
2517 | 204,512 | -3.17
2518 | 204,629 | -2.94

2519 | 204,743 | -2.56
2520 | 204,857 | -2.78
2521 | 204,973 | -3.17
2522 | 205,092 | -3.55
2523 | 205,214 | -3.69
2524 | 205,336 | -3.68
2525 | 205,458 | -3.63
2526 | 205,582 | -3.95
2527 | 205,708 | -4.25
2528 | 205,836 | -4.35
2529 | 205,965 | -4.45
2530 | 206,097 | -4.95
2531 | 206,232 | -5.05
2532 | 206,366 | -4.67
2533 | 206,496 | -4.39
2534 | 206,627 | -4.62
2535 | 206,759 | -4.65
2536 | 206,890 | -4.27
2537 | 207,019 | -4.37
2538 | 207,149 | -4.29
2539 | 207,278 | -4.31
2540 | 207,409 | -4.49
2541 | 207,540 | -4.37
2542 | 207,671 | -4.47
2543 | 207,803 | -4.51
2544 | 207,933 | -4.11
2545 | 208,061 | -4.02
2546 | 208,191 | -4.35
2547 | 208,319 | -3.70
2548 | 208,445 | -3.81

2549 | 208,571 | -3.68
2550 | 208,696 | -3.59
2551 | 208,821 | -3.60
2552 | 208,946 | -3.55
2553 | 209,072 | -3.70
2554 | 209,195 | -3.06
2555 | 209,315 | -2.88
2556 | 209,436 | -3.17
2557 | 209,560 | -3.55
2558 | 209,686 | -3.73
2559 | 209,812 | -3.48
2560 | 209,938 | -3.58
2561 | 210,062 | -3.07
2562 | 210,182 | -2.51
2563 | 210,300 | -2.64
2564 | 210,418 | -2.67
2565 | 210,538 | -2.91
2566 | 210,659 | -2.79
2567 | 210,777 | -2.33
2568 | 210,893 | -2.14
2569 | 211,009 | -2.21
2570 | 211,127 | -2.64
2571 | 211,246 | -2.67
2572 | 211,366 | -2.56
2573 | 211,485 | -2.53
2574 | 211,603 | -2.28
2575 | 211,721 | -2.61
2576 | 211,842 | -2.80
2577 | 211,960 | -2.00
2578 | 212,075 | -2.03

2579 | 212,190 | -1.81
2580 | 212,305 | -2.17
2581 | 212,422 | -2.07
2582 | 212,538 | -2.03
2583 | 212,653 | -1.65
2584 | 212,766 | -1.63
2585 | 212,881 | -2.07
2586 | 212,996 | -1.65
2587 | 213,109 | -1.50
2588 | 213,223 | -1.96
2589 | 213,338 | -1.76
2590 | 213,453 | -1.67
2591 | 213,567 | -1.82
2592 | 213,684 | -2.06
2593 | 213,799 | -1.45
2594 | 213,911 | -1.35
2595 | 214,024 | -1.58
2596 | 214,138 | -1.50
2597 | 214,251 | -1.40
2598 | 214,364 | -1.41
2599 | 214,478 | -1.61
2600 | 214,592 | -1.58
2601 | 214,706 | -1.36
2602 | 214,819 | -1.29
2603 | 214,932 | -1.39
2604 | 215,047 | -1.67
2605 | 215,161 | -1.40
2606 | 215,274 | -1.24
2607 | 215,386 | -1.02
2608 | 215,497 | -0.91

2609 | 215,608 | -1.22
2610 | 215,723 | -1.74
2611 | 215,839 | -1.45
2612 | 215,955 | -1.79
2613 | 216,075 | -2.28
2614 | 216,197 | -2.39
2615 | 216,319 | -2.11
2616 | 216,439 | -2.14
2617 | 216,560 | -2.10
2618 | 216,678 | -1.39
2619 | 216,792 | -1.05
2620 | 216,904 | -0.97
2621 | 217,018 | -1.23
2622 | 217,131 | -1.07
2623 | 217,244 | -0.86
2624 | 217,356 | -1.02
2625 | 217,470 | -1.13
2626 | 217,584 | -1.20
2627 | 217,699 | -1.18
2628 | 217,817 | -2.18
2629 | 217,939 | -2.04
2630 | 218,060 | -1.85
2631 | 218,181 | -1.97
2632 | 218,303 | -2.08
2633 | 218,426 | -2.20
2634 | 218,551 | -2.62
2635 | 218,681 | -3.23
2636 | 218,814 | -3.20
2637 | 218,946 | -2.94
2638 | 219,076 | -2.90

2639 | 219,207 | -2.98
2640 | 219,339 | -3.17
2641 | 219,472 | -3.35
2642 | 219,605 | -2.84
2643 | 219,738 | -3.42
2644 | 219,878 | -4.37
2645 | 220,023 | -4.31
2646 | 220,167 | -4.23
2647 | 220,312 | -4.43
2648 | 220,461 | -4.99
2649 | 220,613 | -4.84
2650 | 220,761 | -4.35
2651 | 220,907 | -4.49
2652 | 221,058 | -5.03
2653 | 221,210 | -4.89
2654 | 221,363 | -4.95
2655 | 221,520 | -5.78
2656 | 221,684 | -6.19
2657 | 221,850 | -6.02
2658 | 222,015 | -6.06
2659 | 222,182 | -6.21
2660 | 222,355 | -7.07
2661 | 222,535 | -7.29
2662 | 222,711 | -6.49
2663 | 222,885 | -6.78
2664 | 223,062 | -7.02
2665 | 223,240 | -6.89
2666 | 223,420 | -7.30
2667 | 223,605 | -7.63
2668 | 223,792 | -7.60

2669 | 223,979 | -7.55
2670 | 224,164 | -7.36
2671 | 224,351 | -7.64
2672 | 224,536 | -7.19
2673 | 224,716 | -6.66
2674 | 224,888 | -5.89
2675 | 225,058 | -6.22
2676 | 225,229 | -6.20
2677 | 225,399 | -5.88
2678 | 225,562 | -4.99
2679 | 225,726 | -5.96
2680 | 225,900 | -6.59
2681 | 226,072 | -5.72
2682 | 226,238 | -5.58
2683 | 226,408 | -6.17
2684 | 226,582 | -6.33
2685 | 226,758 | -6.53
2686 | 226,939 | -7.04
2687 | 227,128 | -7.70
2688 | 227,321 | -7.62
2689 | 227,512 | -7.19
2690 | 227,699 | -7.12
2691 | 227,883 | -6.81
2692 | 228,064 | -6.50
2693 | 228,241 | -6.01
2694 | 228,416 | -6.19
2695 | 228,596 | -6.81
2696 | 228,783 | -7.33
2697 | 228,974 | -7.34
2698 | 229,165 | -7.19

2699	229,354	-7.01
2700	229,541	-6.89
2701	229,726	-6.71
2702	229,909	-6.34
2703	230,088	-6.15
2704	230,264	-5.85
2705	230,440	-6.17
2706	230,618	-5.94
2707	230,794	-5.96
2708	230,972	-6.09
2709	231,146	-5.37
2710	231,315	-5.15
2711	231,482	-5.07
2712	231,649	-4.98
2713	231,815	-4.93
2714	231,979	-4.54
2715	232,140	-4.38
2716	232,299	-4.21
2717	232,458	-4.29
2718	232,617	-4.16
2719	232,774	-3.91
2720	232,930	-3.80
2721	233,083	-3.44
2722	233,236	-3.66
2723	233,390	-3.71
2724	233,542	-3.26
2725	233,691	-2.87
2726	233,836	-2.59
2727	233,979	-2.33
2728	234,122	-2.78

2729	234,268	-2.66
2730	234,411	-2.22
2731	234,552	-2.22
2732	234,694	-2.25
2733	234,837	-2.65
2734	234,984	-2.77
2735	235,129	-2.44
2736	235,271	-1.93
2737	235,411	-1.92
2738	235,552	-2.24
2739	235,694	-2.13
2740	235,836	-2.11
2741	235,977	-1.89
2742	236,114	-1.45
2743	236,250	-1.40
2744	236,385	-1.11
2745	236,518	-1.17
2746	236,652	-1.15
2747	236,785	-0.94
2748	236,916	-0.60
2749	237,043	0.10
2750	237,166	0.16
2751	237,288	0.48
2752	237,410	0.36
2753	237,529	1.03
2754	237,643	1.73
2755	237,755	1.74
2756	237,866	1.99
2757	237,975	2.20
2758	238,084	2.09

2759	238,194	1.90
2760	238,306	1.75
2761	238,419	1.66
2762	238,532	1.55
2763	238,647	1.30
2764	238,764	1.12
2765	238,883	0.99
2766	239,003	0.69
2767	239,125	0.56
2768	239,248	0.46
2769	239,375	-0.18
2770	239,506	-0.54
2771	239,642	-1.29
2772	239,784	-1.76
2773	239,931	-2.22
2774	240,080	-2.12
2775	240,231	-2.59
2776	240,387	-3.09
2777	240,548	-3.43
2778	240,711	-3.55
2779	240,877	-3.85
2780	241,048	-4.17
2781	241,224	-4.82
2782	241,406	-5.23
2783	241,596	-5.80
2784	241,792	-6.21
2785	241,991	-6.10
2786	242,187	-5.86
2787	242,382	-5.80
2788	242,576	-5.74
2789	242,770	-5.65
2790	242,959	-5.02
2791	243,144	-4.93
2792	243,330	-5.18
2793	243,519	-5.46
2794	243,714	-5.93
2795	243,916	-6.38
2796	244,124	-6.82
2797	244,337	-6.94
2798	244,551	-7.02
2799	244,767	-7.07
2800	244,983	-6.97
2801	245,196	-6.72
2802	245,407	-6.58
2803	245,618	-6.71
2804	245,834	-7.19
2805	246,053	-7.07
2806	246,269	-6.86
2807	246,485	-6.83
2808	246,700	-6.78
2809	246,917	-7.07
2810	247,135	-6.92
2811	247,351	-6.60
2812	247,566	-6.78
2813	247,785	-7.13
2814	248,007	-7.13
2815	248,229	-7.10
2816	248,453	-7.36
2817	248,680	-7.41
2818	248,907	-7.24

2819	249,128	-6.76	2849	255,620	-6.39
2820	249,345	-6.51	2850	255,844	-6.43
2821	249,560	-6.48	2851	256,068	-6.36
2822	249,772	-6.11	2852	256,294	-6.69
2823	249,984	-6.39	2853	256,525	-6.95
2824	250,199	-6.52	2854	256,759	-6.93
2825	250,413	-6.17	2855	256,994	-7.13
2826	250,618	-5.31	2856	257,234	-7.42
2827	250,818	-5.29	2857	257,479	-7.58
2828	251,021	-5.72	2858	257,726	-7.70
2829	251,231	-6.34	2859	257,975	-7.85
2830	251,448	-6.48	2860	258,224	-7.50
2831	251,666	-6.50	2861	258,471	-7.67
2832	251,883	-6.25	2862	258,723	-8.01
2833	252,102	-6.79	2863	258,978	-7.98
2834	252,329	-7.16	2864	259,235	-8.09
2835	252,553	-6.49	2865	259,492	-8.05
2836	252,769	-6.01	2866	259,750	-7.99
2837	252,986	-6.55	2867	260,009	-8.30
2838	253,213	-7.13	2868	260,272	-8.31
2839	253,442	-6.93	2869	260,536	-8.26
2840	253,670	-6.88	2870	260,801	-8.52
2841	253,896	-6.63	2871	261,069	-8.43
2842	254,121	-6.75	2872	261,336	-8.43
2843	254,345	-6.45	2873	261,602	-8.19
2844	254,560	-5.60	2874	261,865	-8.16
2845	254,767	-5.14	2875	262,131	-8.42
2846	254,972	-5.41	2876	262,399	-8.30
2847	255,183	-5.77	2877	262,665	-8.21
2848	255,399	-6.16	2878	262,933	-8.42

2879 | 263,201 | -8.13
2880 | 263,466 | -8.08
2881 | 263,732 | -8.30
2882 | 264,004 | -8.53
2883 | 264,273 | -7.93
2884 | 264,533 | -7.52
2885 | 264,792 | -7.86
2886 | 265,055 | -7.85
2887 | 265,321 | -8.13
2888 | 265,595 | -8.57
2889 | 265,865 | -7.78
2890 | 266,123 | -7.18
2891 | 266,380 | -7.61
2892 | 266,643 | -7.75
2893 | 266,905 | -7.54
2894 | 267,167 | -7.66
2895 | 267,431 | -7.65
2896 | 267,692 | -7.28
2897 | 267,946 | -6.86
2898 | 268,194 | -6.59
2899 | 268,435 | -5.96
2900 | 268,677 | -6.61
2901 | 268,926 | -6.77
2902 | 269,178 | -6.84
2903 | 269,429 | -6.65
2904 | 269,678 | -6.59
2905 | 269,926 | -6.47
2906 | 270,169 | -5.95
2907 | 270,409 | -6.00
2908 | 270,648 | -5.80

2909 | 270,883 | -5.44
2910 | 271,114 | -5.30
2911 | 271,347 | -5.65
2912 | 271,584 | -5.74
2913 | 271,822 | -5.65
2914 | 272,053 | -4.80
2915 | 272,276 | -4.60
2916 | 272,500 | -4.90
2917 | 272,724 | -4.65
2918 | 272,946 | -4.43
2919 | 273,166 | -4.35
2920 | 273,385 | -4.33
2921 | 273,605 | -4.33
2922 | 273,827 | -4.55
2923 | 274,048 | -4.10
2924 | 274,263 | -3.84
2925 | 274,477 | -3.80
2926 | 274,692 | -3.95
2927 | 274,909 | -3.95
2928 | 275,127 | -4.13
2929 | 275,344 | -3.73
2930 | 275,559 | -3.71
2931 | 275,772 | -3.53
2932 | 275,983 | -3.38
2933 | 276,195 | -3.48
2934 | 276,409 | -3.62
2935 | 276,622 | -3.35
2936 | 276,830 | -2.88
2937 | 277,036 | -3.01
2938 | 277,246 | -3.33

2939	277,460	-3.61		2969	285,102	-7.45
2940	277,680	-3.95		2970	285,398	-6.99
2941	277,903	-4.00		2971	285,686	-6.57
2942	278,130	-4.44		2972	285,968	-6.40
2943	278,361	-4.46		2973	286,248	-6.27
2944	278,593	-4.44		2974	286,525	-6.10
2945	278,825	-4.43		2975	286,802	-6.15
2946	279,057	-4.44		2976	287,079	-5.96
2947	279,292	-4.67		2977	287,351	-5.62
2948	279,533	-5.10		2978	287,617	-5.28
2949	279,781	-5.49		2979	287,879	-5.09
2950	280,038	-6.17		2980	288,145	-5.68
2951	280,303	-6.18		2981	288,418	-5.80
2952	280,564	-5.69		2982	288,698	-6.34
2953	280,817	-5.30		2983	288,986	-6.50
2954	281,071	-5.66		2984	289,271	-6.06
2955	281,332	-6.11		2985	289,549	-5.67
2956	281,602	-6.50		2986	289,823	-5.56
2957	281,875	-6.39		2987	290,100	-6.00
2958	282,143	-5.96		2988	290,382	-6.00
2959	282,405	-5.67		2989	290,660	-5.39
2960	282,661	-5.15		2990	290,929	-5.15
2961	282,910	-4.85		2991	291,193	-4.65
2962	283,160	-5.22		2992	291,455	-4.93
2963	283,417	-5.55		2993	291,721	-5.07
2964	283,677	-5.54		2994	291,986	-4.62
2965	283,941	-5.95		2995	292,242	-4.13
2966	284,218	-6.83		2996	292,490	-3.52
2967	284,508	-7.27		2997	292,732	-3.46
2968	284,804	-7.30		2998	292,976	-3.67

2999 | 293,232 | -4.76
3000 | 293,502 | -5.22
3001 | 293,779 | -5.34
3002 | 294,060 | -5.60
3003 | 294,348 | -6.06
3004 | 294,637 | -5.56
3005 | 294,919 | -5.34
3006 | 295,198 | -5.13
3007 | 295,474 | -4.96
3008 | 295,749 | -4.94
3009 | 296,020 | -4.57
3010 | 296,289 | -4.53
3011 | 296,562 | -5.04
3012 | 296,840 | -4.92
3013 | 297,116 | -4.81
3014 | 297,393 | -4.90
3015 | 297,669 | -4.70
3016 | 297,943 | -4.57
3017 | 298,215 | -4.48
3018 | 298,485 | -4.20
3019 | 298,749 | -3.81
3020 | 299,012 | -4.04
3021 | 299,275 | -3.69
3022 | 299,533 | -3.45
3023 | 299,788 | -3.19
3024 | 300,039 | -3.08
3025 | 300,293 | -3.26
3026 | 300,547 | -3.13
3027 | 300,799 | -2.97
3028 | 301,052 | -3.17

3029 | 301,305 | -2.92
3030 | 301,560 | -3.32
3031 | 301,813 | -2.52
3032 | 302,061 | -2.76
3033 | 302,320 | -3.64
3034 | 302,583 | -3.19
3035 | 302,834 | -2.28
3036 | 303,077 | -2.04
3037 | 303,316 | -1.74
3038 | 303,558 | -2.34
3039 | 303,810 | -2.88
3040 | 304,068 | -3.03
3041 | 304,333 | -3.46
3042 | 304,606 | -3.88
3043 | 304,886 | -4.13
3044 | 305,171 | -4.32
3045 | 305,463 | -4.82
3046 | 305,764 | -5.14
3047 | 306,073 | -5.42
3048 | 306,386 | -5.54
3049 | 306,698 | -5.14
3050 | 307,005 | -5.09
3051 | 307,309 | -4.72
3052 | 307,608 | -4.58
3053 | 307,908 | -4.69
3054 | 308,206 | -4.41
3055 | 308,500 | -4.09
3056 | 308,789 | -3.95
3057 | 309,077 | -3.86
3058 | 309,365 | -3.83

3059 | 309,651 | -3.69
3060 | 309,934 | -3.39
3061 | 310,215 | -3.32
3062 | 310,493 | -3.06
3063 | 310,766 | -2.78
3064 | 311,035 | -2.54
3065 | 311,302 | -2.39
3066 | 311,568 | -2.33
3067 | 311,832 | -2.22
3068 | 312,099 | -2.45
3069 | 312,367 | -2.35
3070 | 312,634 | -2.29
3071 | 312,899 | -1.92
3072 | 313,159 | -1.79
3073 | 313,418 | -1.61
3074 | 313,674 | -1.34
3075 | 313,928 | -1.38
3076 | 314,183 | -1.36
3077 | 314,438 | -1.32
3078 | 314,692 | -1.10
3079 | 314,943 | -0.99
3080 | 315,194 | -0.97
3081 | 315,443 | -0.64
3082 | 315,689 | -0.48
3083 | 315,934 | -0.46
3084 | 316,179 | -0.51
3085 | 316,424 | -0.18
3086 | 316,666 | -0.18
3087 | 316,910 | -0.33
3088 | 317,155 | -0.20

3089 | 317,401 | -0.39
3090 | 317,649 | -0.42
3091 | 317,897 | -0.14
3092 | 318,141 | -0.03
3093 | 318,385 | 0.02
3094 | 318,628 | 0.20
3095 | 318,872 | 0.01
3096 | 319,118 | 0.02
3097 | 319,364 | 0.16
3098 | 319,608 | 0.24
3099 | 319,853 | 0.15
3100 | 320,102 | -0.12
3101 | 320,354 | -0.18
3102 | 320,607 | -0.09
3103 | 320,856 | 0.38
3104 | 321,095 | 1.14
3105 | 321,329 | 1.26
3106 | 321,559 | 1.66
3107 | 321,783 | 2.18
3108 | 322,000 | 2.65
3109 | 322,214 | 2.85
3110 | 322,426 | 2.82
3111 | 322,638 | 3.19
3112 | 322,847 | 3.18
3113 | 323,057 | 3.08
3114 | 323,269 | 3.08
3115 | 323,482 | 3.14
3116 | 323,695 | 2.99
3117 | 323,911 | 2.84
3118 | 324,129 | 2.86

3119 | 324,349 | 2.65
3120 | 324,574 | 2.34
3121 | 324,804 | 2.00
3122 | 325,039 | 1.74
3123 | 325,278 | 1.59
3124 | 325,527 | 0.68
3125 | 325,789 | 0.05
3126 | 326,061 | -0.25
3127 | 326,342 | -0.96
3128 | 326,639 | -1.73
3129 | 326,952 | -2.39
3130 | 327,273 | -2.41
3131 | 327,597 | -2.55
3132 | 327,924 | -2.64
3133 | 328,258 | -3.03
3134 | 328,602 | -3.41
3135 | 328,955 | -3.68
3136 | 329,318 | -4.21
3137 | 329,694 | -4.56
3138 | 330,081 | -4.90
3139 | 330,476 | -5.11
3140 | 330,883 | -5.67
3141 | 331,300 | -5.76
3142 | 331,725 | -6.06
3143 | 332,164 | -6.63
3144 | 332,624 | -7.30
3145 | 333,106 | -7.86
3146 | 333,602 | -8.04
3147 | 334,101 | -7.94
3148 | 334,600 | -7.93

3149 | 335,100 | -7.87
3150 | 335,595 | -7.58
3151 | 336,081 | -7.28
3152 | 336,567 | -7.43
3153 | 337,055 | -7.30
3154 | 337,544 | -7.42
3155 | 338,033 | -7.16
3156 | 338,521 | -7.30
3157 | 339,015 | -7.40
3158 | 339,505 | -6.98
3159 | 339,993 | -7.13
3160 | 340,490 | -7.44
3161 | 340,984 | -6.88
3162 | 341,462 | -6.31
3163 | 341,930 | -6.26
3164 | 342,401 | -6.32
3165 | 342,879 | -6.59
3166 | 343,364 | -6.64
3167 | 343,853 | -6.76
3168 | 344,348 | -6.84
3169 | 344,839 | -6.43
3170 | 345,327 | -6.58
3171 | 345,821 | -6.65
3172 | 346,319 | -6.76
3173 | 346,823 | -6.84
3174 | 347,330 | -6.83
3175 | 347,841 | -6.92
3176 | 348,357 | -7.05
3177 | 348,877 | -7.02
3178 | 349,404 | -7.32

3179	349,945	-7.61	3209	366,267	-5.13
3180	350,493	-7.64	3210	366,784	-4.84
3181	351,044	-7.64	3211	367,297	-4.86
3182	351,596	-7.57	3212	367,816	-4.97
3183	352,150	-7.64	3213	368,329	-4.46
3184	352,710	-7.77	3214	368,829	-4.03
3185	353,273	-7.71	3215	369,331	-4.54
3186	353,838	-7.71	3216	369,858	-5.23
3187	354,400	-7.46	3217	370,403	-5.42
3188	354,958	-7.41	3218	370,960	-5.74
3189	355,512	-7.13	3219	371,545	-6.67
3190	356,060	-6.94	3220	372,149	-6.58
3191	356,612	-7.24	3221	372,746	-6.15
3192	357,167	-7.01	3222	373,331	-5.90
3193	357,712	-6.58	3223	373,909	-5.66
3194	358,247	-6.39	3224	374,478	-5.33
3195	358,786	-6.64	3225	375,042	-5.26
3196	359,327	-6.41	3226	375,604	-5.08
3197	359,855	-5.80	3227	376,155	-4.55
3198	360,376	-5.89	3228	376,690	-4.07
3199	360,904	-6.05	3229	377,204	-3.27
3200	361,440	-6.24	3230	377,702	-2.99
3201	361,973	-5.79	3231	378,210	-3.72
3202	362,503	-5.90	3232	378,743	-4.29
3203	363,042	-6.20	3233	379,297	-4.63
3204	363,580	-5.72	3234	379,864	-4.88
3205	364,114	-5.84	3235	380,435	-4.68
3206	364,659	-6.21	3236	380,997	-4.32
3207	365,208	-5.88	3237	381,551	-4.05
3208	365,743	-5.40	3238	382,095	-3.72

3239 | 382,622 | -2.92
3240 | 383,134 | -2.72
3241 | 383,641 | -2.50
3242 | 384,159 | -3.24
3243 | 384,710 | -4.10
3244 | 385,288 | -4.54
3245 | 385,894 | -5.36
3246 | 386,528 | -5.65
3247 | 387,167 | -5.46
3248 | 387,804 | -5.40
3249 | 388,435 | -5.02
3250 | 389,061 | -5.07
3251 | 389,700 | -5.44
3252 | 390,349 | -5.34
3253 | 390,999 | -5.36
3254 | 391,658 | -5.60
3255 | 392,317 | -5.17
3256 | 392,972 | -5.31
3257 | 393,636 | -5.40
3258 | 394,297 | -5.02
3259 | 394,935 | -4.23
3260 | 395,543 | -3.43
3261 | 396,126 | -2.80
3262 | 396,694 | -2.54
3263 | 397,253 | -2.20
3264 | 397,808 | -2.13
3265 | 398,361 | -1.92
3266 | 398,909 | -1.68
3267 | 399,455 | -1.65
3268 | 400,003 | -1.64

3269 | 400,553 | -1.58
3270 | 401,097 | -1.17
3271 | 401,636 | -1.09
3272 | 402,179 | -1.26
3273 | 402,728 | -1.23
3274 | 403,276 | -1.01
3275 | 403,815 | -0.58
3276 | 404,346 | -0.37
3277 | 404,874 | -0.28
3278 | 405,401 | -0.10
3279 | 405,923 | 0.18
3280 | 406,441 | 0.31
3281 | 406,957 | 0.49
3282 | 407,469 | 0.71
3283 | 407,977 | 0.85
3284 | 408,485 | 0.94
3285 | 408,992 | 1.12
3286 | 409,496 | 1.28
3287 | 409,995 | 1.64
3288 | 410,483 | 2.13
3289 | 410,979 | 1.27
3290 | 411,492 | 1.34
3291 | 412,009 | 1.23
3292 | 412,533 | 1.15
3293 | 413,062 | 1.08
3294 | 413,601 | 0.80
3295 | 414,147 | 0.85
3296 | 414,692 | 1.07
3297 | 415,235 | 1.16
3298 | 415,781 | 1.13

3299 | 416,327 | 1.32
3300 | 416,872 | 1.36
3301 | 417,419 | 1.43
3302 | 417,969 | 1.40
3303 | 418,526 | 1.27
3304 | 419,095 | 0.94
3305 | 419,682 | 0.51
3306 | 420,281 | 0.54
3307 | 420,888 | 0.32
3308 | 421,507 | 0.15
3309 | 422,135 | 0.08
3310 | 422,766 | 0.23

DATASET 5

Dataset 5 (VostokTempMillBaseDetrend423kyrs). This dataset contains the following Vostok variables: (Age)—ice age in millennia or Kyrs from 423 Kyrs before present (BP) to the present calendar year 1989; (TDv)—detrended millennial-scale atmospheric temperature deviations in degrees Celsius (°C) relative to the Vostok base Kyr1 = 0.00 °C. This dataset is included in appendix: Dataset 5.[374]

Age	TDv
1	0.00
2	-0.21
3	-0.15
4	0.22
5	0.35
6	0.46
7	-0.04
8	0.01
9	0.10
10	0.06
11	-0.08
12	-0.76
13	-3.18
14	-2.83
15	-3.41
16	-4.96
17	-6.58
18	-7.43
19	-7.79
20	-7.80
21	-7.56
22	-7.60
23	-8.06
24	-8.20
25	-7.59
26	-7.34
27	-7.37
28	-8.19
29	-7.45
30	-7.64
31	-7.25
32	-7.35
33	-6.77
34	-7.07
35	-5.72
36	-5.33
37	-6.51
38	-6.41
39	-6.56
40	-5.89
41	-6.31
42	-5.02
43	-5.29
44	-5.91
45	-6.72
46	-7.07
47	-6.83
48	-6.31
49	-6.04
50	-5.48
51	-5.35
52	-5.33
53	-5.81
54	-6.17
55	-6.35
56	-5.35
57	-4.51
58	-4.57
59	-5.07

60 | -5.80
61 | -6.39
62 | -6.70
63 | -7.44
64 | -6.71
65 | -7.46
66 | -7.71
67 | -7.53
68 | -7.59
69 | -7.47
70 | -6.58
71 | -6.45
72 | -5.35
73 | -5.93
74 | -6.59
75 | -5.97
76 | -5.02
77 | -4.86
78 | -6.13
79 | -6.01
80 | -5.03
81 | -4.97
82 | -3.92
83 | -3.74
84 | -3.58
85 | -3.41
86 | -2.94
87 | -3.91
88 | -4.22
89 | -4.87

90 | -5.02
91 | -5.39
92 | -4.66
93 | -4.53
94 | -4.45
95 | -4.16
96 | -4.04
97 | -4.13
98 | -4.62
99 | -4.31
100 | -4.02
101 | -3.60
102 | -2.80
103 | -2.74
104 | -4.05
105 | -4.36
106 | -4.30
107 | -4.25
108 | -6.07
109 | -6.66
110 | -6.34
111 | -5.85
112 | -5.98
113 | -6.00
114 | -5.18
115 | -4.50
116 | -3.14
117 | -2.69
118 | -1.73
119 | -1.27

120 | -1.27
121 | -0.74
122 | -0.46
123 | -0.14
124 | 0.14
125 | 0.40
126 | 0.54
127 | 1.10
128 | 2.12
129 | 2.30
130 | 1.35
131 | 0.31
132 | -1.48
133 | -2.95
134 | -4.07
135 | -6.24
136 | -7.40
137 | -7.77
138 | -8.01
139 | -8.99
140 | -8.78
141 | -8.81
142 | -8.79
143 | -8.68
144 | -8.66
145 | -8.30
146 | -8.56
147 | -8.23
148 | -8.37
149 | -8.14

150 | -7.88
151 | -7.47
152 | -7.61
153 | -8.12
154 | -8.04
155 | -8.39
156 | -8.66
157 | -8.84
158 | -8.41
159 | -7.53
160 | -6.40
161 | -7.16
162 | -7.77
163 | -6.66
164 | -7.57
165 | -8.63
166 | -8.64
167 | -8.46
168 | -8.41
169 | -7.30
170 | -6.58
171 | -7.17
172 | -7.40
173 | -8.19
174 | -7.23
175 | -6.70
176 | -6.94
177 | -7.83
178 | -7.57
179 | -8.30

180	-6.79	210	-4.00
181	-6.39	211	-3.22
182	-6.99	212	-3.11
183	-8.16	213	-2.53
184	-8.54	214	-2.30
185	-8.34	215	-2.08
186	-8.61	216	-2.01
187	-8.44	217	-2.43
188	-8.01	218	-1.98
189	-8.11	219	-3.17
190	-7.15	220	-4.00
191	-5.64	221	-5.19
192	-5.65	222	-6.20
193	-5.98	223	-7.37
194	-6.64	224	-8.00
195	-5.44	225	-7.64
196	-5.91	226	-6.66
197	-5.85	227	-6.91
198	-5.33	228	-7.98
199	-4.11	229	-7.37
200	-3.89	230	-7.56
201	-3.16	231	-6.73
202	-2.67	232	-5.77
203	-2.87	233	-4.86
204	-3.06	234	-3.90
205	-3.51	235	-3.23
206	-4.50	236	-2.86
207	-5.23	237	-1.86
208	-4.95	238	0.29
209	-4.39	239	0.77

240 | -1.43	270 | -7.56
241 | -3.91	271 | -6.72
242 | -6.17	272 | -6.50
243 | -6.44	273 | -5.60
244 | -6.32	274 | -5.32
245 | -7.74	275 | -4.85
246 | -7.60	276 | -4.67
247 | -7.70	277 | -4.26
248 | -7.68	278 | -4.57
249 | -8.06	279 | -5.39
250 | -7.27	280 | -5.97
251 | -6.66	281 | -6.74
252 | -7.12	282 | -7.16
253 | -7.43	283 | -6.41
254 | -7.74	284 | -6.50
255 | -6.76	285 | -8.10
256 | -7.00	286 | -7.87
257 | -7.64	287 | -7.17
258 | -8.44	288 | -6.46
259 | -8.59	289 | -7.03
260 | -8.89	290 | -6.80
261 | -9.22	291 | -6.67
262 | -9.16	292 | -5.85
263 | -9.20	293 | -4.77
264 | -9.06	294 | -6.12
265 | -8.76	295 | -6.69
266 | -8.99	296 | -6.04
267 | -8.41	297 | -5.83
268 | -8.31	298 | -5.80
269 | -7.38	299 | -5.20

300	-4.56		330	-5.59
301	-4.16		331	-6.44
302	-4.01		332	-7.12
303	-4.03		333	-8.26
304	-3.38		334	-9.20
305	-4.74		335	-9.17
306	-5.87		336	-8.92
307	-6.42		337	-8.62
308	-5.83		338	-8.63
309	-5.25		339	-8.51
310	-4.81		340	-8.39
311	-4.16		341	-8.56
312	-3.46		342	-7.56
313	-3.38		343	-7.69
314	-2.67		344	-7.97
315	-2.35		345	-7.96
316	-1.81		346	-7.90
317	-1.45		347	-8.09
318	-1.43		348	-8.17
319	-1.10		349	-8.34
320	-1.01		350	-8.72
321	-1.16		351	-8.95
322	0.51		352	-8.91
323	1.83		353	-9.04
324	1.85		354	-9.04
325	1.25		355	-8.78
326	-0.33		356	-8.47
327	-2.54		357	-8.48
328	-3.71		358	-8.12
329	-4.51		359	-7.87

360	-7.52	390	-6.81
361	-7.31	391	-6.86
362	-7.48	392	-7.12
363	-7.28	393	-6.75
364	-7.25	394	-6.93
365	-7.45	395	-6.25
366	-7.02	396	-4.97
367	-6.39	397	-4.19
368	-6.31	398	-3.72
369	-5.68	399	-3.39
370	-6.23	400	-3.21
371	-6.94	401	-3.16
372	-8.09	402	-2.69
373	-7.77	403	-2.82
374	-7.24	404	-2.40
375	-6.77	405	-1.92
376	-6.58	406	-1.60
377	-5.72	407	-1.23
378	-4.57	408	-0.85
379	-5.47	409	-0.62
380	-6.20	410	-0.24
381	-6.04	411	0.28
382	-5.52	412	-0.28
383	-4.68	413	-0.46
384	-4.07	414	-0.74
385	-5.21	415	-0.66
386	-6.37	416	-0.50
387	-7.15	417	-0.31
388	-6.93	418	-0.23
389	-6.53	419	-0.39

420 | -0.99
421 | -1.22
422 | -1.52
423 | -1.52

DATASET 8

Dataset 8 (VostokWindow10Kyrs). This dataset contains the following Vostok variables: (Age)—in years from 10,000 years before the present (BP) to the present calendar year 1989; (TDv)—semi-annual Vostok atmospheric temperature deviations; and (TDvSW)—Vostok window-scale atmospheric temperature deviations, which are the mean temperature deviations obtained with a sliding window of 50 years. Temperature deviations are in degrees Celsius (°C) relative to Vostok base Kyr1 = 0.00 °C.

Age	TDv	TDvSW
1	0.46	0.46
2	0.46	0.46
3	0.46	0.46
4	0.46	0.46
5	0.46	0.46
6	0.46	0.46
7	0.46	0.46
8	0.46	0.46
9	0.46	0.46
10	0.46	0.46
11	0.46	0.46
12	0.46	0.46
13	0.46	0.46
14	0.46	0.46
15	0.46	0.46
16	0.46	0.46
17	0.46	0.46
18	0.46	0.46
19	0.46	0.46
20	0.46	0.46
21	0.46	0.46
22	0.46	0.46
23	0.46	0.46
24	0.46	0.46
25	0.46	0.46
26	0.46	0.46
27	0.46	0.46
28	0.46	0.46
29	0.46	0.46
30	0.46	0.46
31	0.46	0.46
32	0.46	0.46
33	0.46	0.46
34	0.46	0.46
35	0.46	0.46
36	0.46	0.46
37	0.46	0.46
38	0.46	0.46
39	0.46	0.46
40	0.46	0.46
41	0.46	0.46
42	0.46	0.46
43	0.46	0.46
44	0.46	0.46
45	0.46	0.46
46	0.46	0.46
47	0.46	0.46
48	0.46	0.46
49	0.46	0.46
50	0.46	0.46
51	0.46	0.46
52	0.46	0.46
53	0.46	0.46
54	0.46	0.46
55	0.46	0.46
56	0.46	0.46
57	0.46	0.46
58	0.46	0.46
59	0.46	0.46

60	0.46	0.46		90	0.46	0.46
61	0.46	0.46		91	0.46	0.44
62	0.46	0.46		92	0.46	0.43
63	0.46	0.46		93	0.46	0.41
64	0.46	0.46		94	0.46	0.40
65	0.46	0.46		95	0.46	0.38
66	0.46	0.46		96	0.46	0.36
67	0.46	0.46		97	0.46	0.35
68	0.46	0.46		98	0.46	0.33
69	0.46	0.46		99	0.46	0.31
70	0.46	0.46		100	0.46	0.30
71	0.46	0.46		101	0.46	0.28
72	0.46	0.46		102	0.46	0.27
73	0.46	0.46		103	0.46	0.25
74	0.46	0.46		104	0.46	0.23
75	0.46	0.46		105	0.46	0.22
76	0.46	0.46		106	0.46	0.20
77	0.46	0.46		107	0.46	0.18
78	0.46	0.46		108	0.46	0.17
79	0.46	0.46		109	0.46	0.15
80	0.46	0.46		110	0.46	0.14
81	0.46	0.46		111	0.46	0.12
82	0.46	0.46		112	0.46	0.12
83	0.46	0.46		113	0.46	0.12
84	0.46	0.46		114	0.46	0.12
85	0.46	0.46		115	0.46	0.12
86	0.46	0.46		116	0.46	0.12
87	0.46	0.46		117	0.46	0.12
88	0.46	0.46		118	0.46	0.12
89	0.46	0.46		119	0.46	0.12

120 | 0.46 | 0.12
121 | 0.46 | 0.12
122 | 0.46 | 0.12
123 | 0.46 | 0.12
124 | 0.46 | 0.13
125 | 0.46 | 0.13
126 | 0.46 | 0.13
127 | 0.46 | 0.13
128 | 0.46 | 0.13
129 | 0.46 | 0.13
130 | 0.46 | 0.13
131 | 0.46 | 0.13
132 | 0.46 | 0.14
133 | 0.46 | 0.14
134 | 0.46 | 0.15
135 | 0.46 | 0.16
136 | 0.46 | 0.16
137 | 0.46 | 0.17
138 | 0.46 | 0.18
139 | 0.46 | 0.19
140 | -0.35 | 0.19
141 | -0.35 | 0.22
142 | -0.35 | 0.24
143 | -0.35 | 0.26
144 | -0.35 | 0.29
145 | -0.35 | 0.31
146 | -0.35 | 0.33
147 | -0.35 | 0.36
148 | -0.35 | 0.38
149 | -0.35 | 0.40

150 | -0.35 | 0.43
151 | -0.35 | 0.45
152 | -0.35 | 0.47
153 | -0.35 | 0.47
154 | -0.35 | 0.47
155 | -0.35 | 0.47
156 | -0.35 | 0.46
157 | -0.35 | 0.46
158 | -0.35 | 0.46
159 | -0.35 | 0.45
160 | -0.35 | 0.45
161 | 0.48 | 0.45
162 | 0.48 | 0.43
163 | 0.48 | 0.41
164 | 0.48 | 0.39
165 | 0.48 | 0.37
166 | 0.48 | 0.35
167 | 0.48 | 0.33
168 | 0.48 | 0.31
169 | 0.48 | 0.29
170 | 0.48 | 0.27
171 | 0.48 | 0.25
172 | 0.48 | 0.23
173 | 0.48 | 0.22
174 | 0.48 | 0.20
175 | 0.48 | 0.16
176 | 0.48 | 0.12
177 | 0.48 | 0.08
178 | 0.48 | 0.05
179 | 0.48 | 0.01

295

180 | 0.48 | -0.03
181 | 0.82 | -0.06
182 | 0.82 | -0.11
183 | 0.82 | -0.15
184 | 0.82 | -0.20
185 | 0.82 | -0.24
186 | 0.82 | -0.28
187 | 0.82 | -0.33
188 | 0.82 | -0.37
189 | 0.82 | -0.42
190 | 0.82 | -0.46
191 | 0.82 | -0.50
192 | 0.82 | -0.55
193 | 0.82 | -0.59
194 | 0.82 | -0.64
195 | 0.82 | -0.68
196 | 0.82 | -0.72
197 | 0.82 | -0.77
198 | 0.82 | -0.80
199 | 0.82 | -0.83
200 | 0.82 | -0.86
201 | 0.82 | -0.88
202 | -0.49 | -0.91
203 | -0.49 | -0.92
204 | -0.49 | -0.92
205 | -0.49 | -0.92
206 | -0.49 | -0.92
207 | -0.49 | -0.93
208 | -0.49 | -0.93
209 | -0.49 | -0.93

210 | -0.49 | -0.94
211 | -0.49 | -0.94
212 | -0.49 | -0.94
213 | -0.49 | -0.94
214 | -0.49 | -0.95
215 | -0.49 | -0.95
216 | -0.49 | -0.95
217 | -0.49 | -0.96
218 | -0.49 | -0.96
219 | -0.49 | -0.96
220 | -0.49 | -0.96
221 | -0.49 | -0.97
222 | -0.49 | -0.96
223 | -0.49 | -0.96
224 | -1.38 | -0.95
225 | -1.38 | -0.93
226 | -1.38 | -0.91
227 | -1.38 | -0.89
228 | -1.38 | -0.87
229 | -1.38 | -0.85
230 | -1.38 | -0.82
231 | -1.38 | -0.80
232 | -1.38 | -0.78
233 | -1.38 | -0.76
234 | -1.38 | -0.74
235 | -1.38 | -0.71
236 | -1.38 | -0.69
237 | -1.38 | -0.67
238 | -1.38 | -0.65
239 | -1.38 | -0.63

240	-1.38	-0.61	270	-0.63	-0.04
241	-1.38	-0.58	271	-0.29	-0.02
242	-1.38	-0.56	272	-0.29	-0.02
243	-1.38	-0.54	273	-0.29	-0.01
244	-1.38	-0.52	274	-0.29	-0.01
245	-1.38	-0.49	275	-0.29	0.00
246	-1.38	-0.45	276	-0.29	0.00
247	-0.63	-0.42	277	-0.29	0.01
248	-0.63	-0.40	278	-0.29	0.01
249	-0.63	-0.39	279	-0.29	0.02
250	-0.63	-0.37	280	-0.29	0.02
251	-0.63	-0.35	281	-0.29	0.03
252	-0.63	-0.33	282	-0.29	0.03
253	-0.63	-0.32	283	-0.29	0.04
254	-0.63	-0.30	284	-0.29	0.05
255	-0.63	-0.28	285	-0.29	0.05
256	-0.63	-0.26	286	-0.29	0.06
257	-0.63	-0.25	287	-0.29	0.06
258	-0.63	-0.23	288	-0.29	0.07
259	-0.63	-0.21	289	-0.29	0.07
260	-0.63	-0.20	290	-0.29	0.08
261	-0.63	-0.18	291	-0.29	0.08
262	-0.63	-0.16	292	-0.29	0.08
263	-0.63	-0.14	293	-0.29	0.08
264	-0.63	-0.13	294	0.24	0.08
265	-0.63	-0.11	295	0.24	0.07
266	-0.63	-0.09	296	0.24	0.06
267	-0.63	-0.07	297	0.24	0.05
268	-0.63	-0.06	298	0.24	0.04
269	-0.63	-0.05	299	0.24	0.03

300	0.24	0.01		330	-0.02	0.08
301	0.24	0.00		331	-0.02	0.09
302	0.24	-0.01		332	-0.02	0.11
303	0.24	-0.02		333	-0.02	0.12
304	0.24	-0.03		334	-0.02	0.13
305	0.24	-0.04		335	-0.02	0.15
306	0.24	-0.05		336	-0.02	0.16
307	0.24	-0.06		337	-0.02	0.18
308	0.24	-0.07		338	-0.02	0.21
309	0.24	-0.08		339	-0.02	0.25
310	0.24	-0.09		340	-0.29	0.29
311	0.24	-0.10		341	-0.29	0.33
312	0.24	-0.11		342	-0.29	0.37
313	0.24	-0.12		343	-0.29	0.41
314	0.24	-0.13		344	-0.29	0.45
315	0.24	-0.13		345	-0.29	0.49
316	0.24	-0.12		346	-0.29	0.54
317	-0.02	-0.11		347	-0.29	0.58
318	-0.02	-0.09		348	-0.29	0.62
319	-0.02	-0.08		349	-0.29	0.66
320	-0.02	-0.06		350	-0.29	0.70
321	-0.02	-0.05		351	-0.29	0.74
322	-0.02	-0.04		352	-0.29	0.78
323	-0.02	-0.02		353	-0.29	0.83
324	-0.02	-0.01		354	-0.29	0.87
325	-0.02	0.01		355	-0.29	0.91
326	-0.02	0.02		356	-0.29	0.95
327	-0.02	0.03		357	-0.29	0.99
328	-0.02	0.05		358	-0.29	1.03
329	-0.02	0.06		359	-0.29	1.08

360 | -0.29 | 1.12
361 | -0.29 | 1.14
362 | -0.29 | 1.16
363 | -0.29 | 1.18
364 | 0.69 | 1.21
365 | 0.69 | 1.21
366 | 0.69 | 1.21
367 | 0.69 | 1.21
368 | 0.69 | 1.22
369 | 0.69 | 1.22
370 | 0.69 | 1.22
371 | 0.69 | 1.22
372 | 0.69 | 1.22
373 | 0.69 | 1.23
374 | 0.69 | 1.23
375 | 0.69 | 1.23
376 | 0.69 | 1.23
377 | 0.69 | 1.24
378 | 0.69 | 1.24
379 | 0.69 | 1.24
380 | 0.69 | 1.24
381 | 0.69 | 1.25
382 | 0.69 | 1.25
383 | 0.69 | 1.25
384 | 0.69 | 1.25
385 | 0.69 | 1.25
386 | 0.69 | 1.25
387 | 1.79 | 1.25
388 | 1.79 | 1.22
389 | 1.79 | 1.20

390 | 1.79 | 1.18
391 | 1.79 | 1.16
392 | 1.79 | 1.13
393 | 1.79 | 1.11
394 | 1.79 | 1.09
395 | 1.79 | 1.06
396 | 1.79 | 1.04
397 | 1.79 | 1.02
398 | 1.79 | 0.99
399 | 1.79 | 0.97
400 | 1.79 | 0.95
401 | 1.79 | 0.93
402 | 1.79 | 0.90
403 | 1.79 | 0.88
404 | 1.79 | 0.86
405 | 1.79 | 0.83
406 | 1.79 | 0.81
407 | 1.79 | 0.79
408 | 1.79 | 0.76
409 | 1.79 | 0.74
410 | 0.81 | 0.71
411 | 0.81 | 0.70
412 | 0.81 | 0.69
413 | 0.81 | 0.68
414 | 0.81 | 0.67
415 | 0.81 | 0.66
416 | 0.81 | 0.66
417 | 0.81 | 0.65
418 | 0.81 | 0.64
419 | 0.81 | 0.63

420 | 0.81 | 0.62
421 | 0.81 | 0.61
422 | 0.81 | 0.60
423 | 0.81 | 0.60
424 | 0.81 | 0.59
425 | 0.81 | 0.58
426 | 0.81 | 0.57
427 | 0.81 | 0.56
428 | 0.81 | 0.55
429 | 0.81 | 0.54
430 | 0.81 | 0.54
431 | 0.81 | 0.53
432 | 0.81 | 0.52
433 | 0.64 | 0.51
434 | 0.64 | 0.48
435 | 0.64 | 0.46
436 | 0.64 | 0.43
437 | 0.64 | 0.41
438 | 0.64 | 0.38
439 | 0.64 | 0.36
440 | 0.64 | 0.33
441 | 0.64 | 0.31
442 | 0.64 | 0.28
443 | 0.64 | 0.26
444 | 0.64 | 0.23
445 | 0.64 | 0.21
446 | 0.64 | 0.18
447 | 0.64 | 0.16
448 | 0.64 | 0.13
449 | 0.64 | 0.11

450 | 0.64 | 0.08
451 | 0.64 | 0.06
452 | 0.64 | 0.03
453 | 0.64 | 0.01
454 | 0.64 | -0.02
455 | 0.64 | -0.04
456 | 0.64 | -0.07
457 | 0.64 | -0.09
458 | 0.38 | -0.12
459 | 0.38 | -0.14
460 | 0.38 | -0.16
461 | 0.38 | -0.19
462 | 0.38 | -0.21
463 | 0.38 | -0.24
464 | 0.38 | -0.26
465 | 0.38 | -0.29
466 | 0.38 | -0.32
467 | 0.38 | -0.34
468 | 0.38 | -0.37
469 | 0.38 | -0.40
470 | 0.38 | -0.42
471 | 0.38 | -0.45
472 | 0.38 | -0.47
473 | 0.38 | -0.50
474 | 0.38 | -0.53
475 | 0.38 | -0.55
476 | 0.38 | -0.58
477 | 0.38 | -0.61
478 | 0.38 | -0.63
479 | 0.38 | -0.66

480 | 0.38 | -0.68
481 | 0.38 | -0.71
482 | 0.38 | -0.74
483 | -0.62 | -0.76
484 | -0.62 | -0.77
485 | -0.62 | -0.78
486 | -0.62 | -0.78
487 | -0.62 | -0.79
488 | -0.62 | -0.79
489 | -0.62 | -0.80
490 | -0.62 | -0.81
491 | -0.62 | -0.82
492 | -0.62 | -0.83
493 | -0.62 | -0.84
494 | -0.62 | -0.85
495 | -0.62 | -0.86
496 | -0.62 | -0.87
497 | -0.62 | -0.88
498 | -0.62 | -0.90
499 | -0.62 | -0.91
500 | -0.62 | -0.92
501 | -0.62 | -0.93
502 | -0.62 | -0.94
503 | -0.62 | -0.95
504 | -0.62 | -0.96
505 | -0.62 | -0.97
506 | -0.62 | -0.98
507 | -0.62 | -0.99
508 | -0.62 | -1.00
509 | -0.62 | -1.01

510 | -0.93 | -1.02
511 | -0.93 | -1.03
512 | -0.93 | -1.03
513 | -0.93 | -1.04
514 | -0.93 | -1.04
515 | -0.93 | -1.04
516 | -0.93 | -1.05
517 | -0.93 | -1.05
518 | -0.93 | -1.06
519 | -0.93 | -1.05
520 | -0.93 | -1.04
521 | -0.93 | -1.03
522 | -0.93 | -1.02
523 | -0.93 | -1.01
524 | -0.93 | -1.00
525 | -0.93 | -0.99
526 | -0.93 | -0.98
527 | -0.93 | -0.97
528 | -0.93 | -0.96
529 | -0.93 | -0.95
530 | -0.93 | -0.94
531 | -0.93 | -0.93
532 | -0.93 | -0.92
533 | -0.93 | -0.91
534 | -0.93 | -0.90
535 | -0.93 | -0.89
536 | -0.93 | -0.88
537 | -0.93 | -0.87
538 | -0.93 | -0.86
539 | -1.15 | -0.85

540	-1.15	-0.84	570	-0.44	-0.30
541	-1.15	-0.82	571	-0.44	-0.29
542	-1.15	-0.81	572	-0.44	-0.28
543	-1.15	-0.79	573	-0.44	-0.28
544	-1.15	-0.78	574	-0.44	-0.27
545	-1.15	-0.77	575	-0.44	-0.25
546	-1.15	-0.75	576	-0.44	-0.24
547	-1.15	-0.73	577	-0.44	-0.22
548	-1.15	-0.71	578	-0.44	-0.20
549	-1.15	-0.69	579	-0.44	-0.18
550	-1.15	-0.67	580	-0.44	-0.17
551	-1.15	-0.65	581	-0.44	-0.15
552	-1.15	-0.63	582	-0.44	-0.13
553	-1.15	-0.61	583	-0.44	-0.11
554	-1.15	-0.59	584	-0.44	-0.10
555	-1.15	-0.57	585	-0.44	-0.08
556	-1.15	-0.55	586	-0.44	-0.06
557	-1.15	-0.53	587	-0.44	-0.04
558	-1.15	-0.51	588	-0.44	-0.03
559	-1.15	-0.49	589	-0.44	-0.01
560	-1.15	-0.47	590	-0.44	0.01
561	-1.15	-0.45	591	-0.44	0.03
562	-1.15	-0.43	592	-0.44	0.04
563	-1.15	-0.41	593	-0.44	0.06
564	-1.15	-0.39	594	-0.44	0.08
565	-1.15	-0.37	595	-0.44	0.10
566	-1.15	-0.35	596	-0.14	0.12
567	-1.15	-0.33	597	-0.14	0.13
568	-0.44	-0.31	598	-0.14	0.14
569	-0.44	-0.30	599	-0.14	0.15

600	-0.14	0.16	630	0.44	0.35
601	-0.14	0.17	631	0.44	0.35
602	-0.14	0.18	632	0.44	0.32
603	-0.14	0.19	633	0.44	0.30
604	-0.14	0.20	634	0.44	0.27
605	-0.14	0.21	635	0.44	0.25
606	-0.14	0.22	636	0.44	0.23
607	-0.14	0.23	637	0.44	0.20
608	-0.14	0.24	638	0.44	0.18
609	-0.14	0.24	639	0.44	0.15
610	-0.14	0.25	640	0.44	0.13
611	-0.14	0.26	641	0.44	0.11
612	-0.14	0.27	642	0.44	0.08
613	-0.14	0.28	643	0.44	0.06
614	-0.14	0.29	644	0.44	0.03
615	-0.14	0.29	645	0.44	0.01
616	-0.14	0.30	646	0.44	-0.02
617	-0.14	0.31	647	0.44	-0.04
618	-0.14	0.32	648	0.44	-0.06
619	-0.14	0.33	649	0.44	-0.09
620	-0.14	0.34	650	0.44	-0.11
621	-0.14	0.34	651	0.44	-0.14
622	-0.14	0.35	652	0.28	-0.16
623	-0.14	0.36	653	0.28	-0.18
624	0.44	0.37	654	0.28	-0.20
625	0.44	0.37	655	0.28	-0.22
626	0.44	0.36	656	0.28	-0.25
627	0.44	0.36	657	0.28	-0.27
628	0.44	0.36	658	0.28	-0.29
629	0.44	0.35	659	0.28	-0.31

660	0.28	-0.33
661	0.28	-0.35
662	0.28	-0.37
663	0.28	-0.40
664	0.28	-0.43
665	0.28	-0.45
666	0.28	-0.48
667	0.28	-0.51
668	0.28	-0.53
669	0.28	-0.56
670	0.28	-0.59
671	0.28	-0.62
672	0.28	-0.64
673	0.28	-0.67
674	0.28	-0.70
675	0.28	-0.72
676	0.28	-0.75
677	0.28	-0.78
678	0.28	-0.81
679	0.28	-0.83
680	0.28	-0.86
681	-0.77	-0.89
682	-0.77	-0.89
683	-0.77	-0.90
684	-0.77	-0.91
685	-0.77	-0.91
686	-0.77	-0.92
687	-0.77	-0.92
688	-0.77	-0.93
689	-0.77	-0.94
690	-0.77	-0.94
691	-0.77	-0.95
692	-0.77	-0.96
693	-0.77	-0.96
694	-0.77	-0.95
695	-0.77	-0.95
696	-0.77	-0.94
697	-0.77	-0.93
698	-0.77	-0.92
699	-0.77	-0.92
700	-0.77	-0.91
701	-0.77	-0.90
702	-0.77	-0.89
703	-0.77	-0.89
704	-0.77	-0.88
705	-0.77	-0.87
706	-0.77	-0.86
707	-0.77	-0.86
708	-0.77	-0.85
709	-0.77	-0.84
710	-0.77	-0.83
711	-0.77	-0.83
712	-1.08	-0.82
713	-1.08	-0.80
714	-1.08	-0.79
715	-1.08	-0.78
716	-1.08	-0.76
717	-1.08	-0.75
718	-1.08	-0.74
719	-1.08	-0.72

720	-1.08	-0.71	750	-0.39	0.00
721	-1.08	-0.69	751	-0.39	0.02
722	-1.08	-0.68	752	-0.39	0.03
723	-1.08	-0.67	753	-0.39	0.05
724	-1.08	-0.65	754	-0.39	0.06
725	-1.08	-0.62	755	-0.39	0.07
726	-1.08	-0.59	756	-0.39	0.09
727	-1.08	-0.57	757	-0.39	0.10
728	-1.08	-0.54	758	-0.39	0.11
729	-1.08	-0.51	759	-0.39	0.13
730	-1.08	-0.48	760	-0.39	0.14
731	-1.08	-0.45	761	-0.39	0.16
732	-1.08	-0.42	762	-0.39	0.17
733	-1.08	-0.39	763	-0.39	0.18
734	-1.08	-0.36	764	-0.39	0.20
735	-1.08	-0.34	765	-0.39	0.21
736	-1.08	-0.31	766	-0.39	0.22
737	-1.08	-0.28	767	-0.39	0.24
738	-1.08	-0.25	768	-0.39	0.25
739	-1.08	-0.22	769	-0.39	0.26
740	-1.08	-0.19	770	-0.39	0.28
741	-1.08	-0.16	771	-0.39	0.29
742	-1.08	-0.13	772	-0.39	0.30
743	-0.39	-0.11	773	-0.39	0.32
744	-0.39	-0.09	774	0.36	0.33
745	-0.39	-0.08	775	0.36	0.33
746	-0.39	-0.06	776	0.36	0.33
747	-0.39	-0.05	777	0.36	0.33
748	-0.39	-0.03	778	0.36	0.33
749	-0.39	-0.02	779	0.36	0.33

305

780 | 0.36 | 0.32
781 | 0.36 | 0.32
782 | 0.36 | 0.32
783 | 0.36 | 0.32
784 | 0.36 | 0.32
785 | 0.36 | 0.30
786 | 0.36 | 0.29
787 | 0.36 | 0.28
788 | 0.36 | 0.26
789 | 0.36 | 0.25
790 | 0.36 | 0.24
791 | 0.36 | 0.22
792 | 0.36 | 0.21
793 | 0.36 | 0.20
794 | 0.36 | 0.18
795 | 0.36 | 0.17
796 | 0.36 | 0.15
797 | 0.36 | 0.14
798 | 0.36 | 0.13
799 | 0.36 | 0.11
800 | 0.36 | 0.10
801 | 0.36 | 0.09
802 | 0.36 | 0.07
803 | 0.36 | 0.06
804 | 0.29 | 0.05
805 | 0.29 | 0.03
806 | 0.29 | 0.02
807 | 0.29 | 0.01
808 | 0.29 | 0.00
809 | 0.29 | -0.02

810 | 0.29 | -0.03
811 | 0.29 | -0.04
812 | 0.29 | -0.05
813 | 0.29 | -0.06
814 | 0.29 | -0.08
815 | 0.29 | -0.09
816 | 0.29 | -0.10
817 | 0.29 | -0.11
818 | 0.29 | -0.12
819 | 0.29 | -0.14
820 | 0.29 | -0.15
821 | 0.29 | -0.16
822 | 0.29 | -0.17
823 | 0.29 | -0.19
824 | 0.29 | -0.20
825 | 0.29 | -0.21
826 | 0.29 | -0.22
827 | 0.29 | -0.23
828 | 0.29 | -0.25
829 | 0.29 | -0.26
830 | 0.29 | -0.27
831 | 0.29 | -0.28
832 | 0.29 | -0.30
833 | 0.29 | -0.31
834 | -0.32 | -0.32
835 | -0.32 | -0.32
836 | -0.32 | -0.32
837 | -0.32 | -0.32
838 | -0.32 | -0.32
839 | -0.32 | -0.32

nemonik-thinking.org

840	-0.32	-0.32		870	-0.32	-0.05
841	-0.32	-0.32		871	-0.32	-0.04
842	-0.32	-0.32		872	-0.32	-0.03
843	-0.32	-0.32		873	-0.32	-0.02
844	-0.32	-0.32		874	-0.32	0.00
845	-0.32	-0.32		875	-0.32	0.01
846	-0.32	-0.32		876	-0.32	0.02
847	-0.32	-0.32		877	-0.32	0.03
848	-0.32	-0.32		878	-0.32	0.05
849	-0.32	-0.31		879	-0.32	0.06
850	-0.32	-0.30		880	-0.32	0.07
851	-0.32	-0.28		881	-0.32	0.08
852	-0.32	-0.27		882	-0.32	0.09
853	-0.32	-0.26		883	-0.32	0.10
854	-0.32	-0.25		884	-0.32	0.11
855	-0.32	-0.23		885	-0.32	0.12
856	-0.32	-0.22		886	-0.32	0.13
857	-0.32	-0.21		887	-0.32	0.14
858	-0.32	-0.20		888	-0.32	0.15
859	-0.32	-0.19		889	-0.32	0.16
860	-0.32	-0.17		890	-0.32	0.17
861	-0.32	-0.16		891	-0.32	0.19
862	-0.32	-0.15		892	-0.32	0.20
863	-0.32	-0.14		893	-0.32	0.21
864	-0.32	-0.12		894	-0.32	0.22
865	-0.32	-0.11		895	-0.32	0.23
866	-0.32	-0.10		896	-0.32	0.24
867	-0.32	-0.09		897	-0.32	0.25
868	-0.32	-0.08		898	0.29	0.26
869	-0.32	-0.06		899	0.29	0.26

900 | 0.29 | 0.26
901 | 0.29 | 0.25
902 | 0.29 | 0.25
903 | 0.29 | 0.25
904 | 0.29 | 0.25
905 | 0.29 | 0.25
906 | 0.29 | 0.25
907 | 0.29 | 0.25
908 | 0.29 | 0.24
909 | 0.29 | 0.24
910 | 0.29 | 0.24
911 | 0.29 | 0.24
912 | 0.29 | 0.24
913 | 0.29 | 0.23
914 | 0.29 | 0.23
915 | 0.29 | 0.23
916 | 0.29 | 0.23
917 | 0.29 | 0.23
918 | 0.29 | 0.22
919 | 0.29 | 0.22
920 | 0.29 | 0.22
921 | 0.29 | 0.22
922 | 0.29 | 0.21
923 | 0.29 | 0.21
924 | 0.29 | 0.21
925 | 0.29 | 0.21
926 | 0.29 | 0.21
927 | 0.29 | 0.20
928 | 0.29 | 0.20
929 | 0.21 | 0.20

930 | 0.21 | 0.20
931 | 0.21 | 0.20
932 | 0.21 | 0.20
933 | 0.21 | 0.20
934 | 0.21 | 0.20
935 | 0.21 | 0.20
936 | 0.21 | 0.20
937 | 0.21 | 0.19
938 | 0.21 | 0.19
939 | 0.21 | 0.19
940 | 0.21 | 0.19
941 | 0.21 | 0.19
942 | 0.21 | 0.19
943 | 0.21 | 0.19
944 | 0.21 | 0.19
945 | 0.21 | 0.18
946 | 0.21 | 0.17
947 | 0.21 | 0.16
948 | 0.21 | 0.15
949 | 0.21 | 0.13
950 | 0.21 | 0.12
951 | 0.21 | 0.11
952 | 0.21 | 0.10
953 | 0.21 | 0.09
954 | 0.21 | 0.08
955 | 0.21 | 0.07
956 | 0.21 | 0.06
957 | 0.21 | 0.04
958 | 0.21 | 0.03
959 | 0.21 | 0.02

960 | 0.21 | 0.01
961 | 0.18 | 0.00
962 | 0.18 | -0.01
963 | 0.18 | -0.02
964 | 0.18 | -0.03
965 | 0.18 | -0.04
966 | 0.18 | -0.05
967 | 0.18 | -0.06
968 | 0.18 | -0.07
969 | 0.18 | -0.09
970 | 0.18 | -0.10
971 | 0.18 | -0.11
972 | 0.18 | -0.12
973 | 0.18 | -0.13
974 | 0.18 | -0.14
975 | 0.18 | -0.15
976 | 0.18 | -0.16
977 | 0.18 | -0.17
978 | 0.18 | -0.17
979 | 0.18 | -0.16
980 | 0.18 | -0.16
981 | 0.18 | -0.15
982 | 0.18 | -0.15
983 | 0.18 | -0.14
984 | 0.18 | -0.14
985 | 0.18 | -0.13
986 | 0.18 | -0.13
987 | 0.18 | -0.12
988 | 0.18 | -0.12
989 | 0.18 | -0.11

990 | 0.18 | -0.11
991 | 0.18 | -0.11
992 | 0.18 | -0.10
993 | 0.18 | -0.10
994 | -0.35 | -0.09
995 | -0.35 | -0.08
996 | -0.35 | -0.06
997 | -0.35 | -0.05
998 | -0.35 | -0.03
999 | -0.35 | -0.02
1,000 | -0.35 | 0.00
1,001 | -0.35 | 0.01
1,002 | -0.35 | 0.03
1,003 | -0.35 | 0.05
1,004 | -0.35 | 0.06
1,005 | -0.35 | 0.08
1,006 | -0.35 | 0.09
1,007 | -0.35 | 0.11
1,008 | -0.35 | 0.12
1,009 | -0.35 | 0.14
1,010 | -0.35 | 0.16
1,011 | -0.35 | 0.18
1,012 | -0.35 | 0.20
1,013 | -0.35 | 0.22
1,014 | -0.35 | 0.24
1,015 | -0.35 | 0.27
1,016 | -0.35 | 0.29
1,017 | -0.35 | 0.31
1,018 | -0.35 | 0.33
1,019 | -0.35 | 0.35

1,020	-0.35	0.37	1,050	0.41	0.50
1,021	-0.35	0.40	1,051	0.41	0.48
1,022	-0.35	0.42	1,052	0.41	0.47
1,023	-0.35	0.44	1,053	0.41	0.45
1,024	-0.35	0.46	1,054	0.41	0.43
1,025	-0.35	0.48	1,055	0.41	0.42
1,026	-0.35	0.50	1,056	0.41	0.40
1,027	0.41	0.53	1,057	0.41	0.39
1,028	0.41	0.53	1,058	0.41	0.37
1,029	0.41	0.54	1,059	0.73	0.36
1,030	0.41	0.54	1,060	0.73	0.33
1,031	0.41	0.55	1,061	0.73	0.31
1,032	0.41	0.56	1,062	0.73	0.29
1,033	0.41	0.56	1,063	0.73	0.27
1,034	0.41	0.57	1,064	0.73	0.25
1,035	0.41	0.58	1,065	0.73	0.22
1,036	0.41	0.58	1,066	0.73	0.20
1,037	0.41	0.59	1,067	0.73	0.18
1,038	0.41	0.60	1,068	0.73	0.16
1,039	0.41	0.60	1,069	0.73	0.14
1,040	0.41	0.61	1,070	0.73	0.11
1,041	0.41	0.61	1,071	0.73	0.09
1,042	0.41	0.62	1,072	0.73	0.07
1,043	0.41	0.61	1,073	0.73	0.05
1,044	0.41	0.59	1,074	0.73	0.03
1,045	0.41	0.57	1,075	0.73	0.00
1,046	0.41	0.56	1,076	0.73	-0.02
1,047	0.41	0.54	1,077	0.73	-0.04
1,048	0.41	0.53	1,078	0.73	-0.05
1,049	0.41	0.51	1,079	0.73	-0.07

1,080	0.73	-0.09	1,110	-0.37	-0.25
1,081	0.73	-0.11	1,111	-0.37	-0.24
1,082	0.73	-0.13	1,112	-0.37	-0.23
1,083	0.73	-0.15	1,113	-0.37	-0.21
1,084	0.73	-0.17	1,114	-0.37	-0.20
1,085	0.73	-0.18	1,115	-0.37	-0.19
1,086	0.73	-0.20	1,116	-0.37	-0.18
1,087	0.73	-0.22	1,117	-0.37	-0.17
1,088	0.73	-0.24	1,118	-0.37	-0.16
1,089	0.73	-0.26	1,119	-0.37	-0.15
1,090	0.73	-0.28	1,120	-0.37	-0.14
1,091	0.73	-0.29	1,121	-0.37	-0.12
1,092	-0.37	-0.31	1,122	-0.37	-0.11
1,093	-0.37	-0.31	1,123	-0.37	-0.10
1,094	-0.37	-0.31	1,124	-0.37	-0.09
1,095	-0.37	-0.30	1,125	-0.37	-0.08
1,096	-0.37	-0.30	1,126	-0.19	-0.07
1,097	-0.37	-0.29	1,127	-0.19	-0.06
1,098	-0.37	-0.29	1,128	-0.19	-0.05
1,099	-0.37	-0.29	1,129	-0.19	-0.05
1,100	-0.37	-0.28	1,130	-0.19	-0.04
1,101	-0.37	-0.28	1,131	-0.19	-0.03
1,102	-0.37	-0.28	1,132	-0.19	-0.02
1,103	-0.37	-0.27	1,133	-0.19	-0.02
1,104	-0.37	-0.27	1,134	-0.19	-0.01
1,105	-0.37	-0.27	1,135	-0.19	0.00
1,106	-0.37	-0.26	1,136	-0.19	0.01
1,107	-0.37	-0.26	1,137	-0.19	0.02
1,108	-0.37	-0.25	1,138	-0.19	0.02
1,109	-0.37	-0.25	1,139	-0.19	0.03

1,140	-0.19	0.04		1,170	0.19	0.00
1,141	-0.19	0.05		1,171	0.19	-0.01
1,142	-0.19	0.05		1,172	0.19	-0.02
1,143	-0.19	0.06		1,173	0.19	-0.02
1,144	-0.19	0.07		1,174	0.19	-0.03
1,145	-0.19	0.08		1,175	0.19	-0.04
1,146	-0.19	0.08		1,176	0.19	-0.05
1,147	-0.19	0.08		1,177	0.19	-0.05
1,148	-0.19	0.08		1,178	0.19	-0.06
1,149	-0.19	0.08		1,179	0.19	-0.07
1,150	-0.19	0.08		1,180	0.19	-0.08
1,151	-0.19	0.08		1,181	0.19	-0.10
1,152	-0.19	0.08		1,182	0.19	-0.12
1,153	-0.19	0.08		1,183	0.19	-0.15
1,154	-0.19	0.08		1,184	0.19	-0.17
1,155	-0.19	0.08		1,185	0.19	-0.20
1,156	-0.19	0.08		1,186	0.19	-0.22
1,157	-0.19	0.08		1,187	0.19	-0.25
1,158	-0.19	0.08		1,188	0.19	-0.27
1,159	-0.19	0.08		1,189	0.19	-0.29
1,160	0.19	0.08		1,190	0.19	-0.32
1,161	0.19	0.07		1,191	0.19	-0.34
1,162	0.19	0.06		1,192	0.19	-0.37
1,163	0.19	0.05		1,193	0.19	-0.39
1,164	0.19	0.05		1,194	0.19	-0.41
1,165	0.19	0.04		1,195	-0.19	-0.44
1,166	0.19	0.03		1,196	-0.19	-0.46
1,167	0.19	0.02		1,197	-0.19	-0.47
1,168	0.19	0.02		1,198	-0.19	-0.49
1,169	0.19	0.01		1,199	-0.19	-0.51

1,200	-0.19	-0.52
1,201	-0.19	-0.54
1,202	-0.19	-0.56
1,203	-0.19	-0.57
1,204	-0.19	-0.59
1,205	-0.19	-0.61
1,206	-0.19	-0.62
1,207	-0.19	-0.64
1,208	-0.19	-0.65
1,209	-0.19	-0.67
1,210	-0.19	-0.69
1,211	-0.19	-0.70
1,212	-0.19	-0.72
1,213	-0.19	-0.74
1,214	-0.19	-0.75
1,215	-0.19	-0.77
1,216	-0.19	-0.79
1,217	-0.19	-0.80
1,218	-0.19	-0.82
1,219	-0.19	-0.84
1,220	-0.19	-0.86
1,221	-0.19	-0.88
1,222	-0.19	-0.90
1,223	-0.19	-0.92
1,224	-0.19	-0.94
1,225	-0.19	-0.96
1,226	-0.19	-0.98
1,227	-0.19	-1.00
1,228	-0.19	-1.02
1,229	-0.19	-1.04
1,230	-1.02	-1.06
1,231	-1.02	-1.06
1,232	-1.02	-1.07
1,233	-1.02	-1.07
1,234	-1.02	-1.07
1,235	-1.02	-1.08
1,236	-1.02	-1.08
1,237	-1.02	-1.08
1,238	-1.02	-1.09
1,239	-1.02	-1.09
1,240	-1.02	-1.09
1,241	-1.02	-1.10
1,242	-1.02	-1.10
1,243	-1.02	-1.10
1,244	-1.02	-1.11
1,245	-1.02	-1.11
1,246	-1.02	-1.11
1,247	-1.02	-1.12
1,248	-1.02	-1.12
1,249	-1.02	-1.12
1,250	-1.02	-1.13
1,251	-1.02	-1.13
1,252	-1.02	-1.13
1,253	-1.02	-1.14
1,254	-1.02	-1.14
1,255	-1.02	-1.12
1,256	-1.02	-1.10
1,257	-1.02	-1.07
1,258	-1.02	-1.05
1,259	-1.02	-1.03

1,260	-1.02	-1.01		1,290	-1.18	-0.29
1,261	-1.02	-0.99		1,291	-1.18	-0.25
1,262	-1.02	-0.97		1,292	-1.18	-0.21
1,263	-1.02	-0.94		1,293	-1.18	-0.17
1,264	-1.02	-0.92		1,294	-1.18	-0.13
1,265	-1.02	-0.90		1,295	-1.18	-0.09
1,266	-1.02	-0.88		1,296	-1.18	-0.05
1,267	-1.18	-0.86		1,297	-1.18	-0.01
1,268	-1.18	-0.83		1,298	-1.18	0.03
1,269	-1.18	-0.81		1,299	-1.18	0.07
1,270	-1.18	-0.78		1,300	-1.18	0.11
1,271	-1.18	-0.76		1,301	-1.18	0.15
1,272	-1.18	-0.73		1,302	-1.18	0.19
1,273	-1.18	-0.71		1,303	-1.18	0.23
1,274	-1.18	-0.68		1,304	0.06	0.26
1,275	-1.18	-0.66		1,305	0.06	0.28
1,276	-1.18	-0.63		1,306	0.06	0.29
1,277	-1.18	-0.61		1,307	0.06	0.31
1,278	-1.18	-0.58		1,308	0.06	0.32
1,279	-1.18	-0.56		1,309	0.06	0.34
1,280	-1.18	-0.54		1,310	0.06	0.35
1,281	-1.18	-0.51		1,311	0.06	0.37
1,282	-1.18	-0.49		1,312	0.06	0.38
1,283	-1.18	-0.46		1,313	0.06	0.40
1,284	-1.18	-0.44		1,314	0.06	0.41
1,285	-1.18	-0.41		1,315	0.06	0.43
1,286	-1.18	-0.39		1,316	0.06	0.44
1,287	-1.18	-0.36		1,317	0.06	0.45
1,288	-1.18	-0.34		1,318	0.06	0.47
1,289	-1.18	-0.31		1,319	0.06	0.48

1,320	0.06	0.50	1,350	0.79	0.59
1,321	0.06	0.51	1,351	0.79	0.58
1,322	0.06	0.53	1,352	0.79	0.58
1,323	0.06	0.54	1,353	0.79	0.57
1,324	0.06	0.56	1,354	0.79	0.56
1,325	0.06	0.56	1,355	0.79	0.55
1,326	0.06	0.57	1,356	0.79	0.55
1,327	0.06	0.58	1,357	0.79	0.54
1,328	0.06	0.58	1,358	0.79	0.53
1,329	0.06	0.59	1,359	0.79	0.52
1,330	0.06	0.60	1,360	0.79	0.51
1,331	0.06	0.61	1,361	0.79	0.51
1,332	0.06	0.61	1,362	0.79	0.50
1,333	0.06	0.62	1,363	0.79	0.49
1,334	0.06	0.63	1,364	0.79	0.48
1,335	0.06	0.63	1,365	0.79	0.47
1,336	0.06	0.64	1,366	0.79	0.46
1,337	0.06	0.65	1,367	0.79	0.45
1,338	0.06	0.65	1,368	0.79	0.44
1,339	0.06	0.66	1,369	0.79	0.43
1,340	0.79	0.67	1,370	0.79	0.42
1,341	0.79	0.66	1,371	0.79	0.41
1,342	0.79	0.65	1,372	0.79	0.40
1,343	0.79	0.65	1,373	0.79	0.40
1,344	0.79	0.64	1,374	0.41	0.39
1,345	0.79	0.63	1,375	0.41	0.38
1,346	0.79	0.62	1,376	0.41	0.38
1,347	0.79	0.62	1,377	0.41	0.38
1,348	0.79	0.61	1,378	0.41	0.38
1,349	0.79	0.60	1,379	0.41	0.38

1,380 | 0.41 | 0.38
1,381 | 0.41 | 0.37
1,382 | 0.41 | 0.37
1,383 | 0.41 | 0.37
1,384 | 0.41 | 0.37
1,385 | 0.41 | 0.37
1,386 | 0.41 | 0.36
1,387 | 0.41 | 0.36
1,388 | 0.41 | 0.36
1,389 | 0.41 | 0.36
1,390 | 0.41 | 0.36
1,391 | 0.41 | 0.36
1,392 | 0.41 | 0.35
1,393 | 0.41 | 0.35
1,394 | 0.41 | 0.35
1,395 | 0.41 | 0.35
1,396 | 0.41 | 0.35
1,397 | 0.41 | 0.35
1,398 | 0.41 | 0.35
1,399 | 0.41 | 0.35
1,400 | 0.41 | 0.35
1,401 | 0.41 | 0.35
1,402 | 0.41 | 0.35
1,403 | 0.41 | 0.35
1,404 | 0.41 | 0.35
1,405 | 0.41 | 0.35
1,406 | 0.41 | 0.35
1,407 | 0.41 | 0.35
1,408 | 0.41 | 0.35
1,409 | 0.33 | 0.35

1,410 | 0.33 | 0.35
1,411 | 0.33 | 0.35
1,412 | 0.33 | 0.36
1,413 | 0.33 | 0.36
1,414 | 0.33 | 0.36
1,415 | 0.33 | 0.36
1,416 | 0.33 | 0.36
1,417 | 0.33 | 0.37
1,418 | 0.33 | 0.37
1,419 | 0.33 | 0.37
1,420 | 0.33 | 0.37
1,421 | 0.33 | 0.37
1,422 | 0.33 | 0.37
1,423 | 0.33 | 0.38
1,424 | 0.33 | 0.38
1,425 | 0.32 | 0.38
1,426 | 0.32 | 0.38
1,427 | 0.32 | 0.38
1,428 | 0.32 | 0.39
1,429 | 0.32 | 0.39
1,430 | 0.32 | 0.39
1,431 | 0.32 | 0.38
1,432 | 0.32 | 0.38
1,433 | 0.32 | 0.37
1,434 | 0.32 | 0.36
1,435 | 0.32 | 0.36
1,436 | 0.32 | 0.35
1,437 | 0.32 | 0.34
1,438 | 0.32 | 0.34
1,439 | 0.32 | 0.33

1,440	0.32	0.32	1,470	0.42	0.04
1,441	0.32	0.32	1,471	0.42	0.02
1,442	0.32	0.31	1,472	0.42	0.00
1,443	0.32	0.30	1,473	0.42	-0.01
1,444	0.32	0.30	1,474	0.42	-0.03
1,445	0.42	0.29	1,475	0.42	-0.05
1,446	0.42	0.28	1,476	0.42	-0.07
1,447	0.42	0.27	1,477	0.42	-0.09
1,448	0.42	0.27	1,478	0.42	-0.11
1,449	0.42	0.26	1,479	0.42	-0.13
1,450	0.42	0.25	1,480	-0.01	-0.15
1,451	0.42	0.24	1,481	-0.01	-0.16
1,452	0.42	0.23	1,482	-0.01	-0.17
1,453	0.42	0.22	1,483	-0.01	-0.18
1,454	0.42	0.21	1,484	-0.01	-0.19
1,455	0.42	0.21	1,485	-0.01	-0.20
1,456	0.42	0.20	1,486	-0.01	-0.22
1,457	0.42	0.19	1,487	-0.01	-0.23
1,458	0.42	0.18	1,488	-0.01	-0.24
1,459	0.42	0.17	1,489	-0.01	-0.25
1,460	0.42	0.16	1,490	-0.01	-0.26
1,461	0.42	0.15	1,491	-0.01	-0.27
1,462	0.42	0.14	1,492	-0.01	-0.28
1,463	0.42	0.14	1,493	-0.01	-0.29
1,464	0.42	0.13	1,494	-0.01	-0.30
1,465	0.42	0.12	1,495	-0.01	-0.31
1,466	0.42	0.11	1,496	-0.01	-0.32
1,467	0.42	0.10	1,497	-0.01	-0.33
1,468	0.42	0.08	1,498	-0.01	-0.34
1,469	0.42	0.06	1,499	-0.01	-0.36

1,500	-0.01	-0.37	1,530	-0.55	-0.53
1,501	-0.01	-0.38	1,531	-0.55	-0.53
1,502	-0.01	-0.39	1,532	-0.55	-0.53
1,503	-0.01	-0.40	1,533	-0.55	-0.53
1,504	-0.01	-0.41	1,534	-0.55	-0.53
1,505	-0.01	-0.42	1,535	-0.55	-0.53
1,506	-0.01	-0.43	1,536	-0.55	-0.53
1,507	-0.01	-0.44	1,537	-0.55	-0.52
1,508	-0.01	-0.45	1,538	-0.55	-0.52
1,509	-0.01	-0.46	1,539	-0.55	-0.52
1,510	-0.01	-0.47	1,540	-0.55	-0.52
1,511	-0.01	-0.48	1,541	-0.55	-0.52
1,512	-0.01	-0.49	1,542	-0.55	-0.52
1,513	-0.01	-0.50	1,543	-0.55	-0.52
1,514	-0.01	-0.51	1,544	-0.55	-0.52
1,515	-0.01	-0.52	1,545	-0.55	-0.53
1,516	-0.01	-0.53	1,546	-0.55	-0.53
1,517	-0.55	-0.54	1,547	-0.55	-0.54
1,518	-0.55	-0.54	1,548	-0.55	-0.55
1,519	-0.55	-0.54	1,549	-0.55	-0.55
1,520	-0.55	-0.54	1,550	-0.55	-0.56
1,521	-0.55	-0.54	1,551	-0.55	-0.57
1,522	-0.55	-0.54	1,552	-0.55	-0.57
1,523	-0.55	-0.54	1,553	-0.55	-0.58
1,524	-0.55	-0.53	1,554	-0.55	-0.58
1,525	-0.55	-0.53	1,555	-0.51	-0.59
1,526	-0.55	-0.53	1,556	-0.51	-0.60
1,527	-0.55	-0.53	1,557	-0.51	-0.61
1,528	-0.55	-0.53	1,558	-0.51	-0.61
1,529	-0.55	-0.53	1,559	-0.51	-0.62

1,560	-0.51	-0.63		1,590	-0.51	-0.88
1,561	-0.51	-0.64		1,591	-0.51	-0.89
1,562	-0.51	-0.64		1,592	-0.51	-0.90
1,563	-0.51	-0.65		1,593	-0.51	-0.91
1,564	-0.51	-0.66		1,594	-0.88	-0.92
1,565	-0.51	-0.67		1,595	-0.88	-0.92
1,566	-0.51	-0.67		1,596	-0.88	-0.93
1,567	-0.51	-0.68		1,597	-0.88	-0.93
1,568	-0.51	-0.69		1,598	-0.88	-0.93
1,569	-0.51	-0.70		1,599	-0.88	-0.94
1,570	-0.51	-0.70		1,600	-0.88	-0.94
1,571	-0.51	-0.71		1,601	-0.88	-0.94
1,572	-0.51	-0.72		1,602	-0.88	-0.95
1,573	-0.51	-0.72		1,603	-0.88	-0.95
1,574	-0.51	-0.73		1,604	-0.88	-0.96
1,575	-0.51	-0.74		1,605	-0.88	-0.96
1,576	-0.51	-0.75		1,606	-0.88	-0.96
1,577	-0.51	-0.75		1,607	-0.88	-0.97
1,578	-0.51	-0.76		1,608	-0.88	-0.97
1,579	-0.51	-0.77		1,609	-0.88	-0.97
1,580	-0.51	-0.78		1,610	-0.88	-0.98
1,581	-0.51	-0.78		1,611	-0.88	-0.98
1,582	-0.51	-0.79		1,612	-0.88	-0.98
1,583	-0.51	-0.80		1,613	-0.88	-0.99
1,584	-0.51	-0.81		1,614	-0.88	-0.99
1,585	-0.51	-0.82		1,615	-0.88	-1.00
1,586	-0.51	-0.83		1,616	-0.88	-1.00
1,587	-0.51	-0.84		1,617	-0.88	-1.00
1,588	-0.51	-0.85		1,618	-0.88	-1.01
1,589	-0.51	-0.86		1,619	-0.88	-1.01

1,620 | -0.88 | -1.01
1,621 | -0.88 | -1.02
1,622 | -0.88 | -1.02
1,623 | -0.88 | -1.02
1,624 | -0.88 | -1.02
1,625 | -0.88 | -1.01
1,626 | -0.88 | -1.00
1,627 | -0.88 | -0.99
1,628 | -0.88 | -0.98
1,629 | -0.88 | -0.97
1,630 | -0.88 | -0.96
1,631 | -0.88 | -0.95
1,632 | -0.88 | -0.94
1,633 | -1.06 | -0.93
1,634 | -1.06 | -0.92
1,635 | -1.06 | -0.91
1,636 | -1.06 | -0.90
1,637 | -1.06 | -0.88
1,638 | -1.06 | -0.87
1,639 | -1.06 | -0.86
1,640 | -1.06 | -0.85
1,641 | -1.06 | -0.83
1,642 | -1.06 | -0.82
1,643 | -1.06 | -0.81
1,644 | -1.06 | -0.80
1,645 | -1.06 | -0.78
1,646 | -1.06 | -0.77
1,647 | -1.06 | -0.76
1,648 | -1.06 | -0.75
1,649 | -1.06 | -0.73

1,650 | -1.06 | -0.72
1,651 | -1.06 | -0.71
1,652 | -1.06 | -0.69
1,653 | -1.06 | -0.68
1,654 | -1.06 | -0.67
1,655 | -1.06 | -0.66
1,656 | -1.06 | -0.64
1,657 | -1.06 | -0.63
1,658 | -1.06 | -0.62
1,659 | -1.06 | -0.61
1,660 | -1.06 | -0.59
1,661 | -1.06 | -0.58
1,662 | -1.06 | -0.57
1,663 | -1.06 | -0.56
1,664 | -1.06 | -0.55
1,665 | -1.06 | -0.54
1,666 | -1.06 | -0.53
1,667 | -1.06 | -0.52
1,668 | -1.06 | -0.51
1,669 | -1.06 | -0.50
1,670 | -1.06 | -0.49
1,671 | -1.06 | -0.48
1,672 | -1.06 | -0.48
1,673 | -0.43 | -0.47
1,674 | -0.43 | -0.47
1,675 | -0.43 | -0.47
1,676 | -0.43 | -0.48
1,677 | -0.43 | -0.48
1,678 | -0.43 | -0.48
1,679 | -0.43 | -0.49

1,680	-0.43	-0.49	1,710	-0.43	-0.62
1,681	-0.43	-0.49	1,711	-0.43	-0.63
1,682	-0.43	-0.50	1,712	-0.43	-0.63
1,683	-0.43	-0.50	1,713	-0.61	-0.64
1,684	-0.43	-0.51	1,714	-0.61	-0.64
1,685	-0.43	-0.51	1,715	-0.61	-0.65
1,686	-0.43	-0.51	1,716	-0.61	-0.65
1,687	-0.43	-0.52	1,717	-0.61	-0.65
1,688	-0.43	-0.52	1,718	-0.61	-0.66
1,689	-0.43	-0.52	1,719	-0.61	-0.66
1,690	-0.43	-0.53	1,720	-0.61	-0.66
1,691	-0.43	-0.53	1,721	-0.61	-0.66
1,692	-0.43	-0.53	1,722	-0.61	-0.67
1,693	-0.43	-0.54	1,723	-0.61	-0.67
1,694	-0.43	-0.54	1,724	-0.61	-0.67
1,695	-0.43	-0.55	1,725	-0.61	-0.68
1,696	-0.43	-0.55	1,726	-0.61	-0.68
1,697	-0.43	-0.55	1,727	-0.61	-0.68
1,698	-0.43	-0.56	1,728	-0.61	-0.69
1,699	-0.43	-0.56	1,729	-0.61	-0.69
1,700	-0.43	-0.56	1,730	-0.61	-0.69
1,701	-0.43	-0.57	1,731	-0.61	-0.69
1,702	-0.43	-0.57	1,732	-0.61	-0.70
1,703	-0.43	-0.57	1,733	-0.61	-0.70
1,704	-0.43	-0.58	1,734	-0.61	-0.70
1,705	-0.43	-0.59	1,735	-0.61	-0.71
1,706	-0.43	-0.59	1,736	-0.61	-0.71
1,707	-0.43	-0.60	1,737	-0.61	-0.71
1,708	-0.43	-0.61	1,738	-0.61	-0.72
1,709	-0.43	-0.61	1,739	-0.61	-0.72

1,740	-0.61	-0.72	1,770	-0.76	-0.78
1,741	-0.61	-0.72	1,771	-0.76	-0.78
1,742	-0.61	-0.73	1,772	-0.76	-0.78
1,743	-0.61	-0.73	1,773	-0.76	-0.78
1,744	-0.61	-0.73	1,774	-0.76	-0.78
1,745	-0.61	-0.74	1,775	-0.76	-0.78
1,746	-0.61	-0.74	1,776	-0.76	-0.78
1,747	-0.61	-0.74	1,777	-0.76	-0.78
1,748	-0.61	-0.75	1,778	-0.76	-0.78
1,749	-0.61	-0.75	1,779	-0.76	-0.78
1,750	-0.61	-0.76	1,780	-0.76	-0.78
1,751	-0.61	-0.76	1,781	-0.76	-0.78
1,752	-0.61	-0.76	1,782	-0.76	-0.78
1,753	-0.76	-0.77	1,783	-0.76	-0.78
1,754	-0.76	-0.77	1,784	-0.76	-0.78
1,755	-0.76	-0.77	1,785	-0.76	-0.78
1,756	-0.76	-0.77	1,786	-0.76	-0.77
1,757	-0.76	-0.77	1,787	-0.76	-0.77
1,758	-0.76	-0.77	1,788	-0.76	-0.76
1,759	-0.76	-0.77	1,789	-0.76	-0.76
1,760	-0.76	-0.77	1,790	-0.76	-0.75
1,761	-0.76	-0.77	1,791	-0.76	-0.75
1,762	-0.76	-0.77	1,792	-0.76	-0.74
1,763	-0.76	-0.77	1,793	-0.79	-0.74
1,764	-0.76	-0.77	1,794	-0.79	-0.73
1,765	-0.76	-0.77	1,795	-0.79	-0.73
1,766	-0.76	-0.77	1,796	-0.79	-0.72
1,767	-0.76	-0.77	1,797	-0.79	-0.72
1,768	-0.76	-0.78	1,798	-0.79	-0.71
1,769	-0.76	-0.78	1,799	-0.79	-0.71

1,800 | -0.79 | -0.70
1,801 | -0.79 | -0.69
1,802 | -0.79 | -0.69
1,803 | -0.79 | -0.68
1,804 | -0.79 | -0.68
1,805 | -0.79 | -0.67
1,806 | -0.79 | -0.67
1,807 | -0.79 | -0.66
1,808 | -0.79 | -0.66
1,809 | -0.79 | -0.65
1,810 | -0.79 | -0.64
1,811 | -0.79 | -0.64
1,812 | -0.79 | -0.63
1,813 | -0.79 | -0.63
1,814 | -0.79 | -0.62
1,815 | -0.79 | -0.62
1,816 | -0.79 | -0.61
1,817 | -0.79 | -0.61
1,818 | -0.79 | -0.60
1,819 | -0.79 | -0.59
1,820 | -0.79 | -0.59
1,821 | -0.79 | -0.58
1,822 | -0.79 | -0.58
1,823 | -0.79 | -0.57
1,824 | -0.79 | -0.57
1,825 | -0.79 | -0.55
1,826 | -0.79 | -0.54
1,827 | -0.79 | -0.52
1,828 | -0.79 | -0.51
1,829 | -0.79 | -0.50

1,830 | -0.79 | -0.48
1,831 | -0.79 | -0.47
1,832 | -0.79 | -0.45
1,833 | -0.79 | -0.44
1,834 | -0.51 | -0.42
1,835 | -0.51 | -0.42
1,836 | -0.51 | -0.41
1,837 | -0.51 | -0.40
1,838 | -0.51 | -0.39
1,839 | -0.51 | -0.38
1,840 | -0.51 | -0.37
1,841 | -0.51 | -0.36
1,842 | -0.51 | -0.36
1,843 | -0.51 | -0.35
1,844 | -0.51 | -0.34
1,845 | -0.51 | -0.33
1,846 | -0.51 | -0.32
1,847 | -0.51 | -0.31
1,848 | -0.51 | -0.30
1,849 | -0.51 | -0.30
1,850 | -0.51 | -0.29
1,851 | -0.51 | -0.28
1,852 | -0.51 | -0.27
1,853 | -0.51 | -0.26
1,854 | -0.51 | -0.25
1,855 | -0.51 | -0.24
1,856 | -0.51 | -0.23
1,857 | -0.51 | -0.23
1,858 | -0.51 | -0.22
1,859 | -0.51 | -0.21

1,860 | -0.51 | -0.20
1,861 | -0.51 | -0.19
1,862 | -0.51 | -0.18
1,863 | -0.51 | -0.17
1,864 | -0.51 | -0.15
1,865 | -0.51 | -0.13
1,866 | -0.51 | -0.11
1,867 | -0.51 | -0.09
1,868 | -0.51 | -0.06
1,869 | -0.51 | -0.04
1,870 | -0.51 | -0.02
1,871 | -0.51 | 0.00
1,872 | -0.51 | 0.03
1,873 | -0.51 | 0.05
1,874 | -0.08 | 0.07
1,875 | -0.08 | 0.08
1,876 | -0.08 | 0.10
1,877 | -0.08 | 0.11
1,878 | -0.08 | 0.12
1,879 | -0.08 | 0.14
1,880 | -0.08 | 0.15
1,881 | -0.08 | 0.16
1,882 | -0.08 | 0.18
1,883 | -0.08 | 0.19
1,884 | -0.08 | 0.21
1,885 | -0.08 | 0.22
1,886 | -0.08 | 0.23
1,887 | -0.08 | 0.25
1,888 | -0.08 | 0.26
1,889 | -0.08 | 0.27

1,890 | -0.08 | 0.29
1,891 | -0.08 | 0.30
1,892 | -0.08 | 0.31
1,893 | -0.08 | 0.33
1,894 | -0.08 | 0.34
1,895 | -0.08 | 0.36
1,896 | -0.08 | 0.37
1,897 | -0.08 | 0.38
1,898 | -0.08 | 0.40
1,899 | -0.08 | 0.41
1,900 | -0.08 | 0.42
1,901 | -0.08 | 0.44
1,902 | -0.08 | 0.45
1,903 | -0.08 | 0.46
1,904 | -0.08 | 0.46
1,905 | -0.08 | 0.47
1,906 | -0.08 | 0.47
1,907 | -0.08 | 0.48
1,908 | -0.08 | 0.48
1,909 | -0.08 | 0.49
1,910 | -0.08 | 0.49
1,911 | -0.08 | 0.50
1,912 | -0.08 | 0.50
1,913 | 0.60 | 0.51
1,914 | 0.60 | 0.50
1,915 | 0.60 | 0.49
1,916 | 0.60 | 0.48
1,917 | 0.60 | 0.47
1,918 | 0.60 | 0.46
1,919 | 0.60 | 0.45

1,920	0.60	0.45	1,950	0.60	0.16
1,921	0.60	0.44	1,951	0.60	0.14
1,922	0.60	0.43	1,952	0.17	0.13
1,923	0.60	0.42	1,953	0.17	0.13
1,924	0.60	0.41	1,954	0.17	0.13
1,925	0.60	0.40	1,955	0.17	0.12
1,926	0.60	0.39	1,956	0.17	0.12
1,927	0.60	0.39	1,957	0.17	0.12
1,928	0.60	0.38	1,958	0.17	0.11
1,929	0.60	0.37	1,959	0.17	0.11
1,930	0.60	0.36	1,960	0.17	0.11
1,931	0.60	0.35	1,961	0.17	0.10
1,932	0.60	0.34	1,962	0.17	0.10
1,933	0.60	0.33	1,963	0.17	0.10
1,934	0.60	0.32	1,964	0.17	0.09
1,935	0.60	0.32	1,965	0.17	0.09
1,936	0.60	0.31	1,966	0.17	0.09
1,937	0.60	0.30	1,967	0.17	0.08
1,938	0.60	0.29	1,968	0.17	0.08
1,939	0.60	0.28	1,969	0.17	0.07
1,940	0.60	0.27	1,970	0.17	0.07
1,941	0.60	0.26	1,971	0.17	0.07
1,942	0.60	0.25	1,972	0.17	0.06
1,943	0.60	0.24	1,973	0.17	0.06
1,944	0.60	0.23	1,974	0.17	0.06
1,945	0.60	0.22	1,975	0.17	0.05
1,946	0.60	0.20	1,976	0.17	0.05
1,947	0.60	0.19	1,977	0.17	0.05
1,948	0.60	0.18	1,978	0.17	0.04
1,949	0.60	0.17	1,979	0.17	0.04

1,980	0.17	0.04		2,010	0.00	0.04
1,981	0.17	0.04		2,011	0.00	0.04
1,982	0.17	0.03		2,012	0.00	0.04
1,983	0.17	0.03		2,013	0.00	0.05
1,984	0.17	0.03		2,014	0.00	0.05
1,985	0.17	0.03		2,015	0.00	0.05
1,986	0.17	0.03		2,016	0.00	0.05
1,987	0.17	0.02		2,017	0.00	0.05
1,988	0.17	0.02		2,018	0.00	0.05
1,989	0.17	0.02		2,019	0.00	0.05
1,990	0.17	0.02		2,020	0.00	0.06
1,991	0.00	0.02		2,021	0.00	0.05
1,992	0.00	0.02		2,022	0.00	0.05
1,993	0.00	0.02		2,023	0.00	0.05
1,994	0.00	0.02		2,024	0.00	0.05
1,995	0.00	0.02		2,025	0.00	0.05
1,996	0.00	0.02		2,026	0.00	0.05
1,997	0.00	0.02		2,027	0.00	0.04
1,998	0.00	0.03		2,028	0.00	0.04
1,999	0.00	0.03		2,029	0.00	0.04
2,000	0.00	0.03		2,030	0.07	0.04
2,001	0.00	0.03		2,031	0.07	0.04
2,002	0.00	0.03		2,032	0.07	0.03
2,003	0.00	0.03		2,033	0.07	0.03
2,004	0.00	0.03		2,034	0.07	0.03
2,005	0.00	0.04		2,035	0.07	0.03
2,006	0.00	0.04		2,036	0.07	0.02
2,007	0.00	0.04		2,037	0.07	0.02
2,008	0.00	0.04		2,038	0.07	0.02
2,009	0.00	0.04		2,039	0.07	0.01

2,040	0.07	0.01		2,070	-0.08	-0.10
2,041	0.07	0.01		2,071	-0.08	-0.10
2,042	0.07	0.00		2,072	-0.08	-0.10
2,043	0.07	0.00		2,073	-0.08	-0.10
2,044	0.07	0.00		2,074	-0.08	-0.10
2,045	0.07	0.00		2,075	-0.08	-0.10
2,046	0.07	-0.01		2,076	-0.08	-0.11
2,047	0.07	-0.01		2,077	-0.08	-0.11
2,048	0.07	-0.01		2,078	-0.08	-0.11
2,049	0.07	-0.02		2,079	-0.08	-0.11
2,050	0.07	-0.02		2,080	-0.08	-0.11
2,051	0.07	-0.02		2,081	-0.08	-0.11
2,052	0.07	-0.03		2,082	-0.08	-0.12
2,053	0.07	-0.03		2,083	-0.08	-0.12
2,054	0.07	-0.03		2,084	-0.08	-0.12
2,055	0.07	-0.04		2,085	-0.08	-0.12
2,056	0.07	-0.04		2,086	-0.08	-0.12
2,057	0.07	-0.04		2,087	-0.08	-0.12
2,058	0.07	-0.04		2,088	-0.08	-0.12
2,059	0.07	-0.05		2,089	-0.08	-0.13
2,060	0.07	-0.05		2,090	-0.08	-0.13
2,061	0.07	-0.05		2,091	-0.08	-0.13
2,062	0.07	-0.06		2,092	-0.08	-0.13
2,063	0.07	-0.06		2,093	-0.08	-0.13
2,064	0.07	-0.07		2,094	-0.08	-0.13
2,065	0.07	-0.07		2,095	-0.08	-0.14
2,066	0.07	-0.08		2,096	-0.08	-0.14
2,067	0.07	-0.08		2,097	-0.08	-0.14
2,068	0.07	-0.09		2,098	-0.08	-0.14
2,069	0.07	-0.09		2,099	-0.08	-0.14

2,100	-0.08	-0.14	2,130	-0.16	-0.36
2,101	-0.08	-0.15	2,131	-0.16	-0.36
2,102	-0.08	-0.15	2,132	-0.16	-0.37
2,103	-0.08	-0.16	2,133	-0.16	-0.38
2,104	-0.08	-0.17	2,134	-0.16	-0.38
2,105	-0.08	-0.18	2,135	-0.16	-0.39
2,106	-0.08	-0.19	2,136	-0.16	-0.40
2,107	-0.08	-0.20	2,137	-0.16	-0.40
2,108	-0.08	-0.20	2,138	-0.16	-0.41
2,109	-0.08	-0.21	2,139	-0.16	-0.42
2,110	-0.16	-0.22	2,140	-0.16	-0.43
2,111	-0.16	-0.23	2,141	-0.16	-0.43
2,112	-0.16	-0.23	2,142	-0.16	-0.44
2,113	-0.16	-0.24	2,143	-0.16	-0.45
2,114	-0.16	-0.25	2,144	-0.16	-0.45
2,115	-0.16	-0.26	2,145	-0.16	-0.45
2,116	-0.16	-0.26	2,146	-0.16	-0.45
2,117	-0.16	-0.27	2,147	-0.16	-0.46
2,118	-0.16	-0.28	2,148	-0.16	-0.46
2,119	-0.16	-0.28	2,149	-0.16	-0.46
2,120	-0.16	-0.29	2,150	-0.16	-0.46
2,121	-0.16	-0.30	2,151	-0.50	-0.46
2,122	-0.16	-0.30	2,152	-0.50	-0.46
2,123	-0.16	-0.31	2,153	-0.50	-0.46
2,124	-0.16	-0.32	2,154	-0.50	-0.45
2,125	-0.16	-0.32	2,155	-0.50	-0.45
2,126	-0.16	-0.33	2,156	-0.50	-0.44
2,127	-0.16	-0.34	2,157	-0.50	-0.44
2,128	-0.16	-0.34	2,158	-0.50	-0.43
2,129	-0.16	-0.35	2,159	-0.50	-0.43

2,160 | -0.50 | -0.43
2,161 | -0.50 | -0.42
2,162 | -0.50 | -0.42
2,163 | -0.50 | -0.41
2,164 | -0.50 | -0.41
2,165 | -0.50 | -0.40
2,166 | -0.50 | -0.40
2,167 | -0.50 | -0.39
2,168 | -0.50 | -0.39
2,169 | -0.50 | -0.39
2,170 | -0.50 | -0.38
2,171 | -0.50 | -0.38
2,172 | -0.50 | -0.37
2,173 | -0.50 | -0.37
2,174 | -0.50 | -0.36
2,175 | -0.50 | -0.36
2,176 | -0.50 | -0.35
2,177 | -0.50 | -0.35
2,178 | -0.50 | -0.35
2,179 | -0.50 | -0.34
2,180 | -0.50 | -0.34
2,181 | -0.50 | -0.33
2,182 | -0.50 | -0.33
2,183 | -0.50 | -0.32
2,184 | -0.50 | -0.32
2,185 | -0.50 | -0.30
2,186 | -0.50 | -0.28
2,187 | -0.50 | -0.26
2,188 | -0.50 | -0.23
2,189 | -0.50 | -0.21

2,190 | -0.50 | -0.19
2,191 | -0.50 | -0.17
2,192 | -0.50 | -0.15
2,193 | -0.28 | -0.13
2,194 | -0.28 | -0.11
2,195 | -0.28 | -0.09
2,196 | -0.28 | -0.08
2,197 | -0.28 | -0.06
2,198 | -0.28 | -0.04
2,199 | -0.28 | -0.03
2,200 | -0.28 | -0.01
2,201 | -0.28 | 0.01
2,202 | -0.28 | 0.03
2,203 | -0.28 | 0.04
2,204 | -0.28 | 0.06
2,205 | -0.28 | 0.08
2,206 | -0.28 | 0.09
2,207 | -0.28 | 0.11
2,208 | -0.28 | 0.13
2,209 | -0.28 | 0.15
2,210 | -0.28 | 0.16
2,211 | -0.28 | 0.18
2,212 | -0.28 | 0.20
2,213 | -0.28 | 0.21
2,214 | -0.28 | 0.23
2,215 | -0.28 | 0.25
2,216 | -0.28 | 0.26
2,217 | -0.28 | 0.28
2,218 | -0.28 | 0.30
2,219 | -0.28 | 0.32

2,220	-0.28	0.33	2,250	0.57	1.13
2,221	-0.28	0.35	2,251	0.57	1.15
2,222	-0.28	0.37	2,252	0.57	1.17
2,223	-0.28	0.38	2,253	0.57	1.19
2,224	-0.28	0.42	2,254	0.57	1.21
2,225	-0.28	0.46	2,255	0.57	1.24
2,226	-0.28	0.50	2,256	0.57	1.26
2,227	-0.28	0.53	2,257	0.57	1.28
2,228	-0.28	0.57	2,258	0.57	1.30
2,229	-0.28	0.61	2,259	0.57	1.32
2,230	-0.28	0.65	2,260	0.57	1.34
2,231	-0.28	0.69	2,261	0.57	1.36
2,232	-0.28	0.72	2,262	0.57	1.38
2,233	-0.28	0.76	2,263	0.57	1.36
2,234	0.57	0.80	2,264	0.57	1.34
2,235	0.57	0.82	2,265	0.57	1.32
2,236	0.57	0.84	2,266	0.57	1.29
2,237	0.57	0.86	2,267	0.57	1.27
2,238	0.57	0.88	2,268	0.57	1.25
2,239	0.57	0.90	2,269	0.57	1.23
2,240	0.57	0.92	2,270	0.57	1.21
2,241	0.57	0.94	2,271	0.57	1.18
2,242	0.57	0.97	2,272	0.57	1.16
2,243	0.57	0.99	2,273	1.61	1.14
2,244	0.57	1.01	2,274	1.61	1.10
2,245	0.57	1.03	2,275	1.61	1.05
2,246	0.57	1.05	2,276	1.61	1.01
2,247	0.57	1.07	2,277	1.61	0.97
2,248	0.57	1.09	2,278	1.61	0.93
2,249	0.57	1.11	2,279	1.61	0.88

2,280 | 1.61 | 0.84
2,281 | 1.61 | 0.80
2,282 | 1.61 | 0.75
2,283 | 1.61 | 0.71
2,284 | 1.61 | 0.67
2,285 | 1.61 | 0.63
2,286 | 1.61 | 0.58
2,287 | 1.61 | 0.54
2,288 | 1.61 | 0.50
2,289 | 1.61 | 0.45
2,290 | 1.61 | 0.41
2,291 | 1.61 | 0.37
2,292 | 1.61 | 0.33
2,293 | 1.61 | 0.28
2,294 | 1.61 | 0.24
2,295 | 1.61 | 0.20
2,296 | 1.61 | 0.15
2,297 | 1.61 | 0.11
2,298 | 1.61 | 0.07
2,299 | 1.61 | 0.03
2,300 | 1.61 | -0.02
2,301 | 1.61 | -0.06
2,302 | 1.61 | -0.10
2,303 | 1.61 | -0.14
2,304 | 1.61 | -0.19
2,305 | 1.61 | -0.24
2,306 | 1.61 | -0.29
2,307 | 1.61 | -0.34
2,308 | 1.61 | -0.40
2,309 | 1.61 | -0.45

2,310 | 1.61 | -0.50
2,311 | 1.61 | -0.55
2,312 | -0.53 | -0.60
2,313 | -0.53 | -0.61
2,314 | -0.53 | -0.62
2,315 | -0.53 | -0.63
2,316 | -0.53 | -0.64
2,317 | -0.53 | -0.65
2,318 | -0.53 | -0.66
2,319 | -0.53 | -0.67
2,320 | -0.53 | -0.68
2,321 | -0.53 | -0.69
2,322 | -0.53 | -0.70
2,323 | -0.53 | -0.70
2,324 | -0.53 | -0.71
2,325 | -0.53 | -0.72
2,326 | -0.53 | -0.73
2,327 | -0.53 | -0.74
2,328 | -0.53 | -0.75
2,329 | -0.53 | -0.76
2,330 | -0.53 | -0.77
2,331 | -0.53 | -0.78
2,332 | -0.53 | -0.79
2,333 | -0.53 | -0.80
2,334 | -0.53 | -0.81
2,335 | -0.53 | -0.82
2,336 | -0.53 | -0.82
2,337 | -0.53 | -0.83
2,338 | -0.53 | -0.84
2,339 | -0.53 | -0.85

2,340	-0.53	-0.86	2,370	-0.99	-0.66
2,341	-0.53	-0.87	2,371	-0.99	-0.65
2,342	-0.53	-0.88	2,372	-0.99	-0.64
2,343	-0.53	-0.89	2,373	-0.99	-0.62
2,344	-0.53	-0.90	2,374	-0.99	-0.61
2,345	-0.53	-0.91	2,375	-0.99	-0.59
2,346	-0.53	-0.92	2,376	-0.99	-0.58
2,347	-0.53	-0.93	2,377	-0.99	-0.56
2,348	-0.53	-0.92	2,378	-0.99	-0.55
2,349	-0.53	-0.92	2,379	-0.99	-0.54
2,350	-0.53	-0.91	2,380	-0.99	-0.52
2,351	-0.53	-0.91	2,381	-0.99	-0.51
2,352	-0.53	-0.90	2,382	-0.99	-0.49
2,353	-0.53	-0.90	2,383	-0.99	-0.48
2,354	-0.99	-0.89	2,384	-0.99	-0.46
2,355	-0.99	-0.88	2,385	-0.99	-0.45
2,356	-0.99	-0.86	2,386	-0.99	-0.44
2,357	-0.99	-0.85	2,387	-0.99	-0.42
2,358	-0.99	-0.83	2,388	-0.99	-0.41
2,359	-0.99	-0.82	2,389	-0.99	-0.39
2,360	-0.99	-0.81	2,390	-0.99	-0.38
2,361	-0.99	-0.79	2,391	-0.99	-0.36
2,362	-0.99	-0.78	2,392	-0.99	-0.35
2,363	-0.99	-0.76	2,393	-0.99	-0.33
2,364	-0.99	-0.75	2,394	-0.99	-0.31
2,365	-0.99	-0.73	2,395	-0.99	-0.30
2,366	-0.99	-0.72	2,396	-0.99	-0.28
2,367	-0.99	-0.71	2,397	-0.28	-0.27
2,368	-0.99	-0.69	2,398	-0.28	-0.26
2,369	-0.99	-0.68	2,399	-0.28	-0.26

2,400	-0.28	-0.26	2,430	-0.28	-0.20
2,401	-0.28	-0.26	2,431	-0.28	-0.20
2,402	-0.28	-0.26	2,432	-0.28	-0.20
2,403	-0.28	-0.25	2,433	-0.28	-0.18
2,404	-0.28	-0.25	2,434	-0.28	-0.15
2,405	-0.28	-0.25	2,435	-0.28	-0.13
2,406	-0.28	-0.25	2,436	-0.28	-0.11
2,407	-0.28	-0.25	2,437	-0.28	-0.09
2,408	-0.28	-0.24	2,438	-0.28	-0.07
2,409	-0.28	-0.24	2,439	-0.28	-0.05
2,410	-0.28	-0.24	2,440	-0.18	-0.03
2,411	-0.28	-0.24	2,441	-0.18	-0.01
2,412	-0.28	-0.24	2,442	-0.18	0.01
2,413	-0.28	-0.23	2,443	-0.18	0.02
2,414	-0.28	-0.23	2,444	-0.18	0.04
2,415	-0.28	-0.23	2,445	-0.18	0.06
2,416	-0.28	-0.23	2,446	-0.18	0.08
2,417	-0.28	-0.23	2,447	-0.18	0.10
2,418	-0.28	-0.22	2,448	-0.18	0.12
2,419	-0.28	-0.22	2,449	-0.18	0.14
2,420	-0.28	-0.22	2,450	-0.18	0.15
2,421	-0.28	-0.22	2,451	-0.18	0.17
2,422	-0.28	-0.22	2,452	-0.18	0.19
2,423	-0.28	-0.21	2,453	-0.18	0.21
2,424	-0.28	-0.21	2,454	-0.18	0.23
2,425	-0.28	-0.21	2,455	-0.18	0.25
2,426	-0.28	-0.21	2,456	-0.18	0.27
2,427	-0.28	-0.21	2,457	-0.18	0.29
2,428	-0.28	-0.20	2,458	-0.18	0.30
2,429	-0.28	-0.20	2,459	-0.18	0.32

2,460 | -0.18 | 0.34
2,461 | -0.18 | 0.36
2,462 | -0.18 | 0.38
2,463 | -0.18 | 0.40
2,464 | -0.18 | 0.42
2,465 | -0.18 | 0.43
2,466 | -0.18 | 0.45
2,467 | -0.18 | 0.47
2,468 | -0.18 | 0.49
2,469 | -0.18 | 0.51
2,470 | -0.18 | 0.53
2,471 | -0.18 | 0.55
2,472 | -0.18 | 0.56
2,473 | -0.18 | 0.58
2,474 | -0.18 | 0.59
2,475 | -0.18 | 0.60
2,476 | -0.18 | 0.60
2,477 | -0.18 | 0.61
2,478 | -0.18 | 0.62
2,479 | -0.18 | 0.63
2,480 | -0.18 | 0.63
2,481 | -0.18 | 0.64
2,482 | 0.75 | 0.65
2,483 | 0.75 | 0.64
2,484 | 0.75 | 0.62
2,485 | 0.75 | 0.61
2,486 | 0.75 | 0.60
2,487 | 0.75 | 0.59
2,488 | 0.75 | 0.58
2,489 | 0.75 | 0.57

2,490 | 0.75 | 0.56
2,491 | 0.75 | 0.54
2,492 | 0.75 | 0.53
2,493 | 0.75 | 0.52
2,494 | 0.75 | 0.51
2,495 | 0.75 | 0.50
2,496 | 0.75 | 0.49
2,497 | 0.75 | 0.48
2,498 | 0.75 | 0.47
2,499 | 0.75 | 0.45
2,500 | 0.75 | 0.44
2,501 | 0.75 | 0.43
2,502 | 0.75 | 0.42
2,503 | 0.75 | 0.41
2,504 | 0.75 | 0.40
2,505 | 0.75 | 0.39
2,506 | 0.75 | 0.37
2,507 | 0.75 | 0.36
2,508 | 0.75 | 0.35
2,509 | 0.75 | 0.34
2,510 | 0.75 | 0.33
2,511 | 0.75 | 0.32
2,512 | 0.75 | 0.31
2,513 | 0.75 | 0.29
2,514 | 0.75 | 0.28
2,515 | 0.75 | 0.27
2,516 | 0.75 | 0.24
2,517 | 0.75 | 0.20
2,518 | 0.75 | 0.17
2,519 | 0.75 | 0.14

2,520	0.75	0.11		2,550	0.18	-0.58
2,521	0.75	0.07		2,551	0.18	-0.60
2,522	0.75	0.04		2,552	0.18	-0.63
2,523	0.18	0.01		2,553	0.18	-0.65
2,524	0.18	-0.02		2,554	0.18	-0.67
2,525	0.18	-0.04		2,555	0.18	-0.69
2,526	0.18	-0.06		2,556	0.18	-0.71
2,527	0.18	-0.08		2,557	0.18	-0.74
2,528	0.18	-0.10		2,558	0.18	-0.76
2,529	0.18	-0.13		2,559	0.18	-0.76
2,530	0.18	-0.15		2,560	0.18	-0.75
2,531	0.18	-0.17		2,561	0.18	-0.75
2,532	0.18	-0.19		2,562	0.18	-0.75
2,533	0.18	-0.21		2,563	0.18	-0.75
2,534	0.18	-0.23		2,564	0.18	-0.75
2,535	0.18	-0.26		2,565	-0.91	-0.74
2,536	0.18	-0.28		2,566	-0.91	-0.72
2,537	0.18	-0.30		2,567	-0.91	-0.70
2,538	0.18	-0.32		2,568	-0.91	-0.67
2,539	0.18	-0.34		2,569	-0.91	-0.65
2,540	0.18	-0.37		2,570	-0.91	-0.62
2,541	0.18	-0.39		2,571	-0.91	-0.60
2,542	0.18	-0.41		2,572	-0.91	-0.58
2,543	0.18	-0.43		2,573	-0.91	-0.55
2,544	0.18	-0.45		2,574	-0.91	-0.53
2,545	0.18	-0.47		2,575	-0.91	-0.51
2,546	0.18	-0.50		2,576	-0.91	-0.48
2,547	0.18	-0.52		2,577	-0.91	-0.46
2,548	0.18	-0.54		2,578	-0.91	-0.43
2,549	0.18	-0.56		2,579	-0.91	-0.41

2,580 | -0.91 | -0.39
2,581 | -0.91 | -0.36
2,582 | -0.91 | -0.34
2,583 | -0.91 | -0.31
2,584 | -0.91 | -0.29
2,585 | -0.91 | -0.27
2,586 | -0.91 | -0.24
2,587 | -0.91 | -0.22
2,588 | -0.91 | -0.20
2,589 | -0.91 | -0.17
2,590 | -0.91 | -0.15
2,591 | -0.91 | -0.12
2,592 | -0.91 | -0.10
2,593 | -0.91 | -0.08
2,594 | -0.91 | -0.05
2,595 | -0.91 | -0.03
2,596 | -0.91 | -0.01
2,597 | -0.91 | 0.02
2,598 | -0.91 | 0.04
2,599 | -0.91 | 0.07
2,600 | -0.91 | 0.09
2,601 | -0.91 | 0.12
2,602 | -0.91 | 0.14
2,603 | -0.91 | 0.17
2,604 | -0.91 | 0.20
2,605 | -0.91 | 0.22
2,606 | -0.91 | 0.25
2,607 | -0.91 | 0.28
2,608 | 0.28 | 0.30
2,609 | 0.28 | 0.31

2,610 | 0.28 | 0.31
2,611 | 0.28 | 0.31
2,612 | 0.28 | 0.32
2,613 | 0.28 | 0.32
2,614 | 0.28 | 0.32
2,615 | 0.28 | 0.33
2,616 | 0.28 | 0.33
2,617 | 0.28 | 0.33
2,618 | 0.28 | 0.33
2,619 | 0.28 | 0.34
2,620 | 0.28 | 0.34
2,621 | 0.28 | 0.34
2,622 | 0.28 | 0.35
2,623 | 0.28 | 0.35
2,624 | 0.28 | 0.35
2,625 | 0.28 | 0.36
2,626 | 0.28 | 0.36
2,627 | 0.28 | 0.36
2,628 | 0.28 | 0.36
2,629 | 0.28 | 0.37
2,630 | 0.28 | 0.37
2,631 | 0.28 | 0.37
2,632 | 0.28 | 0.38
2,633 | 0.28 | 0.38
2,634 | 0.28 | 0.38
2,635 | 0.28 | 0.39
2,636 | 0.28 | 0.39
2,637 | 0.28 | 0.39
2,638 | 0.28 | 0.39
2,639 | 0.28 | 0.40

2,640	0.28	0.40	2,670	0.43	-0.20
2,641	0.28	0.40	2,671	0.43	-0.23
2,642	0.28	0.41	2,672	0.43	-0.25
2,643	0.28	0.41	2,673	0.43	-0.27
2,644	0.28	0.39	2,674	0.43	-0.30
2,645	0.28	0.37	2,675	0.43	-0.32
2,646	0.28	0.35	2,676	0.43	-0.34
2,647	0.28	0.33	2,677	0.43	-0.37
2,648	0.28	0.31	2,678	0.43	-0.39
2,649	0.28	0.29	2,679	0.43	-0.41
2,650	0.43	0.27	2,680	0.43	-0.44
2,651	0.43	0.24	2,681	0.43	-0.46
2,652	0.43	0.22	2,682	0.43	-0.48
2,653	0.43	0.20	2,683	0.43	-0.51
2,654	0.43	0.17	2,684	0.43	-0.53
2,655	0.43	0.15	2,685	0.43	-0.55
2,656	0.43	0.13	2,686	0.43	-0.58
2,657	0.43	0.10	2,687	0.43	-0.60
2,658	0.43	0.08	2,688	0.43	-0.62
2,659	0.43	0.06	2,689	0.43	-0.66
2,660	0.43	0.03	2,690	0.43	-0.70
2,661	0.43	0.01	2,691	0.43	-0.74
2,662	0.43	-0.01	2,692	0.43	-0.78
2,663	0.43	-0.04	2,693	-0.74	-0.82
2,664	0.43	-0.06	2,694	-0.74	-0.83
2,665	0.43	-0.08	2,695	-0.74	-0.85
2,666	0.43	-0.11	2,696	-0.74	-0.86
2,667	0.43	-0.13	2,697	-0.74	-0.88
2,668	0.43	-0.16	2,698	-0.74	-0.90
2,669	0.43	-0.18	2,699	-0.74	-0.91

2,700	-0.74	-0.93	2,730	-0.74	-1.40
2,701	-0.74	-0.94	2,731	-0.74	-1.41
2,702	-0.74	-0.96	2,732	-0.74	-1.43
2,703	-0.74	-0.97	2,733	-0.74	-1.44
2,704	-0.74	-0.99	2,734	-0.74	-1.46
2,705	-0.74	-1.01	2,735	-0.74	-1.45
2,706	-0.74	-1.02	2,736	-0.74	-1.43
2,707	-0.74	-1.04	2,737	-0.74	-1.42
2,708	-0.74	-1.05	2,738	-1.52	-1.41
2,709	-0.74	-1.07	2,739	-1.52	-1.38
2,710	-0.74	-1.08	2,740	-1.52	-1.36
2,711	-0.74	-1.10	2,741	-1.52	-1.33
2,712	-0.74	-1.11	2,742	-1.52	-1.30
2,713	-0.74	-1.13	2,743	-1.52	-1.28
2,714	-0.74	-1.15	2,744	-1.52	-1.25
2,715	-0.74	-1.16	2,745	-1.52	-1.22
2,716	-0.74	-1.18	2,746	-1.52	-1.19
2,717	-0.74	-1.19	2,747	-1.52	-1.17
2,718	-0.74	-1.21	2,748	-1.52	-1.14
2,719	-0.74	-1.22	2,749	-1.52	-1.11
2,720	-0.74	-1.24	2,750	-1.52	-1.08
2,721	-0.74	-1.25	2,751	-1.52	-1.06
2,722	-0.74	-1.27	2,752	-1.52	-1.03
2,723	-0.74	-1.29	2,753	-1.52	-1.00
2,724	-0.74	-1.30	2,754	-1.52	-0.98
2,725	-0.74	-1.32	2,755	-1.52	-0.95
2,726	-0.74	-1.33	2,756	-1.52	-0.92
2,727	-0.74	-1.35	2,757	-1.52	-0.89
2,728	-0.74	-1.36	2,758	-1.52	-0.87
2,729	-0.74	-1.38	2,759	-1.52	-0.84

2,760	-1.52	-0.81
2,761	-1.52	-0.79
2,762	-1.52	-0.76
2,763	-1.52	-0.73
2,764	-1.52	-0.70
2,765	-1.52	-0.68
2,766	-1.52	-0.65
2,767	-1.52	-0.62
2,768	-1.52	-0.60
2,769	-1.52	-0.57
2,770	-1.52	-0.54
2,771	-1.52	-0.51
2,772	-1.52	-0.49
2,773	-1.52	-0.46
2,774	-1.52	-0.43
2,775	-1.52	-0.40
2,776	-1.52	-0.38
2,777	-1.52	-0.35
2,778	-1.52	-0.30
2,779	-1.52	-0.25
2,780	-1.52	-0.20
2,781	-1.52	-0.15
2,782	-1.52	-0.10
2,783	-1.52	-0.04
2,784	-0.16	0.01
2,785	-0.16	0.03
2,786	-0.16	0.05
2,787	-0.16	0.08
2,788	-0.16	0.10
2,789	-0.16	0.13
2,790	-0.16	0.15
2,791	-0.16	0.17
2,792	-0.16	0.20
2,793	-0.16	0.22
2,794	-0.16	0.24
2,795	-0.16	0.27
2,796	-0.16	0.29
2,797	-0.16	0.32
2,798	-0.16	0.34
2,799	-0.16	0.36
2,800	-0.16	0.39
2,801	-0.16	0.41
2,802	-0.16	0.44
2,803	-0.16	0.46
2,804	-0.16	0.48
2,805	-0.16	0.51
2,806	-0.16	0.53
2,807	-0.16	0.55
2,808	-0.16	0.58
2,809	-0.16	0.60
2,810	-0.16	0.63
2,811	-0.16	0.65
2,812	-0.16	0.67
2,813	-0.16	0.70
2,814	-0.16	0.72
2,815	-0.16	0.74
2,816	-0.16	0.77
2,817	-0.16	0.79
2,818	-0.16	0.82
2,819	-0.16	0.84

2,820	-0.16	0.84	2,850	1.03	0.43
2,821	-0.16	0.85	2,851	1.03	0.42
2,822	-0.16	0.85	2,852	1.03	0.40
2,823	-0.16	0.86	2,853	1.03	0.38
2,824	-0.16	0.86	2,854	1.03	0.36
2,825	-0.16	0.87	2,855	1.03	0.34
2,826	-0.16	0.87	2,856	1.03	0.32
2,827	1.03	0.88	2,857	1.03	0.30
2,828	1.03	0.86	2,858	1.03	0.28
2,829	1.03	0.84	2,859	1.03	0.26
2,830	1.03	0.82	2,860	1.03	0.24
2,831	1.03	0.80	2,861	1.03	0.22
2,832	1.03	0.78	2,862	1.03	0.20
2,833	1.03	0.76	2,863	1.03	0.19
2,834	1.03	0.74	2,864	1.03	0.15
2,835	1.03	0.72	2,865	1.03	0.11
2,836	1.03	0.70	2,866	1.03	0.07
2,837	1.03	0.68	2,867	1.03	0.03
2,838	1.03	0.67	2,868	1.03	-0.01
2,839	1.03	0.65	2,869	0.07	-0.05
2,840	1.03	0.63	2,870	0.07	-0.06
2,841	1.03	0.61	2,871	0.07	-0.08
2,842	1.03	0.59	2,872	0.07	-0.10
2,843	1.03	0.57	2,873	0.07	-0.12
2,844	1.03	0.55	2,874	0.07	-0.14
2,845	1.03	0.53	2,875	0.07	-0.16
2,846	1.03	0.51	2,876	0.07	-0.18
2,847	1.03	0.49	2,877	0.07	-0.20
2,848	1.03	0.47	2,878	0.07	-0.22
2,849	1.03	0.45	2,879	0.07	-0.24

2,880 | 0.07 | -0.26
2,881 | 0.07 | -0.28
2,882 | 0.07 | -0.29
2,883 | 0.07 | -0.31
2,884 | 0.07 | -0.33
2,885 | 0.07 | -0.35
2,886 | 0.07 | -0.37
2,887 | 0.07 | -0.39
2,888 | 0.07 | -0.41
2,889 | 0.07 | -0.43
2,890 | 0.07 | -0.45
2,891 | 0.07 | -0.47
2,892 | 0.07 | -0.49
2,893 | 0.07 | -0.51
2,894 | 0.07 | -0.53
2,895 | 0.07 | -0.54
2,896 | 0.07 | -0.56
2,897 | 0.07 | -0.58
2,898 | 0.07 | -0.60
2,899 | 0.07 | -0.62
2,900 | 0.07 | -0.64
2,901 | 0.07 | -0.66
2,902 | 0.07 | -0.68
2,903 | 0.07 | -0.70
2,904 | 0.07 | -0.72
2,905 | 0.07 | -0.74
2,906 | 0.07 | -0.76
2,907 | 0.07 | -0.77
2,908 | 0.07 | -0.79
2,909 | 0.07 | -0.81

2,910 | 0.07 | -0.84
2,911 | 0.07 | -0.86
2,912 | 0.07 | -0.88
2,913 | -0.89 | -0.90
2,914 | -0.89 | -0.90
2,915 | -0.89 | -0.90
2,916 | -0.89 | -0.90
2,917 | -0.89 | -0.90
2,918 | -0.89 | -0.90
2,919 | -0.89 | -0.91
2,920 | -0.89 | -0.91
2,921 | -0.89 | -0.91
2,922 | -0.89 | -0.91
2,923 | -0.89 | -0.91
2,924 | -0.89 | -0.91
2,925 | -0.89 | -0.91
2,926 | -0.89 | -0.92
2,927 | -0.89 | -0.92
2,928 | -0.89 | -0.92
2,929 | -0.89 | -0.92
2,930 | -0.89 | -0.92
2,931 | -0.89 | -0.92
2,932 | -0.89 | -0.92
2,933 | -0.89 | -0.93
2,934 | -0.89 | -0.93
2,935 | -0.89 | -0.93
2,936 | -0.89 | -0.93
2,937 | -0.89 | -0.93
2,938 | -0.89 | -0.93
2,939 | -0.89 | -0.93

2,940 | -0.89 | -0.93
2,941 | -0.89 | -0.94
2,942 | -0.89 | -0.94
2,943 | -0.89 | -0.94
2,944 | -0.89 | -0.94
2,945 | -0.89 | -0.94
2,946 | -0.89 | -0.94
2,947 | -0.89 | -0.94
2,948 | -0.89 | -0.95
2,949 | -0.89 | -0.95
2,950 | -0.89 | -0.95
2,951 | -0.89 | -0.95
2,952 | -0.89 | -0.95
2,953 | -0.89 | -0.95
2,954 | -0.89 | -0.95
2,955 | -0.89 | -0.94
2,956 | -0.89 | -0.93
2,957 | -0.89 | -0.92
2,958 | -0.96 | -0.91
2,959 | -0.96 | -0.89
2,960 | -0.96 | -0.88
2,961 | -0.96 | -0.86
2,962 | -0.96 | -0.85
2,963 | -0.96 | -0.84
2,964 | -0.96 | -0.82
2,965 | -0.96 | -0.81
2,966 | -0.96 | -0.80
2,967 | -0.96 | -0.78
2,968 | -0.96 | -0.77
2,969 | -0.96 | -0.76

2,970 | -0.96 | -0.74
2,971 | -0.96 | -0.73
2,972 | -0.96 | -0.72
2,973 | -0.96 | -0.70
2,974 | -0.96 | -0.69
2,975 | -0.96 | -0.67
2,976 | -0.96 | -0.66
2,977 | -0.96 | -0.65
2,978 | -0.96 | -0.63
2,979 | -0.96 | -0.62
2,980 | -0.96 | -0.61
2,981 | -0.96 | -0.59
2,982 | -0.96 | -0.58
2,983 | -0.96 | -0.57
2,984 | -0.96 | -0.55
2,985 | -0.96 | -0.54
2,986 | -0.96 | -0.52
2,987 | -0.96 | -0.51
2,988 | -0.96 | -0.50
2,989 | -0.96 | -0.48
2,990 | -0.96 | -0.47
2,991 | -0.96 | -0.46
2,992 | -0.96 | -0.44
2,993 | -0.96 | -0.43
2,994 | -0.96 | -0.42
2,995 | -0.96 | -0.40
2,996 | -0.96 | -0.39
2,997 | -0.96 | -0.38
2,998 | -0.96 | -0.36
2,999 | -0.96 | -0.35

3,000 | -0.96 | -0.32
3,001 | -0.96 | -0.30
3,002 | -0.96 | -0.28
3,003 | -0.96 | -0.26
3,004 | -0.28 | -0.23
3,005 | -0.28 | -0.22
3,006 | -0.28 | -0.21
3,007 | -0.28 | -0.20
3,008 | -0.28 | -0.19
3,009 | -0.28 | -0.18
3,010 | -0.28 | -0.17
3,011 | -0.28 | -0.16
3,012 | -0.28 | -0.16
3,013 | -0.28 | -0.15
3,014 | -0.28 | -0.14
3,015 | -0.28 | -0.13
3,016 | -0.28 | -0.12
3,017 | -0.28 | -0.11
3,018 | -0.28 | -0.10
3,019 | -0.28 | -0.09
3,020 | -0.28 | -0.08
3,021 | -0.28 | -0.07
3,022 | -0.28 | -0.06
3,023 | -0.28 | -0.05
3,024 | -0.28 | -0.04
3,025 | -0.28 | -0.03
3,026 | -0.28 | -0.02
3,027 | -0.28 | -0.01
3,028 | -0.28 | 0.00
3,029 | -0.28 | 0.01

3,030 | -0.28 | 0.02
3,031 | -0.28 | 0.03
3,032 | -0.28 | 0.04
3,033 | -0.28 | 0.05
3,034 | -0.28 | 0.06
3,035 | -0.28 | 0.07
3,036 | -0.28 | 0.08
3,037 | -0.28 | 0.08
3,038 | -0.28 | 0.09
3,039 | -0.28 | 0.10
3,040 | -0.28 | 0.11
3,041 | -0.28 | 0.12
3,042 | -0.28 | 0.13
3,043 | -0.28 | 0.14
3,044 | -0.28 | 0.15
3,045 | -0.28 | 0.15
3,046 | -0.28 | 0.15
3,047 | -0.28 | 0.16
3,048 | -0.28 | 0.16
3,049 | 0.20 | 0.17
3,050 | 0.20 | 0.16
3,051 | 0.20 | 0.16
3,052 | 0.20 | 0.15
3,053 | 0.20 | 0.14
3,054 | 0.20 | 0.14
3,055 | 0.20 | 0.13
3,056 | 0.20 | 0.13
3,057 | 0.20 | 0.12
3,058 | 0.20 | 0.12
3,059 | 0.20 | 0.11

3,060 | 0.20 | 0.10
3,061 | 0.20 | 0.10
3,062 | 0.20 | 0.09
3,063 | 0.20 | 0.09
3,064 | 0.20 | 0.08
3,065 | 0.20 | 0.08
3,066 | 0.20 | 0.07
3,067 | 0.20 | 0.07
3,068 | 0.20 | 0.06
3,069 | 0.20 | 0.05
3,070 | 0.20 | 0.05
3,071 | 0.20 | 0.04
3,072 | 0.20 | 0.04
3,073 | 0.20 | 0.03
3,074 | 0.20 | 0.03
3,075 | 0.20 | 0.02
3,076 | 0.20 | 0.02
3,077 | 0.20 | 0.01
3,078 | 0.20 | 0.00
3,079 | 0.20 | 0.00
3,080 | 0.20 | -0.01
3,081 | 0.20 | -0.01
3,082 | 0.20 | -0.02
3,083 | 0.20 | -0.02
3,084 | 0.20 | -0.03
3,085 | 0.20 | -0.04
3,086 | 0.20 | -0.04
3,087 | 0.20 | -0.05
3,088 | 0.20 | -0.04
3,089 | 0.20 | -0.04

3,090 | 0.20 | -0.04
3,091 | 0.20 | -0.03
3,092 | 0.20 | -0.03
3,093 | -0.08 | -0.03
3,094 | -0.08 | -0.02
3,095 | -0.08 | -0.01
3,096 | -0.08 | 0.00
3,097 | -0.08 | 0.01
3,098 | -0.08 | 0.02
3,099 | -0.08 | 0.03
3,100 | -0.08 | 0.04
3,101 | -0.08 | 0.05
3,102 | -0.08 | 0.06
3,103 | -0.08 | 0.06
3,104 | -0.08 | 0.07
3,105 | -0.08 | 0.08
3,106 | -0.08 | 0.09
3,107 | -0.08 | 0.10
3,108 | -0.08 | 0.11
3,109 | -0.08 | 0.12
3,110 | -0.08 | 0.13
3,111 | -0.08 | 0.14
3,112 | -0.08 | 0.15
3,113 | -0.08 | 0.15
3,114 | -0.08 | 0.16
3,115 | -0.08 | 0.17
3,116 | -0.08 | 0.18
3,117 | -0.08 | 0.19
3,118 | -0.08 | 0.20
3,119 | -0.08 | 0.21

3,120	-0.08	0.22		3,150	0.37	0.46
3,121	-0.08	0.23		3,151	0.37	0.46
3,122	-0.08	0.24		3,152	0.37	0.47
3,123	-0.08	0.24		3,153	0.37	0.47
3,124	-0.08	0.25		3,154	0.37	0.48
3,125	-0.08	0.26		3,155	0.37	0.48
3,126	-0.08	0.27		3,156	0.37	0.49
3,127	-0.08	0.28		3,157	0.37	0.49
3,128	-0.08	0.29		3,158	0.37	0.49
3,129	-0.08	0.30		3,159	0.37	0.50
3,130	-0.08	0.31		3,160	0.37	0.50
3,131	-0.08	0.32		3,161	0.37	0.51
3,132	-0.08	0.33		3,162	0.37	0.51
3,133	-0.08	0.34		3,163	0.37	0.52
3,134	-0.08	0.36		3,164	0.37	0.52
3,135	-0.08	0.37		3,165	0.37	0.53
3,136	-0.08	0.38		3,166	0.37	0.53
3,137	0.37	0.40		3,167	0.37	0.54
3,138	0.37	0.40		3,168	0.37	0.54
3,139	0.37	0.41		3,169	0.37	0.54
3,140	0.37	0.41		3,170	0.37	0.55
3,141	0.37	0.42		3,171	0.37	0.55
3,142	0.37	0.42		3,172	0.37	0.56
3,143	0.37	0.43		3,173	0.37	0.56
3,144	0.37	0.43		3,174	0.37	0.57
3,145	0.37	0.43		3,175	0.37	0.56
3,146	0.37	0.44		3,176	0.37	0.55
3,147	0.37	0.44		3,177	0.37	0.55
3,148	0.37	0.45		3,178	0.37	0.54
3,149	0.37	0.45		3,179	0.37	0.53

3,180 | 0.37 | 0.53
3,181 | 0.60 | 0.52
3,182 | 0.60 | 0.51
3,183 | 0.60 | 0.50
3,184 | 0.60 | 0.48
3,185 | 0.60 | 0.47
3,186 | 0.60 | 0.46
3,187 | 0.60 | 0.45
3,188 | 0.60 | 0.44
3,189 | 0.60 | 0.43
3,190 | 0.60 | 0.41
3,191 | 0.60 | 0.40
3,192 | 0.60 | 0.39
3,193 | 0.60 | 0.38
3,194 | 0.60 | 0.37
3,195 | 0.60 | 0.36
3,196 | 0.60 | 0.34
3,197 | 0.60 | 0.33
3,198 | 0.60 | 0.32
3,199 | 0.60 | 0.31
3,200 | 0.60 | 0.30
3,201 | 0.60 | 0.29
3,202 | 0.60 | 0.28
3,203 | 0.60 | 0.26
3,204 | 0.60 | 0.25
3,205 | 0.60 | 0.24
3,206 | 0.60 | 0.23
3,207 | 0.60 | 0.22
3,208 | 0.60 | 0.21
3,209 | 0.60 | 0.19

3,210 | 0.60 | 0.18
3,211 | 0.60 | 0.17
3,212 | 0.60 | 0.16
3,213 | 0.60 | 0.15
3,214 | 0.60 | 0.14
3,215 | 0.60 | 0.12
3,216 | 0.60 | 0.11
3,217 | 0.60 | 0.10
3,218 | 0.60 | 0.09
3,219 | 0.60 | 0.07
3,220 | 0.60 | 0.06
3,221 | 0.60 | 0.04
3,222 | 0.60 | 0.03
3,223 | 0.60 | 0.01
3,224 | 0.02 | 0.00
3,225 | 0.02 | -0.01
3,226 | 0.02 | -0.01
3,227 | 0.02 | -0.01
3,228 | 0.02 | -0.02
3,229 | 0.02 | -0.02
3,230 | 0.02 | -0.02
3,231 | 0.02 | -0.03
3,232 | 0.02 | -0.03
3,233 | 0.02 | -0.03
3,234 | 0.02 | -0.04
3,235 | 0.02 | -0.04
3,236 | 0.02 | -0.04
3,237 | 0.02 | -0.05
3,238 | 0.02 | -0.05
3,239 | 0.02 | -0.06

3,240	0.02	-0.06		3,270	-0.16	-0.12
3,241	0.02	-0.06		3,271	-0.16	-0.11
3,242	0.02	-0.07		3,272	-0.16	-0.10
3,243	0.02	-0.07		3,273	-0.16	-0.10
3,244	0.02	-0.07		3,274	-0.16	-0.09
3,245	0.02	-0.08		3,275	-0.16	-0.09
3,246	0.02	-0.08		3,276	-0.16	-0.08
3,247	0.02	-0.08		3,277	-0.16	-0.07
3,248	0.02	-0.09		3,278	-0.16	-0.07
3,249	0.02	-0.09		3,279	-0.16	-0.06
3,250	0.02	-0.10		3,280	-0.16	-0.05
3,251	0.02	-0.10		3,281	-0.16	-0.05
3,252	0.02	-0.10		3,282	-0.16	-0.04
3,253	0.02	-0.11		3,283	-0.16	-0.04
3,254	0.02	-0.11		3,284	-0.16	-0.03
3,255	0.02	-0.11		3,285	-0.16	-0.02
3,256	0.02	-0.12		3,286	-0.16	-0.02
3,257	0.02	-0.12		3,287	-0.16	-0.01
3,258	0.02	-0.12		3,288	-0.16	-0.01
3,259	0.02	-0.13		3,289	-0.16	0.00
3,260	0.02	-0.13		3,290	-0.16	0.01
3,261	0.02	-0.13		3,291	-0.16	0.01
3,262	0.02	-0.14		3,292	-0.16	0.02
3,263	0.02	-0.14		3,293	-0.16	0.03
3,264	0.02	-0.14		3,294	-0.16	0.03
3,265	0.02	-0.14		3,295	-0.16	0.04
3,266	0.02	-0.13		3,296	-0.16	0.04
3,267	0.02	-0.13		3,297	-0.16	0.05
3,268	-0.16	-0.13		3,298	-0.16	0.06
3,269	-0.16	-0.12		3,299	-0.16	0.06

3,300 | -0.16 | 0.07
3,301 | -0.16 | 0.08
3,302 | -0.16 | 0.08
3,303 | -0.16 | 0.09
3,304 | -0.16 | 0.09
3,305 | -0.16 | 0.10
3,306 | -0.16 | 0.11
3,307 | -0.16 | 0.11
3,308 | -0.16 | 0.12
3,309 | -0.16 | 0.13
3,310 | -0.16 | 0.14
3,311 | -0.16 | 0.15
3,312 | -0.16 | 0.16
3,313 | 0.15 | 0.17
3,314 | 0.15 | 0.17
3,315 | 0.15 | 0.17
3,316 | 0.15 | 0.18
3,317 | 0.15 | 0.18
3,318 | 0.15 | 0.18
3,319 | 0.15 | 0.19
3,320 | 0.15 | 0.19
3,321 | 0.15 | 0.19
3,322 | 0.15 | 0.20
3,323 | 0.15 | 0.20
3,324 | 0.15 | 0.20
3,325 | 0.15 | 0.21
3,326 | 0.15 | 0.21
3,327 | 0.15 | 0.21
3,328 | 0.15 | 0.22
3,329 | 0.15 | 0.22

3,330 | 0.15 | 0.22
3,331 | 0.15 | 0.23
3,332 | 0.15 | 0.23
3,333 | 0.15 | 0.24
3,334 | 0.15 | 0.24
3,335 | 0.15 | 0.24
3,336 | 0.15 | 0.25
3,337 | 0.15 | 0.25
3,338 | 0.15 | 0.25
3,339 | 0.15 | 0.26
3,340 | 0.15 | 0.26
3,341 | 0.15 | 0.26
3,342 | 0.15 | 0.27
3,343 | 0.15 | 0.27
3,344 | 0.15 | 0.27
3,345 | 0.15 | 0.28
3,346 | 0.15 | 0.28
3,347 | 0.15 | 0.28
3,348 | 0.15 | 0.29
3,349 | 0.15 | 0.29
3,350 | 0.15 | 0.29
3,351 | 0.15 | 0.30
3,352 | 0.15 | 0.30
3,353 | 0.15 | 0.31
3,354 | 0.15 | 0.32
3,355 | 0.15 | 0.33
3,356 | 0.15 | 0.34
3,357 | 0.15 | 0.35
3,358 | 0.32 | 0.36
3,359 | 0.32 | 0.37

3,360	0.32	0.37	3,390	0.32	0.58
3,361	0.32	0.38	3,391	0.32	0.58
3,362	0.32	0.39	3,392	0.32	0.59
3,363	0.32	0.39	3,393	0.32	0.60
3,364	0.32	0.40	3,394	0.32	0.60
3,365	0.32	0.41	3,395	0.31	0.61
3,366	0.32	0.41	3,396	0.31	0.62
3,367	0.32	0.42	3,397	0.31	0.62
3,368	0.32	0.43	3,398	0.31	0.63
3,369	0.32	0.43	3,399	0.31	0.64
3,370	0.32	0.44	3,400	0.31	0.64
3,371	0.32	0.45	3,401	0.31	0.65
3,372	0.32	0.45	3,402	0.66	0.65
3,373	0.32	0.46	3,403	0.66	0.65
3,374	0.32	0.47	3,404	0.66	0.65
3,375	0.32	0.48	3,405	0.66	0.65
3,376	0.32	0.48	3,406	0.66	0.65
3,377	0.32	0.49	3,407	0.66	0.65
3,378	0.32	0.50	3,408	0.66	0.65
3,379	0.32	0.50	3,409	0.66	0.65
3,380	0.32	0.51	3,410	0.66	0.65
3,381	0.32	0.52	3,411	0.66	0.64
3,382	0.32	0.52	3,412	0.66	0.64
3,383	0.32	0.53	3,413	0.66	0.64
3,384	0.32	0.54	3,414	0.66	0.64
3,385	0.32	0.54	3,415	0.66	0.64
3,386	0.32	0.55	3,416	0.66	0.64
3,387	0.32	0.56	3,417	0.66	0.64
3,388	0.32	0.56	3,418	0.66	0.64
3,389	0.32	0.57	3,419	0.66	0.64

3,420	0.66	0.64	3,450	0.61	0.39
3,421	0.66	0.63	3,451	0.61	0.37
3,422	0.66	0.63	3,452	0.61	0.35
3,423	0.66	0.63	3,453	0.61	0.33
3,424	0.66	0.63	3,454	0.61	0.31
3,425	0.66	0.63	3,455	0.61	0.29
3,426	0.66	0.63	3,456	0.61	0.26
3,427	0.66	0.63	3,457	0.61	0.24
3,428	0.66	0.63	3,458	0.61	0.22
3,429	0.66	0.63	3,459	0.61	0.20
3,430	0.66	0.63	3,460	0.61	0.18
3,431	0.66	0.62	3,461	0.61	0.16
3,432	0.66	0.62	3,462	0.61	0.13
3,433	0.66	0.62	3,463	0.61	0.11
3,434	0.66	0.62	3,464	0.61	0.09
3,435	0.66	0.62	3,465	0.61	0.07
3,436	0.66	0.62	3,466	0.61	0.05
3,437	0.66	0.62	3,467	0.61	0.03
3,438	0.66	0.62	3,468	0.61	0.01
3,439	0.66	0.62	3,469	0.61	-0.02
3,440	0.66	0.62	3,470	0.61	-0.04
3,441	0.66	0.59	3,471	0.61	-0.06
3,442	0.66	0.57	3,472	0.61	-0.08
3,443	0.66	0.55	3,473	0.61	-0.10
3,444	0.66	0.52	3,474	0.61	-0.12
3,445	0.61	0.50	3,475	0.61	-0.15
3,446	0.61	0.48	3,476	0.61	-0.17
3,447	0.61	0.46	3,477	0.61	-0.19
3,448	0.61	0.44	3,478	0.61	-0.21
3,449	0.61	0.42	3,479	0.61	-0.23

3,480	0.61	-0.25	3,510	-0.47	-0.47
3,481	0.61	-0.28	3,511	-0.47	-0.47
3,482	0.61	-0.30	3,512	-0.47	-0.46
3,483	0.61	-0.32	3,513	-0.47	-0.46
3,484	0.61	-0.34	3,514	-0.47	-0.46
3,485	0.61	-0.36	3,515	-0.47	-0.46
3,486	0.61	-0.38	3,516	-0.47	-0.46
3,487	0.61	-0.41	3,517	-0.47	-0.46
3,488	0.61	-0.43	3,518	-0.47	-0.46
3,489	0.61	-0.45	3,519	-0.47	-0.46
3,490	-0.47	-0.47	3,520	-0.47	-0.46
3,491	-0.47	-0.47	3,521	-0.47	-0.46
3,492	-0.47	-0.47	3,522	-0.47	-0.46
3,493	-0.47	-0.47	3,523	-0.47	-0.46
3,494	-0.47	-0.47	3,524	-0.47	-0.46
3,495	-0.47	-0.47	3,525	-0.47	-0.46
3,496	-0.47	-0.47	3,526	-0.47	-0.46
3,497	-0.47	-0.47	3,527	-0.47	-0.46
3,498	-0.47	-0.47	3,528	-0.47	-0.46
3,499	-0.47	-0.47	3,529	-0.47	-0.46
3,500	-0.47	-0.47	3,530	-0.47	-0.46
3,501	-0.47	-0.47	3,531	-0.47	-0.46
3,502	-0.47	-0.47	3,532	-0.47	-0.46
3,503	-0.47	-0.47	3,533	-0.47	-0.44
3,504	-0.47	-0.47	3,534	-0.47	-0.41
3,505	-0.47	-0.47	3,535	-0.47	-0.39
3,506	-0.47	-0.47	3,536	-0.46	-0.36
3,507	-0.47	-0.47	3,537	-0.46	-0.34
3,508	-0.47	-0.47	3,538	-0.46	-0.32
3,509	-0.47	-0.47	3,539	-0.46	-0.29

3,540 | -0.46 | -0.27
3,541 | -0.46 | -0.24
3,542 | -0.46 | -0.22
3,543 | -0.46 | -0.20
3,544 | -0.46 | -0.17
3,545 | -0.46 | -0.15
3,546 | -0.46 | -0.12
3,547 | -0.46 | -0.10
3,548 | -0.46 | -0.08
3,549 | -0.46 | -0.05
3,550 | -0.46 | -0.03
3,551 | -0.46 | 0.00
3,552 | -0.46 | 0.02
3,553 | -0.46 | 0.04
3,554 | -0.46 | 0.07
3,555 | -0.46 | 0.09
3,556 | -0.46 | 0.12
3,557 | -0.46 | 0.14
3,558 | -0.46 | 0.16
3,559 | -0.46 | 0.19
3,560 | -0.46 | 0.21
3,561 | -0.46 | 0.24
3,562 | -0.46 | 0.26
3,563 | -0.46 | 0.28
3,564 | -0.46 | 0.31
3,565 | -0.46 | 0.33
3,566 | -0.46 | 0.36
3,567 | -0.46 | 0.38
3,568 | -0.46 | 0.40
3,569 | -0.46 | 0.43

3,570 | -0.46 | 0.45
3,571 | -0.46 | 0.48
3,572 | -0.46 | 0.50
3,573 | -0.46 | 0.52
3,574 | -0.46 | 0.55
3,575 | -0.46 | 0.57
3,576 | -0.46 | 0.60
3,577 | -0.46 | 0.63
3,578 | -0.46 | 0.66
3,579 | -0.46 | 0.69
3,580 | -0.46 | 0.72
3,581 | -0.46 | 0.75
3,582 | 0.74 | 0.78
3,583 | 0.74 | 0.79
3,584 | 0.74 | 0.80
3,585 | 0.74 | 0.80
3,586 | 0.74 | 0.81
3,587 | 0.74 | 0.82
3,588 | 0.74 | 0.82
3,589 | 0.74 | 0.83
3,590 | 0.74 | 0.84
3,591 | 0.74 | 0.85
3,592 | 0.74 | 0.85
3,593 | 0.74 | 0.86
3,594 | 0.74 | 0.87
3,595 | 0.74 | 0.87
3,596 | 0.74 | 0.88
3,597 | 0.74 | 0.89
3,598 | 0.74 | 0.89
3,599 | 0.74 | 0.90

3,600 | 0.74 | 0.91
3,601 | 0.74 | 0.92
3,602 | 0.74 | 0.92
3,603 | 0.74 | 0.93
3,604 | 0.74 | 0.94
3,605 | 0.74 | 0.94
3,606 | 0.74 | 0.95
3,607 | 0.74 | 0.96
3,608 | 0.74 | 0.96
3,609 | 0.74 | 0.97
3,610 | 0.74 | 0.98
3,611 | 0.74 | 0.99
3,612 | 0.74 | 0.99
3,613 | 0.74 | 1.00
3,614 | 0.74 | 1.01
3,615 | 0.74 | 1.01
3,616 | 0.74 | 1.02
3,617 | 0.74 | 1.03
3,618 | 0.74 | 1.03
3,619 | 0.74 | 1.04
3,620 | 0.74 | 1.05
3,621 | 0.74 | 1.05
3,622 | 0.74 | 1.06
3,623 | 0.74 | 1.06
3,624 | 0.74 | 1.07
3,625 | 0.74 | 1.07
3,626 | 1.09 | 1.08
3,627 | 1.09 | 1.08
3,628 | 1.09 | 1.08
3,629 | 1.09 | 1.08

3,630 | 1.09 | 1.07
3,631 | 1.09 | 1.07
3,632 | 1.09 | 1.07
3,633 | 1.09 | 1.07
3,634 | 1.09 | 1.07
3,635 | 1.09 | 1.07
3,636 | 1.09 | 1.07
3,637 | 1.09 | 1.06
3,638 | 1.09 | 1.06
3,639 | 1.09 | 1.06
3,640 | 1.09 | 1.06
3,641 | 1.09 | 1.06
3,642 | 1.09 | 1.06
3,643 | 1.09 | 1.06
3,644 | 1.09 | 1.06
3,645 | 1.09 | 1.05
3,646 | 1.09 | 1.05
3,647 | 1.09 | 1.05
3,648 | 1.09 | 1.05
3,649 | 1.09 | 1.05
3,650 | 1.09 | 1.05
3,651 | 1.09 | 1.05
3,652 | 1.09 | 1.04
3,653 | 1.09 | 1.04
3,654 | 1.09 | 1.04
3,655 | 1.09 | 1.04
3,656 | 1.09 | 1.04
3,657 | 1.09 | 1.04
3,658 | 1.09 | 1.04
3,659 | 1.09 | 1.03

3,660 | 1.09 | 1.03
3,661 | 1.09 | 1.03
3,662 | 1.09 | 1.03
3,663 | 1.09 | 1.02
3,664 | 1.09 | 1.01
3,665 | 1.09 | 0.99
3,666 | 1.09 | 0.98
3,667 | 1.09 | 0.97
3,668 | 1.09 | 0.96
3,669 | 1.02 | 0.94
3,670 | 1.02 | 0.93
3,671 | 1.02 | 0.92
3,672 | 1.02 | 0.91
3,673 | 1.02 | 0.90
3,674 | 1.02 | 0.89
3,675 | 1.02 | 0.88
3,676 | 1.02 | 0.87
3,677 | 1.02 | 0.86
3,678 | 1.02 | 0.84
3,679 | 1.02 | 0.83
3,680 | 1.02 | 0.82
3,681 | 1.02 | 0.81
3,682 | 1.02 | 0.80
3,683 | 1.02 | 0.79
3,684 | 1.02 | 0.78
3,685 | 1.02 | 0.77
3,686 | 1.02 | 0.76
3,687 | 1.02 | 0.74
3,688 | 1.02 | 0.73
3,689 | 1.02 | 0.72

3,690 | 1.02 | 0.71
3,691 | 1.02 | 0.70
3,692 | 1.02 | 0.69
3,693 | 1.02 | 0.68
3,694 | 1.02 | 0.67
3,695 | 1.02 | 0.66
3,696 | 1.02 | 0.65
3,697 | 1.02 | 0.63
3,698 | 1.02 | 0.62
3,699 | 1.02 | 0.61
3,700 | 1.02 | 0.60
3,701 | 1.02 | 0.59
3,702 | 1.02 | 0.58
3,703 | 1.02 | 0.57
3,704 | 1.02 | 0.56
3,705 | 1.02 | 0.55
3,706 | 1.02 | 0.54
3,707 | 1.02 | 0.51
3,708 | 1.02 | 0.49
3,709 | 1.02 | 0.46
3,710 | 1.02 | 0.44
3,711 | 1.02 | 0.42
3,712 | 0.47 | 0.39
3,713 | 0.47 | 0.38
3,714 | 0.47 | 0.36
3,715 | 0.47 | 0.35
3,716 | 0.47 | 0.34
3,717 | 0.47 | 0.32
3,718 | 0.47 | 0.31
3,719 | 0.47 | 0.30

3,720	0.47	0.29	3,750	0.47	-0.11
3,721	0.47	0.27	3,751	0.47	-0.12
3,722	0.47	0.26	3,752	0.47	-0.14
3,723	0.47	0.25	3,753	0.47	-0.15
3,724	0.47	0.23	3,754	0.47	-0.16
3,725	0.47	0.22	3,755	0.47	-0.17
3,726	0.47	0.21	3,756	-0.19	-0.18
3,727	0.47	0.19	3,757	-0.19	-0.17
3,728	0.47	0.18	3,758	-0.19	-0.17
3,729	0.47	0.17	3,759	-0.19	-0.17
3,730	0.47	0.15	3,760	-0.19	-0.16
3,731	0.47	0.14	3,761	-0.19	-0.16
3,732	0.47	0.13	3,762	-0.19	-0.16
3,733	0.47	0.11	3,763	-0.19	-0.15
3,734	0.47	0.10	3,764	-0.19	-0.15
3,735	0.47	0.09	3,765	-0.19	-0.15
3,736	0.47	0.07	3,766	-0.19	-0.14
3,737	0.47	0.06	3,767	-0.19	-0.14
3,738	0.47	0.05	3,768	-0.19	-0.14
3,739	0.47	0.03	3,769	-0.19	-0.13
3,740	0.47	0.02	3,770	-0.19	-0.13
3,741	0.47	0.01	3,771	-0.19	-0.13
3,742	0.47	-0.01	3,772	-0.19	-0.12
3,743	0.47	-0.02	3,773	-0.19	-0.12
3,744	0.47	-0.03	3,774	-0.19	-0.12
3,745	0.47	-0.04	3,775	-0.19	-0.11
3,746	0.47	-0.06	3,776	-0.19	-0.11
3,747	0.47	-0.07	3,777	-0.19	-0.11
3,748	0.47	-0.08	3,778	-0.19	-0.10
3,749	0.47	-0.10	3,779	-0.19	-0.10

3,780	-0.19	-0.09	3,810	-0.02	0.09
3,781	-0.19	-0.09	3,811	-0.02	0.10
3,782	-0.19	-0.09	3,812	-0.02	0.11
3,783	-0.19	-0.08	3,813	-0.02	0.12
3,784	-0.19	-0.08	3,814	-0.02	0.13
3,785	-0.19	-0.08	3,815	-0.02	0.14
3,786	-0.19	-0.07	3,816	-0.02	0.15
3,787	-0.19	-0.07	3,817	-0.02	0.15
3,788	-0.19	-0.07	3,818	-0.02	0.16
3,789	-0.19	-0.06	3,819	-0.02	0.17
3,790	-0.19	-0.06	3,820	-0.02	0.18
3,791	-0.19	-0.06	3,821	-0.02	0.19
3,792	-0.19	-0.05	3,822	-0.02	0.20
3,793	-0.19	-0.05	3,823	-0.02	0.21
3,794	-0.19	-0.05	3,824	-0.02	0.22
3,795	-0.19	-0.04	3,825	-0.02	0.23
3,796	-0.19	-0.04	3,826	-0.02	0.24
3,797	-0.19	-0.04	3,827	-0.02	0.25
3,798	-0.19	-0.03	3,828	-0.02	0.26
3,799	-0.19	-0.02	3,829	-0.02	0.27
3,800	-0.19	-0.01	3,830	-0.02	0.27
3,801	-0.19	0.00	3,831	-0.02	0.28
3,802	-0.02	0.02	3,832	-0.02	0.29
3,803	-0.02	0.03	3,833	-0.02	0.30
3,804	-0.02	0.04	3,834	-0.02	0.31
3,805	-0.02	0.04	3,835	-0.02	0.32
3,806	-0.02	0.05	3,836	-0.02	0.33
3,807	-0.02	0.06	3,837	-0.02	0.34
3,808	-0.02	0.07	3,838	-0.02	0.35
3,809	-0.02	0.08	3,839	-0.02	0.36

3,840	-0.02	0.37	3,870	0.44	0.32
3,841	-0.02	0.38	3,871	0.44	0.32
3,842	-0.02	0.38	3,872	0.44	0.31
3,843	-0.02	0.39	3,873	0.44	0.31
3,844	-0.02	0.40	3,874	0.44	0.30
3,845	-0.02	0.41	3,875	0.44	0.30
3,846	-0.02	0.41	3,876	0.44	0.29
3,847	-0.02	0.42	3,877	0.44	0.29
3,848	0.44	0.42	3,878	0.44	0.28
3,849	0.44	0.42	3,879	0.44	0.28
3,850	0.44	0.41	3,880	0.44	0.27
3,851	0.44	0.41	3,881	0.44	0.27
3,852	0.44	0.40	3,882	0.44	0.27
3,853	0.44	0.40	3,883	0.44	0.26
3,854	0.44	0.39	3,884	0.44	0.26
3,855	0.44	0.39	3,885	0.44	0.25
3,856	0.44	0.38	3,886	0.44	0.25
3,857	0.44	0.38	3,887	0.44	0.24
3,858	0.44	0.38	3,888	0.44	0.24
3,859	0.44	0.37	3,889	0.44	0.23
3,860	0.44	0.37	3,890	0.44	0.23
3,861	0.44	0.36	3,891	0.44	0.22
3,862	0.44	0.36	3,892	0.44	0.20
3,863	0.44	0.35	3,893	0.44	0.19
3,864	0.44	0.35	3,894	0.21	0.18
3,865	0.44	0.34	3,895	0.21	0.17
3,866	0.44	0.34	3,896	0.21	0.16
3,867	0.44	0.33	3,897	0.21	0.15
3,868	0.44	0.33	3,898	0.21	0.14
3,869	0.44	0.33	3,899	0.21	0.13

3,900 | 0.21 | 0.12
3,901 | 0.21 | 0.12
3,902 | 0.21 | 0.11
3,903 | 0.21 | 0.10
3,904 | 0.21 | 0.09
3,905 | 0.21 | 0.08
3,906 | 0.21 | 0.07
3,907 | 0.21 | 0.06
3,908 | 0.21 | 0.06
3,909 | 0.21 | 0.05
3,910 | 0.21 | 0.04
3,911 | 0.21 | 0.03
3,912 | 0.21 | 0.02
3,913 | 0.21 | 0.01
3,914 | 0.21 | 0.00
3,915 | 0.21 | -0.01
3,916 | 0.21 | -0.01
3,917 | 0.21 | -0.02
3,918 | 0.21 | -0.03
3,919 | 0.21 | -0.04
3,920 | 0.21 | -0.05
3,921 | 0.21 | -0.06
3,922 | 0.21 | -0.07
3,923 | 0.21 | -0.07
3,924 | 0.21 | -0.08
3,925 | 0.21 | -0.09
3,926 | 0.21 | -0.10
3,927 | 0.21 | -0.11
3,928 | 0.21 | -0.12
3,929 | 0.21 | -0.13

3,930 | 0.21 | -0.13
3,931 | 0.21 | -0.14
3,932 | 0.21 | -0.15
3,933 | 0.21 | -0.16
3,934 | 0.21 | -0.17
3,935 | 0.21 | -0.18
3,936 | 0.21 | -0.19
3,937 | 0.21 | -0.19
3,938 | 0.21 | -0.20
3,939 | 0.21 | -0.21
3,940 | -0.22 | -0.22
3,941 | -0.22 | -0.23
3,942 | -0.22 | -0.23
3,943 | -0.22 | -0.23
3,944 | -0.22 | -0.23
3,945 | -0.22 | -0.23
3,946 | -0.22 | -0.23
3,947 | -0.22 | -0.23
3,948 | -0.22 | -0.24
3,949 | -0.22 | -0.24
3,950 | -0.22 | -0.24
3,951 | -0.22 | -0.24
3,952 | -0.22 | -0.24
3,953 | -0.22 | -0.24
3,954 | -0.22 | -0.24
3,955 | -0.22 | -0.25
3,956 | -0.22 | -0.25
3,957 | -0.22 | -0.25
3,958 | -0.22 | -0.25
3,959 | -0.22 | -0.25

3,960	-0.22	-0.25		3,990	-0.29	-0.29
3,961	-0.22	-0.25		3,991	-0.29	-0.29
3,962	-0.22	-0.26		3,992	-0.29	-0.29
3,963	-0.22	-0.26		3,993	-0.29	-0.29
3,964	-0.22	-0.26		3,994	-0.29	-0.29
3,965	-0.22	-0.26		3,995	-0.29	-0.29
3,966	-0.22	-0.26		3,996	-0.29	-0.29
3,967	-0.22	-0.26		3,997	-0.29	-0.30
3,968	-0.22	-0.26		3,998	-0.29	-0.30
3,969	-0.22	-0.26		3,999	-0.29	-0.30
3,970	-0.22	-0.27		4,000	-0.29	-0.30
3,971	-0.22	-0.27		4,001	-0.29	-0.30
3,972	-0.22	-0.27		4,002	-0.29	-0.30
3,973	-0.22	-0.27		4,003	-0.29	-0.30
3,974	-0.22	-0.27		4,004	-0.29	-0.30
3,975	-0.22	-0.27		4,005	-0.29	-0.30
3,976	-0.22	-0.27		4,006	-0.29	-0.30
3,977	-0.22	-0.28		4,007	-0.29	-0.30
3,978	-0.22	-0.28		4,008	-0.29	-0.30
3,979	-0.22	-0.28		4,009	-0.29	-0.30
3,980	-0.22	-0.28		4,010	-0.29	-0.30
3,981	-0.22	-0.28		4,011	-0.29	-0.30
3,982	-0.22	-0.28		4,012	-0.29	-0.30
3,983	-0.22	-0.28		4,013	-0.29	-0.30
3,984	-0.22	-0.29		4,014	-0.29	-0.30
3,985	-0.22	-0.29		4,015	-0.29	-0.30
3,986	-0.22	-0.29		4,016	-0.29	-0.30
3,987	-0.29	-0.29		4,017	-0.29	-0.30
3,988	-0.29	-0.29		4,018	-0.29	-0.30
3,989	-0.29	-0.29		4,019	-0.29	-0.30

4,020	-0.29	-0.30		4,050	-0.31	-0.37
4,021	-0.29	-0.30		4,051	-0.31	-0.37
4,022	-0.29	-0.31		4,052	-0.31	-0.37
4,023	-0.29	-0.31		4,053	-0.31	-0.38
4,024	-0.29	-0.31		4,054	-0.31	-0.38
4,025	-0.29	-0.31		4,055	-0.31	-0.38
4,026	-0.29	-0.31		4,056	-0.31	-0.39
4,027	-0.29	-0.31		4,057	-0.31	-0.39
4,028	-0.29	-0.31		4,058	-0.31	-0.39
4,029	-0.29	-0.31		4,059	-0.31	-0.40
4,030	-0.29	-0.31		4,060	-0.31	-0.40
4,031	-0.29	-0.31		4,061	-0.31	-0.40
4,032	-0.29	-0.31		4,062	-0.31	-0.41
4,033	-0.29	-0.31		4,063	-0.31	-0.41
4,034	-0.31	-0.32		4,064	-0.31	-0.41
4,035	-0.31	-0.32		4,065	-0.31	-0.42
4,036	-0.31	-0.32		4,066	-0.31	-0.42
4,037	-0.31	-0.33		4,067	-0.31	-0.42
4,038	-0.31	-0.33		4,068	-0.31	-0.43
4,039	-0.31	-0.33		4,069	-0.31	-0.43
4,040	-0.31	-0.34		4,070	-0.31	-0.43
4,041	-0.31	-0.34		4,071	-0.31	-0.43
4,042	-0.31	-0.34		4,072	-0.31	-0.44
4,043	-0.31	-0.35		4,073	-0.31	-0.44
4,044	-0.31	-0.35		4,074	-0.31	-0.44
4,045	-0.31	-0.35		4,075	-0.31	-0.45
4,046	-0.31	-0.35		4,076	-0.31	-0.45
4,047	-0.31	-0.36		4,077	-0.31	-0.45
4,048	-0.31	-0.36		4,078	-0.31	-0.46
4,049	-0.31	-0.36		4,079	-0.31	-0.46

4,080	-0.31	-0.46	4,110	-0.47	-0.58
4,081	-0.31	-0.47	4,111	-0.47	-0.58
4,082	-0.47	-0.48	4,112	-0.47	-0.59
4,083	-0.47	-0.48	4,113	-0.47	-0.59
4,084	-0.47	-0.48	4,114	-0.47	-0.59
4,085	-0.47	-0.49	4,115	-0.47	-0.60
4,086	-0.47	-0.49	4,116	-0.47	-0.60
4,087	-0.47	-0.50	4,117	-0.47	-0.60
4,088	-0.47	-0.50	4,118	-0.47	-0.61
4,089	-0.47	-0.50	4,119	-0.47	-0.61
4,090	-0.47	-0.51	4,120	-0.47	-0.61
4,091	-0.47	-0.51	4,121	-0.47	-0.62
4,092	-0.47	-0.51	4,122	-0.47	-0.62
4,093	-0.47	-0.52	4,123	-0.47	-0.62
4,094	-0.47	-0.52	4,124	-0.47	-0.63
4,095	-0.47	-0.52	4,125	-0.47	-0.63
4,096	-0.47	-0.53	4,126	-0.47	-0.64
4,097	-0.47	-0.53	4,127	-0.47	-0.64
4,098	-0.47	-0.53	4,128	-0.47	-0.64
4,099	-0.47	-0.54	4,129	-0.47	-0.65
4,100	-0.47	-0.54	4,130	-0.65	-0.65
4,101	-0.47	-0.55	4,131	-0.65	-0.65
4,102	-0.47	-0.55	4,132	-0.65	-0.64
4,103	-0.47	-0.55	4,133	-0.65	-0.64
4,104	-0.47	-0.56	4,134	-0.65	-0.64
4,105	-0.47	-0.56	4,135	-0.65	-0.64
4,106	-0.47	-0.56	4,136	-0.65	-0.64
4,107	-0.47	-0.57	4,137	-0.65	-0.64
4,108	-0.47	-0.57	4,138	-0.65	-0.63
4,109	-0.47	-0.57	4,139	-0.65	-0.63

4,140	-0.65	-0.63	4,170	-0.65	-0.58
4,141	-0.65	-0.63	4,171	-0.65	-0.57
4,142	-0.65	-0.63	4,172	-0.65	-0.57
4,143	-0.65	-0.62	4,173	-0.65	-0.57
4,144	-0.65	-0.62	4,174	-0.65	-0.57
4,145	-0.65	-0.62	4,175	-0.65	-0.57
4,146	-0.65	-0.62	4,176	-0.65	-0.57
4,147	-0.65	-0.62	4,177	-0.65	-0.56
4,148	-0.65	-0.62	4,178	-0.65	-0.55
4,149	-0.65	-0.61	4,179	-0.56	-0.54
4,150	-0.65	-0.61	4,180	-0.56	-0.53
4,151	-0.65	-0.61	4,181	-0.56	-0.52
4,152	-0.65	-0.61	4,182	-0.56	-0.51
4,153	-0.65	-0.61	4,183	-0.56	-0.50
4,154	-0.65	-0.60	4,184	-0.56	-0.49
4,155	-0.65	-0.60	4,185	-0.56	-0.48
4,156	-0.65	-0.60	4,186	-0.56	-0.48
4,157	-0.65	-0.60	4,187	-0.56	-0.47
4,158	-0.65	-0.60	4,188	-0.56	-0.46
4,159	-0.65	-0.60	4,189	-0.56	-0.45
4,160	-0.65	-0.59	4,190	-0.56	-0.44
4,161	-0.65	-0.59	4,191	-0.56	-0.43
4,162	-0.65	-0.59	4,192	-0.56	-0.42
4,163	-0.65	-0.59	4,193	-0.56	-0.41
4,164	-0.65	-0.59	4,194	-0.56	-0.40
4,165	-0.65	-0.59	4,195	-0.56	-0.39
4,166	-0.65	-0.58	4,196	-0.56	-0.38
4,167	-0.65	-0.58	4,197	-0.56	-0.37
4,168	-0.65	-0.58	4,198	-0.56	-0.36
4,169	-0.65	-0.58	4,199	-0.56	-0.35

4,200 | -0.56 | -0.34
4,201 | -0.56 | -0.33
4,202 | -0.56 | -0.33
4,203 | -0.56 | -0.32
4,204 | -0.56 | -0.31
4,205 | -0.56 | -0.30
4,206 | -0.56 | -0.29
4,207 | -0.56 | -0.28
4,208 | -0.56 | -0.27
4,209 | -0.56 | -0.26
4,210 | -0.56 | -0.25
4,211 | -0.56 | -0.24
4,212 | -0.56 | -0.23
4,213 | -0.56 | -0.22
4,214 | -0.56 | -0.21
4,215 | -0.56 | -0.20
4,216 | -0.56 | -0.19
4,217 | -0.56 | -0.18
4,218 | -0.56 | -0.17
4,219 | -0.56 | -0.17
4,220 | -0.56 | -0.16
4,221 | -0.56 | -0.15
4,222 | -0.56 | -0.14
4,223 | -0.56 | -0.13
4,224 | -0.56 | -0.12
4,225 | -0.56 | -0.09
4,226 | -0.56 | -0.06
4,227 | -0.09 | -0.03
4,228 | -0.09 | 0.00
4,229 | -0.09 | 0.02

4,230 | -0.09 | 0.04
4,231 | -0.09 | 0.06
4,232 | -0.09 | 0.08
4,233 | -0.09 | 0.10
4,234 | -0.09 | 0.13
4,235 | -0.09 | 0.15
4,236 | -0.09 | 0.17
4,237 | -0.09 | 0.19
4,238 | -0.09 | 0.21
4,239 | -0.09 | 0.23
4,240 | -0.09 | 0.26
4,241 | -0.09 | 0.28
4,242 | -0.09 | 0.30
4,243 | -0.09 | 0.32
4,244 | -0.09 | 0.34
4,245 | -0.09 | 0.36
4,246 | -0.09 | 0.39
4,247 | -0.09 | 0.41
4,248 | -0.09 | 0.43
4,249 | -0.09 | 0.45
4,250 | -0.09 | 0.47
4,251 | -0.09 | 0.49
4,252 | -0.09 | 0.51
4,253 | -0.09 | 0.54
4,254 | -0.09 | 0.56
4,255 | -0.09 | 0.58
4,256 | -0.09 | 0.60
4,257 | -0.09 | 0.62
4,258 | -0.09 | 0.64
4,259 | -0.09 | 0.67

4,260 | -0.09 | 0.69
4,261 | -0.09 | 0.71
4,262 | -0.09 | 0.73
4,263 | -0.09 | 0.75
4,264 | -0.09 | 0.77
4,265 | -0.09 | 0.80
4,266 | -0.09 | 0.82
4,267 | -0.09 | 0.84
4,268 | -0.09 | 0.86
4,269 | -0.09 | 0.89
4,270 | -0.09 | 0.93
4,271 | -0.09 | 0.96
4,272 | -0.09 | 0.99
4,273 | -0.09 | 1.03
4,274 | 0.99 | 1.06
4,275 | 0.99 | 1.07
4,276 | 0.99 | 1.08
4,277 | 0.99 | 1.09
4,278 | 0.99 | 1.11
4,279 | 0.99 | 1.12
4,280 | 0.99 | 1.13
4,281 | 0.99 | 1.14
4,282 | 0.99 | 1.15
4,283 | 0.99 | 1.16
4,284 | 0.99 | 1.18
4,285 | 0.99 | 1.19
4,286 | 0.99 | 1.20
4,287 | 0.99 | 1.21
4,288 | 0.99 | 1.22
4,289 | 0.99 | 1.23
4,290 | 0.99 | 1.25
4,291 | 0.99 | 1.26
4,292 | 0.99 | 1.27
4,293 | 0.99 | 1.28
4,294 | 0.99 | 1.29
4,295 | 0.99 | 1.30
4,296 | 0.99 | 1.31
4,297 | 0.99 | 1.33
4,298 | 0.99 | 1.34
4,299 | 0.99 | 1.35
4,300 | 0.99 | 1.36
4,301 | 0.99 | 1.37
4,302 | 0.99 | 1.38
4,303 | 0.99 | 1.40
4,304 | 0.99 | 1.41
4,305 | 0.99 | 1.42
4,306 | 0.99 | 1.43
4,307 | 0.99 | 1.44
4,308 | 0.99 | 1.45
4,309 | 0.99 | 1.47
4,310 | 0.99 | 1.48
4,311 | 0.99 | 1.49
4,312 | 0.99 | 1.50
4,313 | 0.99 | 1.51
4,314 | 0.99 | 1.52
4,315 | 0.99 | 1.54
4,316 | 0.99 | 1.55
4,317 | 0.99 | 1.56
4,318 | 1.57 | 1.57
4,319 | 1.57 | 1.57

4,320	1.57	1.57	4,350	1.57	1.57
4,321	1.57	1.57	4,351	1.57	1.57
4,322	1.57	1.57	4,352	1.57	1.57
4,323	1.57	1.57	4,353	1.57	1.57
4,324	1.57	1.57	4,354	1.57	1.57
4,325	1.57	1.57	4,355	1.57	1.57
4,326	1.57	1.57	4,356	1.57	1.57
4,327	1.57	1.57	4,357	1.57	1.57
4,328	1.57	1.57	4,358	1.57	1.57
4,329	1.57	1.57	4,359	1.57	1.57
4,330	1.57	1.57	4,360	1.57	1.57
4,331	1.57	1.57	4,361	1.57	1.57
4,332	1.57	1.57	4,362	1.57	1.57
4,333	1.57	1.57	4,363	1.57	1.57
4,334	1.57	1.57	4,364	1.57	1.57
4,335	1.57	1.57	4,365	1.57	1.57
4,336	1.57	1.57	4,366	1.57	1.57
4,337	1.57	1.57	4,367	1.57	1.57
4,338	1.57	1.57	4,368	1.57	1.57
4,339	1.57	1.57	4,369	1.57	1.57
4,340	1.57	1.57	4,370	1.57	1.57
4,341	1.57	1.57	4,371	1.57	1.57
4,342	1.57	1.57	4,372	1.57	1.57
4,343	1.57	1.57	4,373	1.57	1.57
4,344	1.57	1.57	4,374	1.57	1.57
4,345	1.57	1.57	4,375	1.57	1.57
4,346	1.57	1.57	4,376	1.57	1.57
4,347	1.57	1.57	4,377	1.57	1.57
4,348	1.57	1.57	4,378	1.57	1.57
4,349	1.57	1.57	4,379	1.57	1.57

4,380 | 1.57 | 1.57
4,381 | 1.57 | 1.57
4,382 | 1.57 | 1.57
4,383 | 1.57 | 1.57
4,384 | 1.57 | 1.57
4,385 | 1.57 | 1.57
4,386 | 1.57 | 1.57
4,387 | 1.57 | 1.57
4,388 | 1.57 | 1.57
4,389 | 1.57 | 1.57
4,390 | 1.57 | 1.57
4,391 | 1.57 | 1.57
4,392 | 1.57 | 1.57
4,393 | 1.57 | 1.57
4,394 | 1.57 | 1.57
4,395 | 1.57 | 1.57
4,396 | 1.57 | 1.57
4,397 | 1.57 | 1.57
4,398 | 1.57 | 1.57
4,399 | 1.57 | 1.57
4,400 | 1.57 | 1.57
4,401 | 1.57 | 1.57
4,402 | 1.57 | 1.57
4,403 | 1.57 | 1.57
4,404 | 1.57 | 1.57
4,405 | 1.57 | 1.57
4,406 | 1.57 | 1.57
4,407 | 1.57 | 1.57
4,408 | 1.57 | 1.57
4,409 | 1.57 | 1.57

4,410 | 1.57 | 1.57
4,411 | 1.57 | 1.57
4,412 | 1.57 | 1.57
4,413 | 1.57 | 1.57
4,414 | 1.57 | 1.57
4,415 | 1.57 | 1.57
4,416 | 1.57 | 1.57
4,417 | 1.57 | 1.57
4,418 | 1.57 | 1.57
4,419 | 1.57 | 1.57
4,420 | 1.57 | 1.57
4,421 | 1.57 | 1.57
4,422 | 1.57 | 1.57
4,423 | 1.57 | 1.57
4,424 | 1.57 | 1.57
4,425 | 1.57 | 1.57
4,426 | 1.57 | 1.57
4,427 | 1.57 | 1.57
4,428 | 1.57 | 1.57
4,429 | 1.57 | 1.57
4,430 | 1.57 | 1.57
4,431 | 1.57 | 1.57
4,432 | 1.57 | 1.57
4,433 | 1.57 | 1.57
4,434 | 1.57 | 1.57
4,435 | 1.57 | 1.57
4,436 | 1.57 | 1.57
4,437 | 1.57 | 1.57
4,438 | 1.57 | 1.57
4,439 | 1.57 | 1.57

4,440	1.57	1.57	4,470	1.57	1.57
4,441	1.57	1.57	4,471	1.57	1.57
4,442	1.57	1.57	4,472	1.57	1.57
4,443	1.57	1.57	4,473	1.57	1.57
4,444	1.57	1.57	4,474	1.57	1.57
4,445	1.57	1.57	4,475	1.57	1.57
4,446	1.57	1.57	4,476	1.57	1.57
4,447	1.57	1.57	4,477	1.57	1.57
4,448	1.57	1.57	4,478	1.57	1.57
4,449	1.57	1.57	4,479	1.57	1.57
4,450	1.57	1.57	4,480	1.57	1.57
4,451	1.57	1.57	4,481	1.57	1.57
4,452	1.57	1.57	4,482	1.57	1.57
4,453	1.57	1.57	4,483	1.57	1.57
4,454	1.57	1.57	4,484	1.57	1.56
4,455	1.57	1.57	4,485	1.57	1.56
4,456	1.57	1.57	4,486	1.57	1.55
4,457	1.57	1.57	4,487	1.57	1.55
4,458	1.57	1.57	4,488	1.57	1.54
4,459	1.57	1.57	4,489	1.57	1.54
4,460	1.57	1.57	4,490	1.57	1.53
4,461	1.57	1.57	4,491	1.57	1.53
4,462	1.57	1.57	4,492	1.57	1.52
4,463	1.57	1.57	4,493	1.57	1.52
4,464	1.57	1.57	4,494	1.57	1.51
4,465	1.57	1.57	4,495	1.57	1.51
4,466	1.57	1.57	4,496	1.57	1.50
4,467	1.57	1.57	4,497	1.57	1.50
4,468	1.57	1.57	4,498	1.57	1.49
4,469	1.57	1.57	4,499	1.57	1.49

4,500 | 1.57 | 1.48
4,501 | 1.57 | 1.48
4,502 | 1.57 | 1.47
4,503 | 1.57 | 1.47
4,504 | 1.57 | 1.46
4,505 | 1.57 | 1.46
4,506 | 1.57 | 1.45
4,507 | 1.57 | 1.45
4,508 | 1.57 | 1.44
4,509 | 1.57 | 1.44
4,510 | 1.57 | 1.43
4,511 | 1.57 | 1.43
4,512 | 1.57 | 1.42
4,513 | 1.57 | 1.42
4,514 | 1.57 | 1.41
4,515 | 1.57 | 1.41
4,516 | 1.57 | 1.40
4,517 | 1.57 | 1.40
4,518 | 1.57 | 1.39
4,519 | 1.57 | 1.39
4,520 | 1.57 | 1.38
4,521 | 1.57 | 1.38
4,522 | 1.57 | 1.37
4,523 | 1.57 | 1.37
4,524 | 1.57 | 1.36
4,525 | 1.57 | 1.36
4,526 | 1.57 | 1.34
4,527 | 1.57 | 1.32
4,528 | 1.57 | 1.30
4,529 | 1.57 | 1.28

4,530 | 1.57 | 1.26
4,531 | 1.57 | 1.24
4,532 | 1.32 | 1.22
4,533 | 1.32 | 1.21
4,534 | 1.32 | 1.20
4,535 | 1.32 | 1.18
4,536 | 1.32 | 1.17
4,537 | 1.32 | 1.16
4,538 | 1.32 | 1.14
4,539 | 1.32 | 1.13
4,540 | 1.32 | 1.12
4,541 | 1.32 | 1.10
4,542 | 1.32 | 1.09
4,543 | 1.32 | 1.08
4,544 | 1.32 | 1.06
4,545 | 1.32 | 1.05
4,546 | 1.32 | 1.03
4,547 | 1.32 | 1.02
4,548 | 1.32 | 1.01
4,549 | 1.32 | 0.99
4,550 | 1.32 | 0.98
4,551 | 1.32 | 0.97
4,552 | 1.32 | 0.95
4,553 | 1.32 | 0.94
4,554 | 1.32 | 0.93
4,555 | 1.32 | 0.91
4,556 | 1.32 | 0.90
4,557 | 1.32 | 0.88
4,558 | 1.32 | 0.87
4,559 | 1.32 | 0.86

4,560	1.32	0.84	4,590	0.64	0.43
4,561	1.32	0.83	4,591	0.64	0.42
4,562	1.32	0.82	4,592	0.64	0.41
4,563	1.32	0.80	4,593	0.64	0.40
4,564	1.32	0.79	4,594	0.64	0.39
4,565	1.32	0.78	4,595	0.64	0.38
4,566	1.32	0.76	4,596	0.64	0.36
4,567	1.32	0.75	4,597	0.64	0.35
4,568	1.32	0.74	4,598	0.64	0.34
4,569	1.32	0.72	4,599	0.64	0.33
4,570	1.32	0.71	4,600	0.64	0.32
4,571	1.32	0.68	4,601	0.64	0.31
4,572	1.32	0.66	4,602	0.64	0.30
4,573	1.32	0.64	4,603	0.64	0.29
4,574	1.32	0.61	4,604	0.64	0.28
4,575	0.64	0.59	4,605	0.64	0.27
4,576	0.64	0.58	4,606	0.64	0.26
4,577	0.64	0.57	4,607	0.64	0.25
4,578	0.64	0.56	4,608	0.64	0.24
4,579	0.64	0.54	4,609	0.64	0.23
4,580	0.64	0.53	4,610	0.64	0.22
4,581	0.64	0.52	4,611	0.64	0.21
4,582	0.64	0.51	4,612	0.64	0.19
4,583	0.64	0.50	4,613	0.64	0.18
4,584	0.64	0.49	4,614	0.64	0.17
4,585	0.64	0.48	4,615	0.64	0.16
4,586	0.64	0.47	4,616	0.64	0.15
4,587	0.64	0.46	4,617	0.64	0.14
4,588	0.64	0.45	4,618	0.64	0.12
4,589	0.64	0.44	4,619	0.64	0.10

4,620 | 0.11 | 0.08
4,621 | 0.11 | 0.07
4,622 | 0.11 | 0.06
4,623 | 0.11 | 0.04
4,624 | 0.11 | 0.03
4,625 | 0.11 | 0.02
4,626 | 0.11 | 0.01
4,627 | 0.11 | 0.00
4,628 | 0.11 | -0.01
4,629 | 0.11 | -0.02
4,630 | 0.11 | -0.03
4,631 | 0.11 | -0.04
4,632 | 0.11 | -0.06
4,633 | 0.11 | -0.07
4,634 | 0.11 | -0.08
4,635 | 0.11 | -0.09
4,636 | 0.11 | -0.10
4,637 | 0.11 | -0.11
4,638 | 0.11 | -0.12
4,639 | 0.11 | -0.13
4,640 | 0.11 | -0.14
4,641 | 0.11 | -0.15
4,642 | 0.11 | -0.17
4,643 | 0.11 | -0.18
4,644 | 0.11 | -0.19
4,645 | 0.11 | -0.20
4,646 | 0.11 | -0.21
4,647 | 0.11 | -0.22
4,648 | 0.11 | -0.23
4,649 | 0.11 | -0.24

4,650 | 0.11 | -0.25
4,651 | 0.11 | -0.26
4,652 | 0.11 | -0.28
4,653 | 0.11 | -0.29
4,654 | 0.11 | -0.30
4,655 | 0.11 | -0.31
4,656 | 0.11 | -0.32
4,657 | 0.11 | -0.33
4,658 | 0.11 | -0.34
4,659 | 0.11 | -0.35
4,660 | 0.11 | -0.36
4,661 | 0.11 | -0.37
4,662 | 0.11 | -0.39
4,663 | 0.11 | -0.40
4,664 | 0.11 | -0.41
4,665 | 0.11 | -0.42
4,666 | 0.11 | -0.43
4,667 | -0.44 | -0.43
4,668 | -0.44 | -0.43
4,669 | -0.44 | -0.42
4,670 | -0.44 | -0.42
4,671 | -0.44 | -0.41
4,672 | -0.44 | -0.40
4,673 | -0.44 | -0.40
4,674 | -0.44 | -0.39
4,675 | -0.44 | -0.39
4,676 | -0.44 | -0.38
4,677 | -0.44 | -0.37
4,678 | -0.44 | -0.37
4,679 | -0.44 | -0.36

4,680 | -0.44 | -0.36
4,681 | -0.44 | -0.35
4,682 | -0.44 | -0.34
4,683 | -0.44 | -0.34
4,684 | -0.44 | -0.33
4,685 | -0.44 | -0.33
4,686 | -0.44 | -0.32
4,687 | -0.44 | -0.31
4,688 | -0.44 | -0.31
4,689 | -0.44 | -0.30
4,690 | -0.44 | -0.30
4,691 | -0.44 | -0.29
4,692 | -0.44 | -0.28
4,693 | -0.44 | -0.28
4,694 | -0.44 | -0.27
4,695 | -0.44 | -0.27
4,696 | -0.44 | -0.26
4,697 | -0.44 | -0.25
4,698 | -0.44 | -0.25
4,699 | -0.44 | -0.24
4,700 | -0.44 | -0.24
4,701 | -0.44 | -0.23
4,702 | -0.44 | -0.22
4,703 | -0.44 | -0.22
4,704 | -0.44 | -0.21
4,705 | -0.44 | -0.21
4,706 | -0.44 | -0.20
4,707 | -0.44 | -0.19
4,708 | -0.44 | -0.19
4,709 | -0.44 | -0.18

4,710 | -0.44 | -0.18
4,711 | -0.44 | -0.17
4,712 | -0.44 | -0.16
4,713 | -0.44 | -0.16
4,714 | -0.44 | -0.15
4,715 | -0.44 | -0.13
4,716 | -0.14 | -0.11
4,717 | -0.14 | -0.09
4,718 | -0.14 | -0.07
4,719 | -0.14 | -0.06
4,720 | -0.14 | -0.04
4,721 | -0.14 | -0.02
4,722 | -0.14 | -0.01
4,723 | -0.14 | 0.01
4,724 | -0.14 | 0.03
4,725 | -0.14 | 0.04
4,726 | -0.14 | 0.06
4,727 | -0.14 | 0.08
4,728 | -0.14 | 0.09
4,729 | -0.14 | 0.11
4,730 | -0.14 | 0.13
4,731 | -0.14 | 0.14
4,732 | -0.14 | 0.16
4,733 | -0.14 | 0.18
4,734 | -0.14 | 0.19
4,735 | -0.14 | 0.21
4,736 | -0.14 | 0.23
4,737 | -0.14 | 0.24
4,738 | -0.14 | 0.26
4,739 | -0.14 | 0.28

4,740	-0.14	0.29		4,770	0.69	0.64
4,741	-0.14	0.31		4,771	0.69	0.64
4,742	-0.14	0.32		4,772	0.69	0.63
4,743	-0.14	0.34		4,773	0.69	0.63
4,744	-0.14	0.36		4,774	0.69	0.62
4,745	-0.14	0.37		4,775	0.69	0.62
4,746	-0.14	0.39		4,776	0.69	0.61
4,747	-0.14	0.41		4,777	0.69	0.61
4,748	-0.14	0.42		4,778	0.69	0.60
4,749	-0.14	0.44		4,779	0.69	0.60
4,750	-0.14	0.46		4,780	0.69	0.59
4,751	-0.14	0.47		4,781	0.69	0.59
4,752	-0.14	0.49		4,782	0.69	0.58
4,753	-0.14	0.51		4,783	0.69	0.58
4,754	-0.14	0.52		4,784	0.69	0.57
4,755	-0.14	0.54		4,785	0.69	0.57
4,756	-0.14	0.56		4,786	0.69	0.56
4,757	-0.14	0.57		4,787	0.69	0.56
4,758	-0.14	0.59		4,788	0.69	0.55
4,759	-0.14	0.61		4,789	0.69	0.55
4,760	-0.14	0.62		4,790	0.69	0.54
4,761	-0.14	0.64		4,791	0.69	0.54
4,762	-0.14	0.65		4,792	0.69	0.53
4,763	-0.14	0.66		4,793	0.69	0.53
4,764	0.69	0.67		4,794	0.69	0.52
4,765	0.69	0.67		4,795	0.69	0.52
4,766	0.69	0.66		4,796	0.69	0.51
4,767	0.69	0.66		4,797	0.69	0.51
4,768	0.69	0.65		4,798	0.69	0.50
4,769	0.69	0.65		4,799	0.69	0.50

4,800	0.69	0.49		4,830	0.44	-0.08
4,801	0.69	0.49		4,831	0.44	-0.11
4,802	0.69	0.48		4,832	0.44	-0.13
4,803	0.69	0.48		4,833	0.44	-0.15
4,804	0.69	0.47		4,834	0.44	-0.18
4,805	0.69	0.47		4,835	0.44	-0.20
4,806	0.69	0.46		4,836	0.44	-0.22
4,807	0.69	0.46		4,837	0.44	-0.24
4,808	0.69	0.43		4,838	0.44	-0.27
4,809	0.69	0.40		4,839	0.44	-0.29
4,810	0.44	0.37		4,840	0.44	-0.31
4,811	0.44	0.35		4,841	0.44	-0.34
4,812	0.44	0.33		4,842	0.44	-0.36
4,813	0.44	0.30		4,843	0.44	-0.38
4,814	0.44	0.28		4,844	0.44	-0.40
4,815	0.44	0.26		4,845	0.44	-0.43
4,816	0.44	0.23		4,846	0.44	-0.45
4,817	0.44	0.21		4,847	0.44	-0.47
4,818	0.44	0.19		4,848	0.44	-0.49
4,819	0.44	0.17		4,849	0.44	-0.52
4,820	0.44	0.14		4,850	0.44	-0.54
4,821	0.44	0.12		4,851	0.44	-0.56
4,822	0.44	0.10		4,852	0.44	-0.59
4,823	0.44	0.08		4,853	0.44	-0.61
4,824	0.44	0.05		4,854	0.44	-0.63
4,825	0.44	0.03		4,855	0.44	-0.65
4,826	0.44	0.01		4,856	0.44	-0.68
4,827	0.44	-0.02		4,857	-0.70	-0.69
4,828	0.44	-0.04		4,858	-0.70	-0.68
4,829	0.44	-0.06		4,859	-0.70	-0.67

4,860	-0.70	-0.66		4,890	-0.70	-0.37
4,861	-0.70	-0.65		4,891	-0.70	-0.36
4,862	-0.70	-0.64		4,892	-0.70	-0.35
4,863	-0.70	-0.63		4,893	-0.70	-0.34
4,864	-0.70	-0.62		4,894	-0.70	-0.34
4,865	-0.70	-0.61		4,895	-0.70	-0.33
4,866	-0.70	-0.60		4,896	-0.70	-0.32
4,867	-0.70	-0.59		4,897	-0.70	-0.31
4,868	-0.70	-0.58		4,898	-0.70	-0.30
4,869	-0.70	-0.58		4,899	-0.70	-0.29
4,870	-0.70	-0.57		4,900	-0.70	-0.28
4,871	-0.70	-0.56		4,901	-0.70	-0.27
4,872	-0.70	-0.55		4,902	-0.70	-0.26
4,873	-0.70	-0.54		4,903	-0.70	-0.25
4,874	-0.70	-0.53		4,904	-0.70	-0.24
4,875	-0.70	-0.52		4,905	-0.70	-0.22
4,876	-0.70	-0.51		4,906	-0.22	-0.21
4,877	-0.70	-0.50		4,907	-0.22	-0.20
4,878	-0.70	-0.49		4,908	-0.22	-0.20
4,879	-0.70	-0.48		4,909	-0.22	-0.19
4,880	-0.70	-0.47		4,910	-0.22	-0.18
4,881	-0.70	-0.46		4,911	-0.22	-0.18
4,882	-0.70	-0.45		4,912	-0.22	-0.17
4,883	-0.70	-0.44		4,913	-0.22	-0.17
4,884	-0.70	-0.43		4,914	-0.22	-0.16
4,885	-0.70	-0.42		4,915	-0.22	-0.15
4,886	-0.70	-0.41		4,916	-0.22	-0.15
4,887	-0.70	-0.40		4,917	-0.22	-0.14
4,888	-0.70	-0.39		4,918	-0.22	-0.14
4,889	-0.70	-0.38		4,919	-0.22	-0.13

4,920 | -0.22 | -0.12
4,921 | -0.22 | -0.12
4,922 | -0.22 | -0.11
4,923 | -0.22 | -0.11
4,924 | -0.22 | -0.10
4,925 | -0.22 | -0.09
4,926 | -0.22 | -0.09
4,927 | -0.22 | -0.08
4,928 | -0.22 | -0.08
4,929 | -0.22 | -0.07
4,930 | -0.22 | -0.06
4,931 | -0.22 | -0.06
4,932 | -0.22 | -0.05
4,933 | -0.22 | -0.05
4,934 | -0.22 | -0.04
4,935 | -0.22 | -0.03
4,936 | -0.22 | -0.03
4,937 | -0.22 | -0.02
4,938 | -0.22 | -0.02
4,939 | -0.22 | -0.01
4,940 | -0.22 | 0.00
4,941 | -0.22 | 0.00
4,942 | -0.22 | 0.01
4,943 | -0.22 | 0.01
4,944 | -0.22 | 0.02
4,945 | -0.22 | 0.03
4,946 | -0.22 | 0.03
4,947 | -0.22 | 0.04
4,948 | -0.22 | 0.04
4,949 | -0.22 | 0.05

4,950 | -0.22 | 0.06
4,951 | -0.22 | 0.06
4,952 | -0.22 | 0.07
4,953 | -0.22 | 0.07
4,954 | 0.08 | 0.07
4,955 | 0.08 | 0.06
4,956 | 0.08 | 0.06
4,957 | 0.08 | 0.05
4,958 | 0.08 | 0.04
4,959 | 0.08 | 0.04
4,960 | 0.08 | 0.03
4,961 | 0.08 | 0.03
4,962 | 0.08 | 0.02
4,963 | 0.08 | 0.01
4,964 | 0.08 | 0.01
4,965 | 0.08 | 0.00
4,966 | 0.08 | 0.00
4,967 | 0.08 | -0.01
4,968 | 0.08 | -0.02
4,969 | 0.08 | -0.02
4,970 | 0.08 | -0.03
4,971 | 0.08 | -0.03
4,972 | 0.08 | -0.04
4,973 | 0.08 | -0.05
4,974 | 0.08 | -0.05
4,975 | 0.08 | -0.06
4,976 | 0.08 | -0.06
4,977 | 0.08 | -0.07
4,978 | 0.08 | -0.08
4,979 | 0.08 | -0.08

4,980	0.08	-0.09
4,981	0.08	-0.09
4,982	0.08	-0.10
4,983	0.08	-0.11
4,984	0.08	-0.11
4,985	0.08	-0.12
4,986	0.08	-0.12
4,987	0.08	-0.13
4,988	0.08	-0.14
4,989	0.08	-0.14
4,990	0.08	-0.15
4,991	0.08	-0.15
4,992	0.08	-0.16
4,993	0.08	-0.17
4,994	0.08	-0.17
4,995	0.08	-0.18
4,996	0.08	-0.18
4,997	0.08	-0.19
4,998	0.08	-0.20
4,999	0.08	-0.20
5,000	0.08	-0.21
5,001	0.08	-0.20
5,002	-0.22	-0.19
5,003	-0.22	-0.17
5,004	-0.22	-0.16
5,005	-0.22	-0.14
5,006	-0.22	-0.13
5,007	-0.22	-0.11
5,008	-0.22	-0.09
5,009	-0.22	-0.08
5,010	-0.22	-0.06
5,011	-0.22	-0.05
5,012	-0.22	-0.03
5,013	-0.22	-0.01
5,014	-0.22	0.00
5,015	-0.22	0.02
5,016	-0.22	0.03
5,017	-0.22	0.05
5,018	-0.22	0.06
5,019	-0.22	0.08
5,020	-0.22	0.10
5,021	-0.22	0.11
5,022	-0.22	0.13
5,023	-0.22	0.14
5,024	-0.22	0.16
5,025	-0.22	0.18
5,026	-0.22	0.19
5,027	-0.22	0.21
5,028	-0.22	0.22
5,029	-0.22	0.24
5,030	-0.22	0.25
5,031	-0.22	0.27
5,032	-0.22	0.29
5,033	-0.22	0.30
5,034	-0.22	0.32
5,035	-0.22	0.33
5,036	-0.22	0.35
5,037	-0.22	0.36
5,038	-0.22	0.38
5,039	-0.22	0.40

5,040	-0.22	0.41	5,070	0.57	0.59
5,041	-0.22	0.43	5,071	0.57	0.60
5,042	-0.22	0.44	5,072	0.57	0.60
5,043	-0.22	0.46	5,073	0.57	0.60
5,044	-0.22	0.48	5,074	0.57	0.60
5,045	-0.22	0.49	5,075	0.57	0.60
5,046	-0.22	0.51	5,076	0.57	0.60
5,047	-0.22	0.52	5,077	0.57	0.60
5,048	-0.22	0.54	5,078	0.57	0.60
5,049	-0.22	0.56	5,079	0.57	0.60
5,050	0.57	0.57	5,080	0.57	0.60
5,051	0.57	0.58	5,081	0.57	0.61
5,052	0.57	0.58	5,082	0.57	0.61
5,053	0.57	0.58	5,083	0.57	0.61
5,054	0.57	0.58	5,084	0.57	0.61
5,055	0.57	0.58	5,085	0.57	0.61
5,056	0.57	0.58	5,086	0.57	0.61
5,057	0.57	0.58	5,087	0.57	0.61
5,058	0.57	0.58	5,088	0.57	0.61
5,059	0.57	0.58	5,089	0.57	0.61
5,060	0.57	0.58	5,090	0.57	0.61
5,061	0.57	0.59	5,091	0.57	0.62
5,062	0.57	0.59	5,092	0.57	0.62
5,063	0.57	0.59	5,093	0.57	0.62
5,064	0.57	0.59	5,094	0.57	0.61
5,065	0.57	0.59	5,095	0.57	0.60
5,066	0.57	0.59	5,096	0.62	0.59
5,067	0.57	0.59	5,097	0.62	0.58
5,068	0.57	0.59	5,098	0.62	0.56
5,069	0.57	0.59	5,099	0.62	0.55

5,100 | 0.62 | 0.54
5,101 | 0.62 | 0.53
5,102 | 0.62 | 0.52
5,103 | 0.62 | 0.51
5,104 | 0.62 | 0.50
5,105 | 0.62 | 0.49
5,106 | 0.62 | 0.47
5,107 | 0.62 | 0.46
5,108 | 0.62 | 0.45
5,109 | 0.62 | 0.44
5,110 | 0.62 | 0.43
5,111 | 0.62 | 0.42
5,112 | 0.62 | 0.41
5,113 | 0.62 | 0.40
5,114 | 0.62 | 0.38
5,115 | 0.62 | 0.37
5,116 | 0.62 | 0.36
5,117 | 0.62 | 0.35
5,118 | 0.62 | 0.34
5,119 | 0.62 | 0.33
5,120 | 0.62 | 0.32
5,121 | 0.62 | 0.31
5,122 | 0.62 | 0.30
5,123 | 0.62 | 0.28
5,124 | 0.62 | 0.27
5,125 | 0.62 | 0.26
5,126 | 0.62 | 0.25
5,127 | 0.62 | 0.24
5,128 | 0.62 | 0.23
5,129 | 0.62 | 0.22

5,130 | 0.62 | 0.21
5,131 | 0.62 | 0.19
5,132 | 0.62 | 0.18
5,133 | 0.62 | 0.17
5,134 | 0.62 | 0.16
5,135 | 0.62 | 0.15
5,136 | 0.62 | 0.14
5,137 | 0.62 | 0.13
5,138 | 0.62 | 0.12
5,139 | 0.62 | 0.10
5,140 | 0.62 | 0.09
5,141 | 0.62 | 0.09
5,142 | 0.62 | 0.08
5,143 | 0.06 | 0.07
5,144 | 0.06 | 0.07
5,145 | 0.06 | 0.08
5,146 | 0.06 | 0.08
5,147 | 0.06 | 0.08
5,148 | 0.06 | 0.09
5,149 | 0.06 | 0.09
5,150 | 0.06 | 0.09
5,151 | 0.06 | 0.10
5,152 | 0.06 | 0.10
5,153 | 0.06 | 0.10
5,154 | 0.06 | 0.10
5,155 | 0.06 | 0.11
5,156 | 0.06 | 0.11
5,157 | 0.06 | 0.11
5,158 | 0.06 | 0.12
5,159 | 0.06 | 0.12

5,160	0.06	0.12	5,190	0.22	0.25
5,161	0.06	0.13	5,191	0.22	0.26
5,162	0.06	0.13	5,192	0.22	0.27
5,163	0.06	0.13	5,193	0.22	0.28
5,164	0.06	0.14	5,194	0.22	0.29
5,165	0.06	0.14	5,195	0.22	0.30
5,166	0.06	0.14	5,196	0.22	0.31
5,167	0.06	0.15	5,197	0.22	0.32
5,168	0.06	0.15	5,198	0.22	0.33
5,169	0.06	0.15	5,199	0.22	0.34
5,170	0.06	0.16	5,200	0.22	0.35
5,171	0.06	0.16	5,201	0.22	0.36
5,172	0.06	0.16	5,202	0.22	0.37
5,173	0.06	0.17	5,203	0.22	0.38
5,174	0.06	0.17	5,204	0.22	0.39
5,175	0.06	0.17	5,205	0.22	0.40
5,176	0.06	0.18	5,206	0.22	0.41
5,177	0.06	0.18	5,207	0.22	0.42
5,178	0.06	0.18	5,208	0.22	0.43
5,179	0.06	0.18	5,209	0.22	0.44
5,180	0.06	0.19	5,210	0.22	0.45
5,181	0.06	0.19	5,211	0.22	0.46
5,182	0.06	0.19	5,212	0.22	0.47
5,183	0.06	0.20	5,213	0.22	0.48
5,184	0.06	0.20	5,214	0.22	0.49
5,185	0.06	0.20	5,215	0.22	0.50
5,186	0.06	0.21	5,216	0.22	0.51
5,187	0.06	0.21	5,217	0.22	0.52
5,188	0.06	0.22	5,218	0.22	0.53
5,189	0.06	0.24	5,219	0.22	0.54

5,220 | 0.22 | 0.55
5,221 | 0.22 | 0.56
5,222 | 0.22 | 0.57
5,223 | 0.22 | 0.58
5,224 | 0.22 | 0.59
5,225 | 0.22 | 0.60
5,226 | 0.22 | 0.61
5,227 | 0.22 | 0.62
5,228 | 0.22 | 0.63
5,229 | 0.22 | 0.64
5,230 | 0.22 | 0.65
5,231 | 0.22 | 0.66
5,232 | 0.22 | 0.67
5,233 | 0.22 | 0.68
5,234 | 0.22 | 0.69
5,235 | 0.22 | 0.69
5,236 | 0.22 | 0.70
5,237 | 0.72 | 0.71
5,238 | 0.72 | 0.70
5,239 | 0.72 | 0.70
5,240 | 0.72 | 0.69
5,241 | 0.72 | 0.69
5,242 | 0.72 | 0.69
5,243 | 0.72 | 0.68
5,244 | 0.72 | 0.68
5,245 | 0.72 | 0.68
5,246 | 0.72 | 0.67
5,247 | 0.72 | 0.67
5,248 | 0.72 | 0.67
5,249 | 0.72 | 0.66

5,250 | 0.72 | 0.66
5,251 | 0.72 | 0.66
5,252 | 0.72 | 0.65
5,253 | 0.72 | 0.65
5,254 | 0.72 | 0.64
5,255 | 0.72 | 0.64
5,256 | 0.72 | 0.64
5,257 | 0.72 | 0.63
5,258 | 0.72 | 0.63
5,259 | 0.72 | 0.63
5,260 | 0.72 | 0.62
5,261 | 0.72 | 0.62
5,262 | 0.72 | 0.62
5,263 | 0.72 | 0.61
5,264 | 0.72 | 0.61
5,265 | 0.72 | 0.60
5,266 | 0.72 | 0.60
5,267 | 0.72 | 0.60
5,268 | 0.72 | 0.59
5,269 | 0.72 | 0.59
5,270 | 0.72 | 0.59
5,271 | 0.72 | 0.58
5,272 | 0.72 | 0.58
5,273 | 0.72 | 0.58
5,274 | 0.72 | 0.57
5,275 | 0.72 | 0.57
5,276 | 0.72 | 0.57
5,277 | 0.72 | 0.56
5,278 | 0.72 | 0.56
5,279 | 0.72 | 0.55

5,280	0.72	0.55	5,310	0.54	0.53
5,281	0.72	0.55	5,311	0.54	0.53
5,282	0.72	0.54	5,312	0.54	0.53
5,283	0.54	0.54	5,313	0.54	0.53
5,284	0.54	0.54	5,314	0.54	0.53
5,285	0.54	0.54	5,315	0.54	0.53
5,286	0.54	0.54	5,316	0.54	0.53
5,287	0.54	0.54	5,317	0.54	0.52
5,288	0.54	0.54	5,318	0.54	0.52
5,289	0.54	0.54	5,319	0.54	0.52
5,290	0.54	0.54	5,320	0.54	0.52
5,291	0.54	0.54	5,321	0.54	0.52
5,292	0.54	0.53	5,322	0.54	0.52
5,293	0.54	0.53	5,323	0.54	0.52
5,294	0.54	0.53	5,324	0.54	0.52
5,295	0.54	0.53	5,325	0.54	0.52
5,296	0.54	0.53	5,326	0.54	0.53
5,297	0.54	0.53	5,327	0.54	0.54
5,298	0.54	0.53	5,328	0.54	0.54
5,299	0.54	0.53	5,329	0.52	0.55
5,300	0.54	0.53	5,330	0.52	0.56
5,301	0.54	0.53	5,331	0.52	0.57
5,302	0.54	0.53	5,332	0.52	0.58
5,303	0.54	0.53	5,333	0.52	0.58
5,304	0.54	0.53	5,334	0.52	0.59
5,305	0.54	0.53	5,335	0.52	0.60
5,306	0.54	0.53	5,336	0.52	0.61
5,307	0.54	0.53	5,337	0.52	0.62
5,308	0.54	0.53	5,338	0.52	0.62
5,309	0.54	0.53	5,339	0.52	0.63

5,340 | 0.52 | 0.64
5,341 | 0.52 | 0.65
5,342 | 0.52 | 0.66
5,343 | 0.52 | 0.67
5,344 | 0.52 | 0.67
5,345 | 0.52 | 0.68
5,346 | 0.52 | 0.69
5,347 | 0.52 | 0.70
5,348 | 0.52 | 0.71
5,349 | 0.52 | 0.71
5,350 | 0.52 | 0.72
5,351 | 0.52 | 0.73
5,352 | 0.52 | 0.74
5,353 | 0.52 | 0.75
5,354 | 0.52 | 0.76
5,355 | 0.52 | 0.76
5,356 | 0.52 | 0.77
5,357 | 0.52 | 0.78
5,358 | 0.52 | 0.79
5,359 | 0.52 | 0.80
5,360 | 0.52 | 0.81
5,361 | 0.52 | 0.81
5,362 | 0.52 | 0.82
5,363 | 0.52 | 0.83
5,364 | 0.52 | 0.84
5,365 | 0.51 | 0.85
5,366 | 0.51 | 0.85
5,367 | 0.51 | 0.86
5,368 | 0.51 | 0.87
5,369 | 0.51 | 0.88

5,370 | 0.51 | 0.89
5,371 | 0.51 | 0.90
5,372 | 0.51 | 0.90
5,373 | 0.51 | 0.91
5,374 | 0.51 | 0.92
5,375 | 0.93 | 0.93
5,376 | 0.93 | 0.93
5,377 | 0.93 | 0.93
5,378 | 0.93 | 0.93
5,379 | 0.93 | 0.93
5,380 | 0.93 | 0.93
5,381 | 0.93 | 0.93
5,382 | 0.93 | 0.93
5,383 | 0.93 | 0.93
5,384 | 0.93 | 0.93
5,385 | 0.93 | 0.93
5,386 | 0.93 | 0.93
5,387 | 0.93 | 0.93
5,388 | 0.93 | 0.93
5,389 | 0.93 | 0.93
5,390 | 0.93 | 0.93
5,391 | 0.93 | 0.93
5,392 | 0.93 | 0.93
5,393 | 0.93 | 0.93
5,394 | 0.93 | 0.93
5,395 | 0.93 | 0.93
5,396 | 0.93 | 0.93
5,397 | 0.93 | 0.93
5,398 | 0.93 | 0.93
5,399 | 0.93 | 0.93

5,400	0.93	0.93		5,430	0.93	0.85
5,401	0.93	0.93		5,431	0.93	0.84
5,402	0.93	0.93		5,432	0.93	0.83
5,403	0.93	0.93		5,433	0.93	0.83
5,404	0.93	0.93		5,434	0.93	0.82
5,405	0.93	0.93		5,435	0.93	0.82
5,406	0.93	0.93		5,436	0.93	0.81
5,407	0.93	0.93		5,437	0.93	0.80
5,408	0.93	0.93		5,438	0.93	0.80
5,409	0.93	0.93		5,439	0.93	0.79
5,410	0.93	0.93		5,440	0.93	0.79
5,411	0.93	0.93		5,441	0.93	0.78
5,412	0.93	0.93		5,442	0.93	0.77
5,413	0.93	0.93		5,443	0.93	0.77
5,414	0.93	0.93		5,444	0.93	0.76
5,415	0.93	0.93		5,445	0.93	0.76
5,416	0.93	0.93		5,446	0.93	0.75
5,417	0.93	0.92		5,447	0.93	0.74
5,418	0.93	0.92		5,448	0.93	0.74
5,419	0.93	0.91		5,449	0.93	0.73
5,420	0.93	0.91		5,450	0.93	0.73
5,421	0.93	0.90		5,451	0.93	0.72
5,422	0.93	0.89		5,452	0.93	0.71
5,423	0.93	0.89		5,453	0.93	0.71
5,424	0.93	0.88		5,454	0.93	0.70
5,425	0.93	0.88		5,455	0.93	0.70
5,426	0.93	0.87		5,456	0.93	0.69
5,427	0.93	0.86		5,457	0.93	0.68
5,428	0.93	0.86		5,458	0.93	0.68
5,429	0.93	0.85		5,459	0.93	0.67

5,460 | 0.93 | 0.67
5,461 | 0.93 | 0.66
5,462 | 0.93 | 0.65
5,463 | 0.93 | 0.64
5,464 | 0.93 | 0.63
5,465 | 0.93 | 0.63
5,466 | 0.63 | 0.62
5,467 | 0.63 | 0.61
5,468 | 0.63 | 0.61
5,469 | 0.63 | 0.60
5,470 | 0.63 | 0.60
5,471 | 0.63 | 0.60
5,472 | 0.63 | 0.59
5,473 | 0.63 | 0.59
5,474 | 0.63 | 0.59
5,475 | 0.63 | 0.58
5,476 | 0.63 | 0.58
5,477 | 0.63 | 0.58
5,478 | 0.63 | 0.57
5,479 | 0.63 | 0.57
5,480 | 0.63 | 0.57
5,481 | 0.63 | 0.56
5,482 | 0.63 | 0.56
5,483 | 0.63 | 0.55
5,484 | 0.63 | 0.55
5,485 | 0.63 | 0.55
5,486 | 0.63 | 0.54
5,487 | 0.63 | 0.54
5,488 | 0.63 | 0.54
5,489 | 0.63 | 0.53

5,490 | 0.63 | 0.53
5,491 | 0.63 | 0.53
5,492 | 0.63 | 0.52
5,493 | 0.63 | 0.52
5,494 | 0.63 | 0.51
5,495 | 0.63 | 0.51
5,496 | 0.63 | 0.51
5,497 | 0.63 | 0.50
5,498 | 0.63 | 0.50
5,499 | 0.63 | 0.50
5,500 | 0.63 | 0.49
5,501 | 0.63 | 0.49
5,502 | 0.63 | 0.49
5,503 | 0.63 | 0.48
5,504 | 0.63 | 0.48
5,505 | 0.63 | 0.48
5,506 | 0.63 | 0.47
5,507 | 0.63 | 0.47
5,508 | 0.63 | 0.46
5,509 | 0.63 | 0.46
5,510 | 0.63 | 0.46
5,511 | 0.63 | 0.46
5,512 | 0.45 | 0.45
5,513 | 0.45 | 0.45
5,514 | 0.45 | 0.46
5,515 | 0.45 | 0.46
5,516 | 0.45 | 0.46
5,517 | 0.45 | 0.46
5,518 | 0.45 | 0.46
5,519 | 0.45 | 0.46

5,520	0.45	0.46	5,550	0.45	0.50
5,521	0.45	0.46	5,551	0.45	0.50
5,522	0.45	0.47	5,552	0.45	0.50
5,523	0.45	0.47	5,553	0.45	0.50
5,524	0.45	0.47	5,554	0.45	0.50
5,525	0.45	0.47	5,555	0.45	0.51
5,526	0.45	0.47	5,556	0.45	0.51
5,527	0.45	0.47	5,557	0.45	0.51
5,528	0.45	0.47	5,558	0.45	0.51
5,529	0.45	0.47	5,559	0.51	0.51
5,530	0.45	0.48	5,560	0.51	0.51
5,531	0.45	0.48	5,561	0.51	0.50
5,532	0.45	0.48	5,562	0.51	0.50
5,533	0.45	0.48	5,563	0.51	0.50
5,534	0.45	0.48	5,564	0.51	0.50
5,535	0.45	0.48	5,565	0.51	0.50
5,536	0.45	0.48	5,566	0.51	0.50
5,537	0.45	0.48	5,567	0.51	0.50
5,538	0.45	0.48	5,568	0.51	0.50
5,539	0.45	0.49	5,569	0.51	0.50
5,540	0.45	0.49	5,570	0.51	0.50
5,541	0.45	0.49	5,571	0.51	0.49
5,542	0.45	0.49	5,572	0.51	0.49
5,543	0.45	0.49	5,573	0.51	0.49
5,544	0.45	0.49	5,574	0.51	0.49
5,545	0.45	0.49	5,575	0.51	0.49
5,546	0.45	0.49	5,576	0.51	0.49
5,547	0.45	0.50	5,577	0.51	0.49
5,548	0.45	0.50	5,578	0.51	0.49
5,549	0.45	0.50	5,579	0.51	0.49

5,580	0.51	0.49		5,610	0.46	0.45
5,581	0.51	0.48		5,611	0.46	0.44
5,582	0.51	0.48		5,612	0.46	0.44
5,583	0.51	0.48		5,613	0.46	0.44
5,584	0.51	0.48		5,614	0.46	0.44
5,585	0.51	0.48		5,615	0.46	0.44
5,586	0.51	0.48		5,616	0.46	0.43
5,587	0.51	0.48		5,617	0.46	0.43
5,588	0.51	0.48		5,618	0.46	0.43
5,589	0.51	0.48		5,619	0.46	0.43
5,590	0.51	0.48		5,620	0.46	0.43
5,591	0.51	0.47		5,621	0.46	0.43
5,592	0.51	0.47		5,622	0.46	0.42
5,593	0.51	0.47		5,623	0.46	0.42
5,594	0.51	0.47		5,624	0.46	0.42
5,595	0.51	0.47		5,625	0.46	0.42
5,596	0.51	0.47		5,626	0.46	0.42
5,597	0.51	0.47		5,627	0.46	0.42
5,598	0.51	0.47		5,628	0.46	0.41
5,599	0.51	0.47		5,629	0.46	0.41
5,600	0.51	0.47		5,630	0.46	0.41
5,601	0.51	0.46		5,631	0.46	0.41
5,602	0.51	0.46		5,632	0.46	0.41
5,603	0.51	0.46		5,633	0.46	0.40
5,604	0.51	0.46		5,634	0.46	0.40
5,605	0.46	0.45		5,635	0.46	0.40
5,606	0.46	0.45		5,636	0.46	0.40
5,607	0.46	0.45		5,637	0.46	0.40
5,608	0.46	0.45		5,638	0.46	0.40
5,609	0.46	0.45		5,639	0.46	0.39

5,640 | 0.46 | 0.39
5,641 | 0.46 | 0.39
5,642 | 0.46 | 0.39
5,643 | 0.46 | 0.39
5,644 | 0.46 | 0.38
5,645 | 0.46 | 0.38
5,646 | 0.46 | 0.38
5,647 | 0.46 | 0.38
5,648 | 0.46 | 0.38
5,649 | 0.46 | 0.38
5,650 | 0.46 | 0.37
5,651 | 0.46 | 0.37
5,652 | 0.37 | 0.36
5,653 | 0.37 | 0.36
5,654 | 0.37 | 0.36
5,655 | 0.37 | 0.36
5,656 | 0.37 | 0.36
5,657 | 0.37 | 0.36
5,658 | 0.37 | 0.35
5,659 | 0.37 | 0.35
5,660 | 0.37 | 0.35
5,661 | 0.37 | 0.35
5,662 | 0.37 | 0.35
5,663 | 0.37 | 0.34
5,664 | 0.37 | 0.34
5,665 | 0.37 | 0.34
5,666 | 0.37 | 0.34
5,667 | 0.37 | 0.34
5,668 | 0.37 | 0.34
5,669 | 0.37 | 0.33

5,670 | 0.37 | 0.33
5,671 | 0.37 | 0.33
5,672 | 0.37 | 0.33
5,673 | 0.37 | 0.33
5,674 | 0.37 | 0.33
5,675 | 0.37 | 0.32
5,676 | 0.37 | 0.32
5,677 | 0.37 | 0.32
5,678 | 0.37 | 0.32
5,679 | 0.37 | 0.32
5,680 | 0.37 | 0.31
5,681 | 0.37 | 0.31
5,682 | 0.37 | 0.31
5,683 | 0.37 | 0.31
5,684 | 0.37 | 0.31
5,685 | 0.37 | 0.31
5,686 | 0.37 | 0.30
5,687 | 0.37 | 0.30
5,688 | 0.37 | 0.30
5,689 | 0.37 | 0.30
5,690 | 0.37 | 0.30
5,691 | 0.37 | 0.29
5,692 | 0.37 | 0.29
5,693 | 0.37 | 0.29
5,694 | 0.37 | 0.29
5,695 | 0.37 | 0.29
5,696 | 0.37 | 0.29
5,697 | 0.37 | 0.28
5,698 | 0.37 | 0.28
5,699 | 0.28 | 0.27

5,700	0.28	0.27	5,730	0.28	0.21
5,701	0.28	0.27	5,731	0.28	0.21
5,702	0.28	0.27	5,732	0.28	0.21
5,703	0.28	0.27	5,733	0.28	0.21
5,704	0.28	0.26	5,734	0.28	0.20
5,705	0.28	0.26	5,735	0.28	0.20
5,706	0.28	0.26	5,736	0.28	0.20
5,707	0.28	0.26	5,737	0.28	0.20
5,708	0.28	0.26	5,738	0.28	0.20
5,709	0.28	0.25	5,739	0.28	0.19
5,710	0.28	0.25	5,740	0.28	0.19
5,711	0.28	0.25	5,741	0.28	0.19
5,712	0.28	0.25	5,742	0.28	0.19
5,713	0.28	0.25	5,743	0.28	0.19
5,714	0.28	0.24	5,744	0.28	0.18
5,715	0.28	0.24	5,745	0.28	0.18
5,716	0.28	0.24	5,746	0.18	0.18
5,717	0.28	0.24	5,747	0.18	0.18
5,718	0.28	0.24	5,748	0.18	0.18
5,719	0.28	0.23	5,749	0.18	0.18
5,720	0.28	0.23	5,750	0.18	0.18
5,721	0.28	0.23	5,751	0.18	0.18
5,722	0.28	0.23	5,752	0.18	0.18
5,723	0.28	0.23	5,753	0.18	0.19
5,724	0.28	0.22	5,754	0.18	0.19
5,725	0.28	0.22	5,755	0.18	0.19
5,726	0.28	0.22	5,756	0.18	0.19
5,727	0.28	0.22	5,757	0.18	0.19
5,728	0.28	0.22	5,758	0.18	0.19
5,729	0.28	0.21	5,759	0.18	0.19

5,760	0.18	0.19		5,790	0.18	0.21
5,761	0.18	0.19		5,791	0.18	0.21
5,762	0.18	0.19		5,792	0.18	0.22
5,763	0.18	0.19		5,793	0.18	0.22
5,764	0.18	0.19		5,794	0.21	0.23
5,765	0.18	0.19		5,795	0.21	0.24
5,766	0.18	0.19		5,796	0.21	0.25
5,767	0.18	0.19		5,797	0.21	0.25
5,768	0.18	0.19		5,798	0.21	0.26
5,769	0.18	0.20		5,799	0.21	0.27
5,770	0.18	0.20		5,800	0.21	0.28
5,771	0.18	0.20		5,801	0.21	0.28
5,772	0.18	0.20		5,802	0.21	0.29
5,773	0.18	0.20		5,803	0.21	0.30
5,774	0.18	0.20		5,804	0.21	0.31
5,775	0.18	0.20		5,805	0.21	0.31
5,776	0.18	0.20		5,806	0.21	0.32
5,777	0.18	0.20		5,807	0.21	0.33
5,778	0.18	0.20		5,808	0.21	0.34
5,779	0.18	0.20		5,809	0.21	0.34
5,780	0.18	0.20		5,810	0.21	0.35
5,781	0.18	0.20		5,811	0.21	0.36
5,782	0.18	0.20		5,812	0.21	0.37
5,783	0.18	0.20		5,813	0.21	0.37
5,784	0.18	0.20		5,814	0.21	0.38
5,785	0.18	0.20		5,815	0.21	0.39
5,786	0.18	0.21		5,816	0.21	0.40
5,787	0.18	0.21		5,817	0.21	0.40
5,788	0.18	0.21		5,818	0.21	0.41
5,789	0.18	0.21		5,819	0.21	0.42

5,820 | 0.21 | 0.42
5,821 | 0.21 | 0.43
5,822 | 0.21 | 0.44
5,823 | 0.21 | 0.45
5,824 | 0.21 | 0.45
5,825 | 0.21 | 0.46
5,826 | 0.21 | 0.47
5,827 | 0.21 | 0.48
5,828 | 0.21 | 0.48
5,829 | 0.21 | 0.49
5,830 | 0.21 | 0.50
5,831 | 0.21 | 0.51
5,832 | 0.21 | 0.51
5,833 | 0.21 | 0.52
5,834 | 0.21 | 0.53
5,835 | 0.21 | 0.54
5,836 | 0.21 | 0.54
5,837 | 0.21 | 0.55
5,838 | 0.21 | 0.56
5,839 | 0.21 | 0.58
5,840 | 0.21 | 0.59
5,841 | 0.58 | 0.60
5,842 | 0.58 | 0.61
5,843 | 0.58 | 0.62
5,844 | 0.58 | 0.62
5,845 | 0.58 | 0.63
5,846 | 0.58 | 0.63
5,847 | 0.58 | 0.64
5,848 | 0.58 | 0.65
5,849 | 0.58 | 0.65

5,850 | 0.58 | 0.66
5,851 | 0.58 | 0.66
5,852 | 0.58 | 0.67
5,853 | 0.58 | 0.68
5,854 | 0.58 | 0.68
5,855 | 0.58 | 0.69
5,856 | 0.58 | 0.69
5,857 | 0.58 | 0.70
5,858 | 0.58 | 0.71
5,859 | 0.58 | 0.71
5,860 | 0.58 | 0.72
5,861 | 0.58 | 0.72
5,862 | 0.58 | 0.73
5,863 | 0.58 | 0.74
5,864 | 0.58 | 0.74
5,865 | 0.58 | 0.75
5,866 | 0.58 | 0.75
5,867 | 0.58 | 0.76
5,868 | 0.58 | 0.77
5,869 | 0.58 | 0.77
5,870 | 0.58 | 0.78
5,871 | 0.58 | 0.78
5,872 | 0.58 | 0.79
5,873 | 0.58 | 0.80
5,874 | 0.58 | 0.80
5,875 | 0.58 | 0.81
5,876 | 0.58 | 0.81
5,877 | 0.58 | 0.82
5,878 | 0.58 | 0.83
5,879 | 0.58 | 0.83

5,880	0.58	0.84		5,910	0.88	0.57
5,881	0.58	0.84		5,911	0.88	0.56
5,882	0.58	0.85		5,912	0.88	0.54
5,883	0.58	0.86		5,913	0.88	0.53
5,884	0.58	0.85		5,914	0.88	0.52
5,885	0.58	0.84		5,915	0.88	0.51
5,886	0.58	0.84		5,916	0.88	0.50
5,887	0.88	0.83		5,917	0.88	0.49
5,888	0.88	0.82		5,918	0.88	0.47
5,889	0.88	0.81		5,919	0.88	0.46
5,890	0.88	0.80		5,920	0.88	0.45
5,891	0.88	0.79		5,921	0.88	0.44
5,892	0.88	0.78		5,922	0.88	0.43
5,893	0.88	0.76		5,923	0.88	0.42
5,894	0.88	0.75		5,924	0.88	0.40
5,895	0.88	0.74		5,925	0.88	0.39
5,896	0.88	0.73		5,926	0.88	0.38
5,897	0.88	0.72		5,927	0.88	0.37
5,898	0.88	0.71		5,928	0.88	0.36
5,899	0.88	0.69		5,929	0.88	0.35
5,900	0.88	0.68		5,930	0.88	0.33
5,901	0.88	0.67		5,931	0.88	0.32
5,902	0.88	0.66		5,932	0.88	0.30
5,903	0.88	0.65		5,933	0.30	0.28
5,904	0.88	0.64		5,934	0.30	0.26
5,905	0.88	0.62		5,935	0.30	0.25
5,906	0.88	0.61		5,936	0.30	0.24
5,907	0.88	0.60		5,937	0.30	0.23
5,908	0.88	0.59		5,938	0.30	0.22
5,909	0.88	0.58		5,939	0.30	0.20

5,940	0.30	0.19	5,970	0.30	-0.17
5,941	0.30	0.18	5,971	0.30	-0.18
5,942	0.30	0.17	5,972	0.30	-0.19
5,943	0.30	0.16	5,973	0.30	-0.20
5,944	0.30	0.14	5,974	0.30	-0.22
5,945	0.30	0.13	5,975	0.30	-0.23
5,946	0.30	0.12	5,976	0.30	-0.24
5,947	0.30	0.11	5,977	0.30	-0.25
5,948	0.30	0.10	5,978	0.30	-0.26
5,949	0.30	0.08	5,979	0.30	-0.28
5,950	0.30	0.07	5,980	0.30	-0.28
5,951	0.30	0.06	5,981	-0.30	-0.28
5,952	0.30	0.05	5,982	-0.30	-0.27
5,953	0.30	0.04	5,983	-0.30	-0.26
5,954	0.30	0.02	5,984	-0.30	-0.25
5,955	0.30	0.01	5,985	-0.30	-0.24
5,956	0.30	0.00	5,986	-0.30	-0.23
5,957	0.30	-0.01	5,987	-0.30	-0.22
5,958	0.30	-0.02	5,988	-0.30	-0.21
5,959	0.30	-0.04	5,989	-0.30	-0.20
5,960	0.30	-0.05	5,990	-0.30	-0.19
5,961	0.30	-0.06	5,991	-0.30	-0.18
5,962	0.30	-0.07	5,992	-0.30	-0.16
5,963	0.30	-0.08	5,993	-0.30	-0.15
5,964	0.30	-0.10	5,994	-0.30	-0.14
5,965	0.30	-0.11	5,995	-0.30	-0.13
5,966	0.30	-0.12	5,996	-0.30	-0.12
5,967	0.30	-0.13	5,997	-0.30	-0.11
5,968	0.30	-0.14	5,998	-0.30	-0.10
5,969	0.30	-0.16	5,999	-0.30	-0.09

6,000 | -0.30 | -0.08
6,001 | -0.30 | -0.07
6,002 | -0.30 | -0.06
6,003 | -0.30 | -0.05
6,004 | -0.30 | -0.04
6,005 | -0.30 | -0.03
6,006 | -0.30 | -0.02
6,007 | -0.30 | -0.01
6,008 | -0.30 | 0.00
6,009 | -0.30 | 0.01
6,010 | -0.30 | 0.02
6,011 | -0.30 | 0.03
6,012 | -0.30 | 0.04
6,013 | -0.30 | 0.05
6,014 | -0.30 | 0.06
6,015 | -0.30 | 0.07
6,016 | -0.30 | 0.08
6,017 | -0.30 | 0.10
6,018 | -0.30 | 0.11
6,019 | -0.30 | 0.12
6,020 | -0.30 | 0.13
6,021 | -0.30 | 0.14
6,022 | -0.30 | 0.15
6,023 | -0.30 | 0.16
6,024 | -0.30 | 0.17
6,025 | -0.30 | 0.18
6,026 | -0.30 | 0.19
6,027 | -0.30 | 0.20
6,028 | -0.30 | 0.22
6,029 | 0.22 | 0.24

6,030 | 0.22 | 0.25
6,031 | 0.22 | 0.26
6,032 | 0.22 | 0.27
6,033 | 0.22 | 0.28
6,034 | 0.22 | 0.29
6,035 | 0.22 | 0.30
6,036 | 0.22 | 0.32
6,037 | 0.22 | 0.33
6,038 | 0.22 | 0.34
6,039 | 0.22 | 0.35
6,040 | 0.22 | 0.36
6,041 | 0.22 | 0.37
6,042 | 0.22 | 0.38
6,043 | 0.22 | 0.39
6,044 | 0.22 | 0.40
6,045 | 0.22 | 0.41
6,046 | 0.22 | 0.42
6,047 | 0.22 | 0.43
6,048 | 0.22 | 0.44
6,049 | 0.22 | 0.45
6,050 | 0.22 | 0.46
6,051 | 0.22 | 0.47
6,052 | 0.22 | 0.49
6,053 | 0.22 | 0.50
6,054 | 0.22 | 0.51
6,055 | 0.22 | 0.52
6,056 | 0.22 | 0.53
6,057 | 0.22 | 0.54
6,058 | 0.22 | 0.55
6,059 | 0.22 | 0.56

6,060	0.22	0.57		6,090	0.75	0.58
6,061	0.22	0.58		6,091	0.75	0.57
6,062	0.22	0.59		6,092	0.75	0.56
6,063	0.22	0.60		6,093	0.75	0.55
6,064	0.22	0.61		6,094	0.75	0.54
6,065	0.22	0.62		6,095	0.75	0.53
6,066	0.22	0.63		6,096	0.75	0.52
6,067	0.22	0.64		6,097	0.75	0.51
6,068	0.22	0.65		6,098	0.75	0.51
6,069	0.22	0.67		6,099	0.75	0.50
6,070	0.22	0.68		6,100	0.75	0.49
6,071	0.22	0.69		6,101	0.75	0.48
6,072	0.22	0.70		6,102	0.75	0.47
6,073	0.22	0.71		6,103	0.75	0.46
6,074	0.22	0.71		6,104	0.75	0.45
6,075	0.22	0.71		6,105	0.75	0.44
6,076	0.22	0.71		6,106	0.75	0.43
6,077	0.75	0.71		6,107	0.75	0.42
6,078	0.75	0.70		6,108	0.75	0.41
6,079	0.75	0.69		6,109	0.75	0.40
6,080	0.75	0.68		6,110	0.75	0.39
6,081	0.75	0.67		6,111	0.75	0.38
6,082	0.75	0.66		6,112	0.75	0.37
6,083	0.75	0.65		6,113	0.75	0.36
6,084	0.75	0.64		6,114	0.75	0.35
6,085	0.75	0.63		6,115	0.75	0.34
6,086	0.75	0.62		6,116	0.75	0.33
6,087	0.75	0.61		6,117	0.75	0.32
6,088	0.75	0.60		6,118	0.75	0.31
6,089	0.75	0.59		6,119	0.75	0.30

6,120 | 0.75 | 0.29
6,121 | 0.75 | 0.28
6,122 | 0.75 | 0.26
6,123 | 0.26 | 0.25
6,124 | 0.26 | 0.25
6,125 | 0.26 | 0.24
6,126 | 0.26 | 0.24
6,127 | 0.26 | 0.24
6,128 | 0.26 | 0.23
6,129 | 0.26 | 0.23
6,130 | 0.26 | 0.23
6,131 | 0.26 | 0.22
6,132 | 0.26 | 0.22
6,133 | 0.26 | 0.22
6,134 | 0.26 | 0.22
6,135 | 0.26 | 0.21
6,136 | 0.26 | 0.21
6,137 | 0.26 | 0.21
6,138 | 0.26 | 0.20
6,139 | 0.26 | 0.20
6,140 | 0.26 | 0.20
6,141 | 0.26 | 0.19
6,142 | 0.26 | 0.19
6,143 | 0.26 | 0.19
6,144 | 0.26 | 0.18
6,145 | 0.26 | 0.18
6,146 | 0.26 | 0.18
6,147 | 0.26 | 0.17
6,148 | 0.26 | 0.17
6,149 | 0.26 | 0.17

6,150 | 0.26 | 0.16
6,151 | 0.26 | 0.16
6,152 | 0.26 | 0.16
6,153 | 0.26 | 0.15
6,154 | 0.26 | 0.15
6,155 | 0.26 | 0.15
6,156 | 0.26 | 0.14
6,157 | 0.26 | 0.14
6,158 | 0.26 | 0.14
6,159 | 0.26 | 0.14
6,160 | 0.26 | 0.13
6,161 | 0.26 | 0.13
6,162 | 0.26 | 0.13
6,163 | 0.26 | 0.12
6,164 | 0.26 | 0.12
6,165 | 0.26 | 0.12
6,166 | 0.26 | 0.11
6,167 | 0.26 | 0.11
6,168 | 0.26 | 0.11
6,169 | 0.26 | 0.11
6,170 | 0.10 | 0.12
6,171 | 0.10 | 0.13
6,172 | 0.10 | 0.13
6,173 | 0.10 | 0.14
6,174 | 0.10 | 0.15
6,175 | 0.10 | 0.16
6,176 | 0.10 | 0.17
6,177 | 0.10 | 0.18
6,178 | 0.10 | 0.18
6,179 | 0.10 | 0.19

6,180 | 0.10 | 0.20
6,181 | 0.10 | 0.21
6,182 | 0.10 | 0.22
6,183 | 0.10 | 0.23
6,184 | 0.10 | 0.23
6,185 | 0.10 | 0.24
6,186 | 0.10 | 0.25
6,187 | 0.10 | 0.26
6,188 | 0.10 | 0.27
6,189 | 0.10 | 0.28
6,190 | 0.10 | 0.28
6,191 | 0.10 | 0.29
6,192 | 0.10 | 0.30
6,193 | 0.10 | 0.31
6,194 | 0.10 | 0.32
6,195 | 0.10 | 0.33
6,196 | 0.10 | 0.34
6,197 | 0.10 | 0.34
6,198 | 0.10 | 0.35
6,199 | 0.10 | 0.36
6,200 | 0.10 | 0.37
6,201 | 0.10 | 0.38
6,202 | 0.10 | 0.39
6,203 | 0.10 | 0.39
6,204 | 0.10 | 0.40
6,205 | 0.10 | 0.41
6,206 | 0.10 | 0.42
6,207 | 0.10 | 0.43
6,208 | 0.10 | 0.44
6,209 | 0.10 | 0.44

6,210 | 0.10 | 0.45
6,211 | 0.10 | 0.46
6,212 | 0.10 | 0.47
6,213 | 0.10 | 0.48
6,214 | 0.10 | 0.49
6,215 | 0.10 | 0.49
6,216 | 0.10 | 0.51
6,217 | 0.10 | 0.52
6,218 | 0.52 | 0.53
6,219 | 0.52 | 0.54
6,220 | 0.52 | 0.54
6,221 | 0.52 | 0.54
6,222 | 0.52 | 0.55
6,223 | 0.52 | 0.55
6,224 | 0.52 | 0.55
6,225 | 0.52 | 0.56
6,226 | 0.52 | 0.56
6,227 | 0.52 | 0.57
6,228 | 0.52 | 0.57
6,229 | 0.52 | 0.57
6,230 | 0.52 | 0.58
6,231 | 0.52 | 0.58
6,232 | 0.52 | 0.58
6,233 | 0.52 | 0.59
6,234 | 0.52 | 0.59
6,235 | 0.52 | 0.60
6,236 | 0.52 | 0.60
6,237 | 0.52 | 0.60
6,238 | 0.52 | 0.61
6,239 | 0.52 | 0.61

6,240 | 0.52 | 0.62
6,241 | 0.52 | 0.62
6,242 | 0.52 | 0.62
6,243 | 0.52 | 0.63
6,244 | 0.52 | 0.63
6,245 | 0.52 | 0.63
6,246 | 0.52 | 0.64
6,247 | 0.52 | 0.64
6,248 | 0.52 | 0.65
6,249 | 0.52 | 0.65
6,250 | 0.52 | 0.65
6,251 | 0.52 | 0.66
6,252 | 0.52 | 0.66
6,253 | 0.52 | 0.66
6,254 | 0.52 | 0.67
6,255 | 0.52 | 0.67
6,256 | 0.52 | 0.68
6,257 | 0.52 | 0.68
6,258 | 0.52 | 0.68
6,259 | 0.52 | 0.69
6,260 | 0.52 | 0.69
6,261 | 0.52 | 0.69
6,262 | 0.52 | 0.70
6,263 | 0.52 | 0.69
6,264 | 0.52 | 0.67
6,265 | 0.71 | 0.66
6,266 | 0.71 | 0.64
6,267 | 0.71 | 0.63
6,268 | 0.71 | 0.61
6,269 | 0.71 | 0.59

6,270 | 0.71 | 0.58
6,271 | 0.71 | 0.56
6,272 | 0.71 | 0.54
6,273 | 0.71 | 0.53
6,274 | 0.71 | 0.51
6,275 | 0.71 | 0.49
6,276 | 0.71 | 0.48
6,277 | 0.71 | 0.46
6,278 | 0.71 | 0.44
6,279 | 0.71 | 0.43
6,280 | 0.71 | 0.41
6,281 | 0.71 | 0.39
6,282 | 0.71 | 0.38
6,283 | 0.71 | 0.36
6,284 | 0.71 | 0.34
6,285 | 0.71 | 0.33
6,286 | 0.71 | 0.31
6,287 | 0.71 | 0.30
6,288 | 0.71 | 0.28
6,289 | 0.71 | 0.26
6,290 | 0.71 | 0.25
6,291 | 0.71 | 0.23
6,292 | 0.71 | 0.21
6,293 | 0.71 | 0.20
6,294 | 0.71 | 0.18
6,295 | 0.71 | 0.16
6,296 | 0.71 | 0.15
6,297 | 0.71 | 0.13
6,298 | 0.71 | 0.11
6,299 | 0.71 | 0.10

6,300 | 0.71 | 0.08
6,301 | 0.71 | 0.06
6,302 | 0.71 | 0.05
6,303 | 0.71 | 0.03
6,304 | 0.71 | 0.01
6,305 | 0.71 | 0.00
6,306 | 0.71 | -0.02
6,307 | 0.71 | -0.04
6,308 | 0.71 | -0.05
6,309 | 0.71 | -0.07
6,310 | 0.71 | -0.09
6,311 | 0.71 | -0.10
6,312 | -0.12 | -0.14
6,313 | -0.12 | -0.16
6,314 | -0.12 | -0.18
6,315 | -0.12 | -0.20
6,316 | -0.12 | -0.22
6,317 | -0.12 | -0.24
6,318 | -0.12 | -0.27
6,319 | -0.12 | -0.29
6,320 | -0.12 | -0.31
6,321 | -0.12 | -0.33
6,322 | -0.12 | -0.35
6,323 | -0.12 | -0.37
6,324 | -0.12 | -0.39
6,325 | -0.12 | -0.41
6,326 | -0.12 | -0.43
6,327 | -0.12 | -0.45
6,328 | -0.12 | -0.47
6,329 | -0.12 | -0.49

6,330 | -0.12 | -0.52
6,331 | -0.12 | -0.54
6,332 | -0.12 | -0.56
6,333 | -0.12 | -0.58
6,334 | -0.12 | -0.60
6,335 | -0.12 | -0.62
6,336 | -0.12 | -0.64
6,337 | -0.12 | -0.66
6,338 | -0.12 | -0.68
6,339 | -0.12 | -0.70
6,340 | -0.12 | -0.72
6,341 | -0.12 | -0.74
6,342 | -0.12 | -0.76
6,343 | -0.12 | -0.79
6,344 | -0.12 | -0.81
6,345 | -0.12 | -0.83
6,346 | -0.12 | -0.85
6,347 | -0.12 | -0.87
6,348 | -0.12 | -0.89
6,349 | -0.12 | -0.91
6,350 | -0.12 | -0.93
6,351 | -0.12 | -0.95
6,352 | -0.12 | -0.97
6,353 | -0.12 | -0.99
6,354 | -0.12 | -1.01
6,355 | -0.12 | -1.04
6,356 | -0.12 | -1.06
6,357 | -0.12 | -1.08
6,358 | -0.12 | -1.10
6,359 | -0.12 | -1.12

6,360	-0.12	-1.14
6,361	-1.16	-1.16
6,362	-1.16	-1.16
6,363	-1.16	-1.15
6,364	-1.16	-1.14
6,365	-1.16	-1.13
6,366	-1.16	-1.12
6,367	-1.16	-1.11
6,368	-1.16	-1.10
6,369	-1.16	-1.09
6,370	-1.16	-1.08
6,371	-1.16	-1.07
6,372	-1.16	-1.06
6,373	-1.16	-1.05
6,374	-1.16	-1.04
6,375	-1.16	-1.03
6,376	-1.16	-1.02
6,377	-1.16	-1.01
6,378	-1.16	-1.00
6,379	-1.16	-0.99
6,380	-1.16	-0.98
6,381	-1.16	-0.97
6,382	-1.16	-0.96
6,383	-1.16	-0.95
6,384	-1.16	-0.94
6,385	-1.16	-0.93
6,386	-1.16	-0.92
6,387	-1.16	-0.91
6,388	-1.16	-0.90
6,389	-1.16	-0.89
6,390	-1.16	-0.88
6,391	-1.16	-0.87
6,392	-1.16	-0.86
6,393	-1.16	-0.85
6,394	-1.16	-0.84
6,395	-1.16	-0.83
6,396	-1.16	-0.82
6,397	-1.16	-0.81
6,398	-1.16	-0.80
6,399	-1.16	-0.79
6,400	-1.16	-0.78
6,401	-1.16	-0.77
6,402	-1.16	-0.76
6,403	-1.16	-0.75
6,404	-1.16	-0.74
6,405	-1.16	-0.73
6,406	-1.16	-0.72
6,407	-1.16	-0.71
6,408	-1.16	-0.70
6,409	-1.16	-0.69
6,410	-1.16	-0.68
6,411	-1.16	-0.67
6,412	-0.66	-0.66
6,413	-0.66	-0.64
6,414	-0.66	-0.63
6,415	-0.66	-0.61
6,416	-0.66	-0.60
6,417	-0.66	-0.58
6,418	-0.66	-0.57
6,419	-0.66	-0.55

6,420 | -0.66 | -0.54
6,421 | -0.66 | -0.52
6,422 | -0.66 | -0.50
6,423 | -0.66 | -0.49
6,424 | -0.66 | -0.47
6,425 | -0.66 | -0.46
6,426 | -0.66 | -0.44
6,427 | -0.66 | -0.43
6,428 | -0.66 | -0.41
6,429 | -0.66 | -0.39
6,430 | -0.66 | -0.38
6,431 | -0.66 | -0.36
6,432 | -0.66 | -0.35
6,433 | -0.66 | -0.33
6,434 | -0.66 | -0.32
6,435 | -0.66 | -0.30
6,436 | -0.66 | -0.29
6,437 | -0.66 | -0.27
6,438 | -0.66 | -0.25
6,439 | -0.66 | -0.24
6,440 | -0.66 | -0.22
6,441 | -0.66 | -0.21
6,442 | -0.66 | -0.19
6,443 | -0.66 | -0.18
6,444 | -0.66 | -0.16
6,445 | -0.66 | -0.15
6,446 | -0.66 | -0.13
6,447 | -0.66 | -0.11
6,448 | -0.66 | -0.10
6,449 | -0.66 | -0.08

6,450 | -0.66 | -0.07
6,451 | -0.66 | -0.05
6,452 | -0.66 | -0.04
6,453 | -0.66 | -0.02
6,454 | -0.66 | 0.00
6,455 | -0.66 | 0.01
6,456 | -0.66 | 0.03
6,457 | -0.66 | 0.04
6,458 | -0.66 | 0.06
6,459 | -0.66 | 0.07
6,460 | -0.66 | 0.09
6,461 | -0.66 | 0.10
6,462 | 0.12 | 0.11
6,463 | 0.12 | 0.11
6,464 | 0.12 | 0.10
6,465 | 0.12 | 0.10
6,466 | 0.12 | 0.09
6,467 | 0.12 | 0.08
6,468 | 0.12 | 0.08
6,469 | 0.12 | 0.07
6,470 | 0.12 | 0.06
6,471 | 0.12 | 0.06
6,472 | 0.12 | 0.05
6,473 | 0.12 | 0.05
6,474 | 0.12 | 0.04
6,475 | 0.12 | 0.03
6,476 | 0.12 | 0.03
6,477 | 0.12 | 0.02
6,478 | 0.12 | 0.01
6,479 | 0.12 | 0.01

6,480	0.12	0.00	6,510	0.12	-0.18
6,481	0.12	0.00	6,511	-0.19	-0.19
6,482	0.12	-0.01	6,512	-0.19	-0.19
6,483	0.12	-0.02	6,513	-0.19	-0.19
6,484	0.12	-0.02	6,514	-0.19	-0.19
6,485	0.12	-0.03	6,515	-0.19	-0.19
6,486	0.12	-0.03	6,516	-0.19	-0.19
6,487	0.12	-0.04	6,517	-0.19	-0.19
6,488	0.12	-0.05	6,518	-0.19	-0.19
6,489	0.12	-0.05	6,519	-0.19	-0.20
6,490	0.12	-0.06	6,520	-0.19	-0.20
6,491	0.12	-0.07	6,521	-0.19	-0.20
6,492	0.12	-0.07	6,522	-0.19	-0.20
6,493	0.12	-0.08	6,523	-0.19	-0.20
6,494	0.12	-0.08	6,524	-0.19	-0.20
6,495	0.12	-0.09	6,525	-0.19	-0.20
6,496	0.12	-0.10	6,526	-0.19	-0.20
6,497	0.12	-0.10	6,527	-0.19	-0.20
6,498	0.12	-0.11	6,528	-0.19	-0.20
6,499	0.12	-0.12	6,529	-0.19	-0.20
6,500	0.12	-0.12	6,530	-0.19	-0.20
6,501	0.12	-0.13	6,531	-0.19	-0.20
6,502	0.12	-0.13	6,532	-0.19	-0.20
6,503	0.12	-0.14	6,533	-0.19	-0.20
6,504	0.12	-0.15	6,534	-0.19	-0.20
6,505	0.12	-0.15	6,535	-0.19	-0.21
6,506	0.12	-0.16	6,536	-0.19	-0.21
6,507	0.12	-0.17	6,537	-0.19	-0.21
6,508	0.12	-0.17	6,538	-0.19	-0.21
6,509	0.12	-0.18	6,539	-0.19	-0.21

6,540 | -0.19 | -0.21
6,541 | -0.19 | -0.21
6,542 | -0.19 | -0.21
6,543 | -0.19 | -0.21
6,544 | -0.19 | -0.21
6,545 | -0.19 | -0.21
6,546 | -0.19 | -0.21
6,547 | -0.19 | -0.21
6,548 | -0.19 | -0.21
6,549 | -0.19 | -0.21
6,550 | -0.19 | -0.21
6,551 | -0.19 | -0.21
6,552 | -0.19 | -0.22
6,553 | -0.19 | -0.22
6,554 | -0.19 | -0.22
6,555 | -0.19 | -0.22
6,556 | -0.19 | -0.22
6,557 | -0.19 | -0.22
6,558 | -0.19 | -0.22
6,559 | -0.19 | -0.20
6,560 | -0.22 | -0.18
6,561 | -0.22 | -0.16
6,562 | -0.22 | -0.14
6,563 | -0.22 | -0.12
6,564 | -0.22 | -0.10
6,565 | -0.22 | -0.08
6,566 | -0.22 | -0.06
6,567 | -0.22 | -0.04
6,568 | -0.22 | -0.02
6,569 | -0.22 | 0.00

6,570 | -0.22 | 0.02
6,571 | -0.22 | 0.05
6,572 | -0.22 | 0.07
6,573 | -0.22 | 0.09
6,574 | -0.22 | 0.11
6,575 | -0.22 | 0.13
6,576 | -0.22 | 0.15
6,577 | -0.22 | 0.17
6,578 | -0.22 | 0.19
6,579 | -0.22 | 0.21
6,580 | -0.22 | 0.23
6,581 | -0.22 | 0.25
6,582 | -0.22 | 0.27
6,583 | -0.22 | 0.29
6,584 | -0.22 | 0.31
6,585 | -0.22 | 0.33
6,586 | -0.22 | 0.35
6,587 | -0.22 | 0.37
6,588 | -0.22 | 0.39
6,589 | -0.22 | 0.41
6,590 | -0.22 | 0.43
6,591 | -0.22 | 0.45
6,592 | -0.22 | 0.47
6,593 | -0.22 | 0.49
6,594 | -0.22 | 0.51
6,595 | -0.22 | 0.53
6,596 | -0.22 | 0.56
6,597 | -0.22 | 0.58
6,598 | -0.22 | 0.60
6,599 | -0.22 | 0.62

6,600 | -0.22 | 0.64
6,601 | -0.22 | 0.66
6,602 | -0.22 | 0.68
6,603 | -0.22 | 0.70
6,604 | -0.22 | 0.72
6,605 | -0.22 | 0.74
6,606 | -0.22 | 0.76
6,607 | -0.22 | 0.78
6,608 | 0.80 | 0.80
6,609 | 0.80 | 0.80
6,610 | 0.80 | 0.80
6,611 | 0.80 | 0.80
6,612 | 0.80 | 0.80
6,613 | 0.80 | 0.80
6,614 | 0.80 | 0.80
6,615 | 0.80 | 0.80
6,616 | 0.80 | 0.80
6,617 | 0.80 | 0.80
6,618 | 0.80 | 0.80
6,619 | 0.80 | 0.80
6,620 | 0.80 | 0.80
6,621 | 0.80 | 0.80
6,622 | 0.80 | 0.80
6,623 | 0.80 | 0.80
6,624 | 0.80 | 0.80
6,625 | 0.80 | 0.80
6,626 | 0.80 | 0.80
6,627 | 0.80 | 0.80
6,628 | 0.80 | 0.80
6,629 | 0.80 | 0.80

6,630 | 0.80 | 0.80
6,631 | 0.80 | 0.80
6,632 | 0.80 | 0.80
6,633 | 0.80 | 0.80
6,634 | 0.80 | 0.80
6,635 | 0.80 | 0.80
6,636 | 0.80 | 0.80
6,637 | 0.80 | 0.80
6,638 | 0.80 | 0.80
6,639 | 0.80 | 0.80
6,640 | 0.80 | 0.80
6,641 | 0.80 | 0.80
6,642 | 0.80 | 0.80
6,643 | 0.80 | 0.80
6,644 | 0.80 | 0.80
6,645 | 0.80 | 0.80
6,646 | 0.80 | 0.80
6,647 | 0.80 | 0.80
6,648 | 0.80 | 0.80
6,649 | 0.80 | 0.80
6,650 | 0.80 | 0.80
6,651 | 0.80 | 0.80
6,652 | 0.80 | 0.80
6,653 | 0.80 | 0.79
6,654 | 0.80 | 0.77
6,655 | 0.80 | 0.76
6,656 | 0.80 | 0.74
6,657 | 0.80 | 0.73
6,658 | 0.80 | 0.71
6,659 | 0.80 | 0.70

6,660 | 0.80 | 0.68
6,661 | 0.80 | 0.67
6,662 | 0.80 | 0.65
6,663 | 0.80 | 0.64
6,664 | 0.80 | 0.62
6,665 | 0.80 | 0.61
6,666 | 0.80 | 0.60
6,667 | 0.80 | 0.58
6,668 | 0.80 | 0.57
6,669 | 0.80 | 0.55
6,670 | 0.80 | 0.54
6,671 | 0.80 | 0.52
6,672 | 0.80 | 0.51
6,673 | 0.80 | 0.49
6,674 | 0.80 | 0.48
6,675 | 0.80 | 0.46
6,676 | 0.80 | 0.45
6,677 | 0.80 | 0.44
6,678 | 0.80 | 0.42
6,679 | 0.80 | 0.41
6,680 | 0.80 | 0.39
6,681 | 0.80 | 0.38
6,682 | 0.80 | 0.36
6,683 | 0.80 | 0.35
6,684 | 0.80 | 0.33
6,685 | 0.80 | 0.32
6,686 | 0.80 | 0.30
6,687 | 0.80 | 0.29
6,688 | 0.80 | 0.27
6,689 | 0.80 | 0.26

6,690 | 0.80 | 0.25
6,691 | 0.80 | 0.23
6,692 | 0.80 | 0.22
6,693 | 0.80 | 0.20
6,694 | 0.80 | 0.19
6,695 | 0.80 | 0.17
6,696 | 0.80 | 0.16
6,697 | 0.80 | 0.14
6,698 | 0.80 | 0.13
6,699 | 0.80 | 0.11
6,700 | 0.80 | 0.10
6,701 | 0.80 | 0.07
6,702 | 0.07 | 0.05
6,703 | 0.07 | 0.04
6,704 | 0.07 | 0.03
6,705 | 0.07 | 0.02
6,706 | 0.07 | 0.01
6,707 | 0.07 | 0.00
6,708 | 0.07 | -0.01
6,709 | 0.07 | -0.02
6,710 | 0.07 | -0.03
6,711 | 0.07 | -0.04
6,712 | 0.07 | -0.05
6,713 | 0.07 | -0.05
6,714 | 0.07 | -0.06
6,715 | 0.07 | -0.07
6,716 | 0.07 | -0.08
6,717 | 0.07 | -0.09
6,718 | 0.07 | -0.10
6,719 | 0.07 | -0.11

6,720 | 0.07 | -0.12
6,721 | 0.07 | -0.13
6,722 | 0.07 | -0.14
6,723 | 0.07 | -0.15
6,724 | 0.07 | -0.16
6,725 | 0.07 | -0.17
6,726 | 0.07 | -0.18
6,727 | 0.07 | -0.19
6,728 | 0.07 | -0.20
6,729 | 0.07 | -0.21
6,730 | 0.07 | -0.22
6,731 | 0.07 | -0.23
6,732 | 0.07 | -0.24
6,733 | 0.07 | -0.25
6,734 | 0.07 | -0.26
6,735 | 0.07 | -0.27
6,736 | 0.07 | -0.28
6,737 | 0.07 | -0.29
6,738 | 0.07 | -0.29
6,739 | 0.07 | -0.30
6,740 | 0.07 | -0.31
6,741 | 0.07 | -0.32
6,742 | 0.07 | -0.33
6,743 | 0.07 | -0.34
6,744 | 0.07 | -0.35
6,745 | 0.07 | -0.36
6,746 | 0.07 | -0.37
6,747 | 0.07 | -0.38
6,748 | 0.07 | -0.39
6,749 | 0.07 | -0.40

6,750 | -0.41 | -0.42
6,751 | -0.41 | -0.43
6,752 | -0.41 | -0.44
6,753 | -0.41 | -0.45
6,754 | -0.41 | -0.45
6,755 | -0.41 | -0.46
6,756 | -0.41 | -0.47
6,757 | -0.41 | -0.48
6,758 | -0.41 | -0.49
6,759 | -0.41 | -0.50
6,760 | -0.41 | -0.51
6,761 | -0.41 | -0.52
6,762 | -0.41 | -0.52
6,763 | -0.41 | -0.53
6,764 | -0.41 | -0.54
6,765 | -0.41 | -0.55
6,766 | -0.41 | -0.56
6,767 | -0.41 | -0.57
6,768 | -0.41 | -0.58
6,769 | -0.41 | -0.59
6,770 | -0.41 | -0.59
6,771 | -0.41 | -0.60
6,772 | -0.41 | -0.61
6,773 | -0.41 | -0.62
6,774 | -0.41 | -0.63
6,775 | -0.41 | -0.64
6,776 | -0.41 | -0.65
6,777 | -0.41 | -0.66
6,778 | -0.41 | -0.67
6,779 | -0.41 | -0.67

6,780 | -0.41 | -0.68
6,781 | -0.41 | -0.69
6,782 | -0.41 | -0.70
6,783 | -0.41 | -0.71
6,784 | -0.41 | -0.72
6,785 | -0.41 | -0.73
6,786 | -0.41 | -0.74
6,787 | -0.41 | -0.74
6,788 | -0.41 | -0.75
6,789 | -0.41 | -0.76
6,790 | -0.41 | -0.77
6,791 | -0.41 | -0.78
6,792 | -0.41 | -0.79
6,793 | -0.41 | -0.80
6,794 | -0.41 | -0.81
6,795 | -0.41 | -0.81
6,796 | -0.41 | -0.82
6,797 | -0.41 | -0.83
6,798 | -0.41 | -0.84
6,799 | -0.85 | -0.85
6,800 | -0.85 | -0.85
6,801 | -0.85 | -0.85
6,802 | -0.85 | -0.84
6,803 | -0.85 | -0.84
6,804 | -0.85 | -0.84
6,805 | -0.85 | -0.84
6,806 | -0.85 | -0.83
6,807 | -0.85 | -0.83
6,808 | -0.85 | -0.83
6,809 | -0.85 | -0.82

6,810 | -0.85 | -0.82
6,811 | -0.85 | -0.82
6,812 | -0.85 | -0.82
6,813 | -0.85 | -0.81
6,814 | -0.85 | -0.81
6,815 | -0.85 | -0.81
6,816 | -0.85 | -0.81
6,817 | -0.85 | -0.80
6,818 | -0.85 | -0.80
6,819 | -0.85 | -0.80
6,820 | -0.85 | -0.79
6,821 | -0.85 | -0.79
6,822 | -0.85 | -0.79
6,823 | -0.85 | -0.79
6,824 | -0.85 | -0.78
6,825 | -0.85 | -0.78
6,826 | -0.85 | -0.78
6,827 | -0.85 | -0.77
6,828 | -0.85 | -0.77
6,829 | -0.85 | -0.77
6,830 | -0.85 | -0.77
6,831 | -0.85 | -0.76
6,832 | -0.85 | -0.76
6,833 | -0.85 | -0.76
6,834 | -0.85 | -0.75
6,835 | -0.85 | -0.75
6,836 | -0.85 | -0.75
6,837 | -0.85 | -0.75
6,838 | -0.85 | -0.74
6,839 | -0.85 | -0.74

6,840	-0.85	-0.74	6,870	-0.71	-0.49
6,841	-0.85	-0.74	6,871	-0.71	-0.47
6,842	-0.85	-0.73	6,872	-0.71	-0.46
6,843	-0.85	-0.73	6,873	-0.71	-0.45
6,844	-0.85	-0.73	6,874	-0.71	-0.44
6,845	-0.85	-0.72	6,875	-0.71	-0.43
6,846	-0.85	-0.72	6,876	-0.71	-0.42
6,847	-0.85	-0.72	6,877	-0.71	-0.41
6,848	-0.85	-0.72	6,878	-0.71	-0.40
6,849	-0.85	-0.71	6,879	-0.71	-0.39
6,850	-0.71	-0.71	6,880	-0.71	-0.37
6,851	-0.71	-0.70	6,881	-0.71	-0.36
6,852	-0.71	-0.69	6,882	-0.71	-0.35
6,853	-0.71	-0.68	6,883	-0.71	-0.34
6,854	-0.71	-0.67	6,884	-0.71	-0.33
6,855	-0.71	-0.65	6,885	-0.71	-0.32
6,856	-0.71	-0.64	6,886	-0.71	-0.31
6,857	-0.71	-0.63	6,887	-0.71	-0.30
6,858	-0.71	-0.62	6,888	-0.71	-0.28
6,859	-0.71	-0.61	6,889	-0.71	-0.27
6,860	-0.71	-0.60	6,890	-0.71	-0.26
6,861	-0.71	-0.59	6,891	-0.71	-0.25
6,862	-0.71	-0.58	6,892	-0.71	-0.24
6,863	-0.71	-0.56	6,893	-0.71	-0.23
6,864	-0.71	-0.55	6,894	-0.71	-0.22
6,865	-0.71	-0.54	6,895	-0.71	-0.21
6,866	-0.71	-0.53	6,896	-0.71	-0.19
6,867	-0.71	-0.52	6,897	-0.71	-0.18
6,868	-0.71	-0.51	6,898	-0.71	-0.17
6,869	-0.71	-0.50	6,899	-0.71	-0.16

6,900 | -0.15 | -0.15
6,901 | -0.15 | -0.15
6,902 | -0.15 | -0.15
6,903 | -0.15 | -0.15
6,904 | -0.15 | -0.15
6,905 | -0.15 | -0.15
6,906 | -0.15 | -0.15
6,907 | -0.15 | -0.15
6,908 | -0.15 | -0.15
6,909 | -0.15 | -0.15
6,910 | -0.15 | -0.15
6,911 | -0.15 | -0.15
6,912 | -0.15 | -0.15
6,913 | -0.15 | -0.15
6,914 | -0.15 | -0.15
6,915 | -0.15 | -0.15
6,916 | -0.15 | -0.15
6,917 | -0.15 | -0.15
6,918 | -0.15 | -0.15
6,919 | -0.15 | -0.15
6,920 | -0.15 | -0.15
6,921 | -0.15 | -0.15
6,922 | -0.15 | -0.15
6,923 | -0.15 | -0.15
6,924 | -0.15 | -0.15
6,925 | -0.15 | -0.16
6,926 | -0.15 | -0.16
6,927 | -0.15 | -0.16
6,928 | -0.15 | -0.16
6,929 | -0.15 | -0.16

6,930 | -0.15 | -0.16
6,931 | -0.15 | -0.16
6,932 | -0.15 | -0.16
6,933 | -0.15 | -0.16
6,934 | -0.15 | -0.16
6,935 | -0.15 | -0.16
6,936 | -0.15 | -0.16
6,937 | -0.15 | -0.16
6,938 | -0.15 | -0.16
6,939 | -0.15 | -0.16
6,940 | -0.15 | -0.16
6,941 | -0.15 | -0.16
6,942 | -0.15 | -0.16
6,943 | -0.15 | -0.16
6,944 | -0.15 | -0.16
6,945 | -0.15 | -0.16
6,946 | -0.15 | -0.16
6,947 | -0.15 | -0.16
6,948 | -0.15 | -0.16
6,949 | -0.15 | -0.16
6,950 | -0.16 | -0.17
6,951 | -0.16 | -0.18
6,952 | -0.16 | -0.19
6,953 | -0.16 | -0.20
6,954 | -0.16 | -0.21
6,955 | -0.16 | -0.22
6,956 | -0.16 | -0.22
6,957 | -0.16 | -0.23
6,958 | -0.16 | -0.24
6,959 | -0.16 | -0.25

6,960 | -0.16 | -0.26
6,961 | -0.16 | -0.27
6,962 | -0.16 | -0.28
6,963 | -0.16 | -0.29
6,964 | -0.16 | -0.30
6,965 | -0.16 | -0.31
6,966 | -0.16 | -0.32
6,967 | -0.16 | -0.33
6,968 | -0.16 | -0.33
6,969 | -0.16 | -0.34
6,970 | -0.16 | -0.35
6,971 | -0.16 | -0.36
6,972 | -0.16 | -0.37
6,973 | -0.16 | -0.38
6,974 | -0.16 | -0.39
6,975 | -0.16 | -0.40
6,976 | -0.16 | -0.41
6,977 | -0.16 | -0.42
6,978 | -0.16 | -0.43
6,979 | -0.16 | -0.44
6,980 | -0.16 | -0.45
6,981 | -0.16 | -0.45
6,982 | -0.16 | -0.46
6,983 | -0.16 | -0.47
6,984 | -0.16 | -0.48
6,985 | -0.16 | -0.49
6,986 | -0.16 | -0.50
6,987 | -0.16 | -0.51
6,988 | -0.16 | -0.52
6,989 | -0.16 | -0.53

6,990 | -0.16 | -0.54
6,991 | -0.16 | -0.55
6,992 | -0.16 | -0.56
6,993 | -0.16 | -0.56
6,994 | -0.16 | -0.57
6,995 | -0.16 | -0.58
6,996 | -0.16 | -0.59
6,997 | -0.16 | -0.60
6,998 | -0.16 | -0.61
6,999 | -0.62 | -0.62
7,000 | -0.62 | -0.62
7,001 | -0.62 | -0.62
7,002 | -0.62 | -0.63
7,003 | -0.62 | -0.63
7,004 | -0.62 | -0.63
7,005 | -0.62 | -0.64
7,006 | -0.62 | -0.64
7,007 | -0.62 | -0.64
7,008 | -0.62 | -0.64
7,009 | -0.62 | -0.65
7,010 | -0.62 | -0.65
7,011 | -0.62 | -0.65
7,012 | -0.62 | -0.66
7,013 | -0.62 | -0.66
7,014 | -0.62 | -0.66
7,015 | -0.62 | -0.67
7,016 | -0.62 | -0.67
7,017 | -0.62 | -0.67
7,018 | -0.62 | -0.67
7,019 | -0.62 | -0.68

7,020 | -0.62 | -0.68
7,021 | -0.62 | -0.68
7,022 | -0.62 | -0.69
7,023 | -0.62 | -0.69
7,024 | -0.62 | -0.69
7,025 | -0.62 | -0.70
7,026 | -0.62 | -0.70
7,027 | -0.62 | -0.70
7,028 | -0.62 | -0.70
7,029 | -0.62 | -0.71
7,030 | -0.62 | -0.71
7,031 | -0.62 | -0.71
7,032 | -0.62 | -0.72
7,033 | -0.62 | -0.72
7,034 | -0.62 | -0.72
7,035 | -0.62 | -0.73
7,036 | -0.62 | -0.73
7,037 | -0.62 | -0.73
7,038 | -0.62 | -0.73
7,039 | -0.62 | -0.74
7,040 | -0.62 | -0.74
7,041 | -0.62 | -0.74
7,042 | -0.62 | -0.75
7,043 | -0.62 | -0.75
7,044 | -0.62 | -0.75
7,045 | -0.62 | -0.76
7,046 | -0.62 | -0.76
7,047 | -0.62 | -0.76
7,048 | -0.62 | -0.76
7,049 | -0.62 | -0.77

7,050 | -0.77 | -0.77
7,051 | -0.77 | -0.76
7,052 | -0.77 | -0.75
7,053 | -0.77 | -0.74
7,054 | -0.77 | -0.72
7,055 | -0.77 | -0.71
7,056 | -0.77 | -0.70
7,057 | -0.77 | -0.69
7,058 | -0.77 | -0.68
7,059 | -0.77 | -0.67
7,060 | -0.77 | -0.66
7,061 | -0.77 | -0.64
7,062 | -0.77 | -0.63
7,063 | -0.77 | -0.62
7,064 | -0.77 | -0.61
7,065 | -0.77 | -0.60
7,066 | -0.77 | -0.59
7,067 | -0.77 | -0.58
7,068 | -0.77 | -0.56
7,069 | -0.77 | -0.55
7,070 | -0.77 | -0.54
7,071 | -0.77 | -0.53
7,072 | -0.77 | -0.52
7,073 | -0.77 | -0.51
7,074 | -0.77 | -0.50
7,075 | -0.77 | -0.48
7,076 | -0.77 | -0.47
7,077 | -0.77 | -0.46
7,078 | -0.77 | -0.45
7,079 | -0.77 | -0.44

7,080 | -0.77 | -0.43
7,081 | -0.77 | -0.42
7,082 | -0.77 | -0.41
7,083 | -0.77 | -0.39
7,084 | -0.77 | -0.38
7,085 | -0.77 | -0.37
7,086 | -0.77 | -0.36
7,087 | -0.77 | -0.35
7,088 | -0.77 | -0.34
7,089 | -0.77 | -0.33
7,090 | -0.77 | -0.31
7,091 | -0.77 | -0.30
7,092 | -0.77 | -0.29
7,093 | -0.77 | -0.28
7,094 | -0.77 | -0.27
7,095 | -0.77 | -0.26
7,096 | -0.77 | -0.25
7,097 | -0.77 | -0.23
7,098 | -0.77 | -0.22
7,099 | -0.77 | -0.21
7,100 | -0.20 | -0.19
7,101 | -0.20 | -0.17
7,102 | -0.20 | -0.16
7,103 | -0.20 | -0.15
7,104 | -0.20 | -0.14
7,105 | -0.20 | -0.12
7,106 | -0.20 | -0.11
7,107 | -0.20 | -0.10
7,108 | -0.20 | -0.08
7,109 | -0.20 | -0.07

7,110 | -0.20 | -0.06
7,111 | -0.20 | -0.04
7,112 | -0.20 | -0.03
7,113 | -0.20 | -0.02
7,114 | -0.20 | -0.01
7,115 | -0.20 | 0.01
7,116 | -0.20 | 0.02
7,117 | -0.20 | 0.03
7,118 | -0.20 | 0.05
7,119 | -0.20 | 0.06
7,120 | -0.20 | 0.07
7,121 | -0.20 | 0.09
7,122 | -0.20 | 0.10
7,123 | -0.20 | 0.11
7,124 | -0.20 | 0.13
7,125 | -0.20 | 0.14
7,126 | -0.20 | 0.15
7,127 | -0.20 | 0.16
7,128 | -0.20 | 0.18
7,129 | -0.20 | 0.19
7,130 | -0.20 | 0.20
7,131 | -0.20 | 0.22
7,132 | -0.20 | 0.23
7,133 | -0.20 | 0.24
7,134 | -0.20 | 0.26
7,135 | -0.20 | 0.27
7,136 | -0.20 | 0.28
7,137 | -0.20 | 0.29
7,138 | -0.20 | 0.31
7,139 | -0.20 | 0.32

7,140	-0.20	0.33		7,170	0.45	0.42
7,141	-0.20	0.35		7,171	0.45	0.42
7,142	-0.20	0.36		7,172	0.45	0.42
7,143	-0.20	0.37		7,173	0.45	0.42
7,144	-0.20	0.39		7,174	0.45	0.42
7,145	-0.20	0.40		7,175	0.45	0.42
7,146	-0.20	0.41		7,176	0.45	0.42
7,147	-0.20	0.42		7,177	0.45	0.41
7,148	-0.20	0.44		7,178	0.45	0.41
7,149	0.45	0.45		7,179	0.45	0.41
7,150	0.45	0.45		7,180	0.45	0.41
7,151	0.45	0.45		7,181	0.45	0.41
7,152	0.45	0.44		7,182	0.45	0.41
7,153	0.45	0.44		7,183	0.45	0.41
7,154	0.45	0.44		7,184	0.45	0.41
7,155	0.45	0.44		7,185	0.45	0.40
7,156	0.45	0.44		7,186	0.45	0.40
7,157	0.45	0.44		7,187	0.45	0.40
7,158	0.45	0.44		7,188	0.45	0.40
7,159	0.45	0.44		7,189	0.45	0.40
7,160	0.45	0.43		7,190	0.45	0.40
7,161	0.45	0.43		7,191	0.45	0.40
7,162	0.45	0.43		7,192	0.45	0.40
7,163	0.45	0.43		7,193	0.45	0.39
7,164	0.45	0.43		7,194	0.45	0.39
7,165	0.45	0.43		7,195	0.45	0.39
7,166	0.45	0.43		7,196	0.45	0.39
7,167	0.45	0.43		7,197	0.39	0.39
7,168	0.45	0.42		7,198	0.39	0.39
7,169	0.45	0.42		7,199	0.39	0.38

7,200 | 0.39 | 0.38
7,201 | 0.39 | 0.38
7,202 | 0.39 | 0.38
7,203 | 0.39 | 0.38
7,204 | 0.39 | 0.38
7,205 | 0.39 | 0.38
7,206 | 0.39 | 0.37
7,207 | 0.39 | 0.37
7,208 | 0.39 | 0.37
7,209 | 0.39 | 0.37
7,210 | 0.39 | 0.37
7,211 | 0.39 | 0.37
7,212 | 0.39 | 0.37
7,213 | 0.39 | 0.36
7,214 | 0.39 | 0.36
7,215 | 0.39 | 0.36
7,216 | 0.39 | 0.36
7,217 | 0.39 | 0.36
7,218 | 0.39 | 0.36
7,219 | 0.39 | 0.36
7,220 | 0.39 | 0.36
7,221 | 0.39 | 0.35
7,222 | 0.39 | 0.35
7,223 | 0.39 | 0.35
7,224 | 0.39 | 0.35
7,225 | 0.39 | 0.35
7,226 | 0.39 | 0.35
7,227 | 0.39 | 0.35
7,228 | 0.39 | 0.34
7,229 | 0.39 | 0.34

7,230 | 0.39 | 0.34
7,231 | 0.39 | 0.34
7,232 | 0.39 | 0.34
7,233 | 0.39 | 0.34
7,234 | 0.39 | 0.34
7,235 | 0.39 | 0.33
7,236 | 0.39 | 0.33
7,237 | 0.39 | 0.33
7,238 | 0.39 | 0.33
7,239 | 0.39 | 0.33
7,240 | 0.39 | 0.33
7,241 | 0.39 | 0.33
7,242 | 0.39 | 0.32
7,243 | 0.39 | 0.32
7,244 | 0.39 | 0.31
7,245 | 0.32 | 0.31
7,246 | 0.32 | 0.31
7,247 | 0.32 | 0.30
7,248 | 0.32 | 0.30
7,249 | 0.32 | 0.29
7,250 | 0.32 | 0.29
7,251 | 0.32 | 0.29
7,252 | 0.32 | 0.28
7,253 | 0.32 | 0.28
7,254 | 0.32 | 0.28
7,255 | 0.32 | 0.27
7,256 | 0.32 | 0.27
7,257 | 0.32 | 0.27
7,258 | 0.32 | 0.26
7,259 | 0.32 | 0.26

7,260 | 0.32 | 0.26
7,261 | 0.32 | 0.25
7,262 | 0.32 | 0.25
7,263 | 0.32 | 0.24
7,264 | 0.32 | 0.24
7,265 | 0.32 | 0.24
7,266 | 0.32 | 0.23
7,267 | 0.32 | 0.23
7,268 | 0.32 | 0.23
7,269 | 0.32 | 0.22
7,270 | 0.32 | 0.22
7,271 | 0.32 | 0.22
7,272 | 0.32 | 0.21
7,273 | 0.32 | 0.21
7,274 | 0.32 | 0.20
7,275 | 0.32 | 0.20
7,276 | 0.32 | 0.20
7,277 | 0.32 | 0.19
7,278 | 0.32 | 0.19
7,279 | 0.32 | 0.19
7,280 | 0.32 | 0.18
7,281 | 0.32 | 0.18
7,282 | 0.32 | 0.18
7,283 | 0.32 | 0.17
7,284 | 0.32 | 0.17
7,285 | 0.32 | 0.17
7,286 | 0.32 | 0.16
7,287 | 0.32 | 0.16
7,288 | 0.32 | 0.15
7,289 | 0.32 | 0.15

7,290 | 0.32 | 0.15
7,291 | 0.32 | 0.14
7,292 | 0.14 | 0.13
7,293 | 0.14 | 0.12
7,294 | 0.14 | 0.11
7,295 | 0.14 | 0.10
7,296 | 0.14 | 0.09
7,297 | 0.14 | 0.08
7,298 | 0.14 | 0.06
7,299 | 0.14 | 0.05
7,300 | 0.14 | 0.04
7,301 | 0.14 | 0.03
7,302 | 0.14 | 0.02
7,303 | 0.14 | 0.01
7,304 | 0.14 | 0.00
7,305 | 0.14 | -0.01
7,306 | 0.14 | -0.02
7,307 | 0.14 | -0.03
7,308 | 0.14 | -0.04
7,309 | 0.14 | -0.05
7,310 | 0.14 | -0.06
7,311 | 0.14 | -0.07
7,312 | 0.14 | -0.08
7,313 | 0.14 | -0.09
7,314 | 0.14 | -0.11
7,315 | 0.14 | -0.12
7,316 | 0.14 | -0.13
7,317 | 0.14 | -0.14
7,318 | 0.14 | -0.15
7,319 | 0.14 | -0.16

7,320	0.14	-0.17	7,350	-0.39	-0.34
7,321	0.14	-0.18	7,351	-0.39	-0.34
7,322	0.14	-0.19	7,352	-0.39	-0.33
7,323	0.14	-0.20	7,353	-0.39	-0.33
7,324	0.14	-0.21	7,354	-0.39	-0.33
7,325	0.14	-0.22	7,355	-0.39	-0.32
7,326	0.14	-0.23	7,356	-0.39	-0.32
7,327	0.14	-0.24	7,357	-0.39	-0.31
7,328	0.14	-0.25	7,358	-0.39	-0.31
7,329	0.14	-0.26	7,359	-0.39	-0.30
7,330	0.14	-0.27	7,360	-0.39	-0.30
7,331	0.14	-0.29	7,361	-0.39	-0.29
7,332	0.14	-0.30	7,362	-0.39	-0.29
7,333	0.14	-0.31	7,363	-0.39	-0.28
7,334	0.14	-0.32	7,364	-0.39	-0.28
7,335	0.13	-0.33	7,365	-0.39	-0.28
7,336	0.13	-0.34	7,366	-0.39	-0.27
7,337	0.13	-0.35	7,367	-0.39	-0.27
7,338	0.13	-0.36	7,368	-0.39	-0.26
7,339	0.13	-0.37	7,369	-0.39	-0.26
7,340	0.13	-0.38	7,370	-0.39	-0.25
7,341	-0.39	-0.39	7,371	-0.39	-0.25
7,342	-0.39	-0.38	7,372	-0.39	-0.24
7,343	-0.39	-0.38	7,373	-0.39	-0.24
7,344	-0.39	-0.37	7,374	-0.39	-0.23
7,345	-0.39	-0.37	7,375	-0.39	-0.23
7,346	-0.39	-0.36	7,376	-0.39	-0.22
7,347	-0.39	-0.36	7,377	-0.39	-0.22
7,348	-0.39	-0.35	7,378	-0.39	-0.22
7,349	-0.39	-0.35	7,379	-0.39	-0.21

7,380 | -0.39 | -0.21
7,381 | -0.39 | -0.20
7,382 | -0.39 | -0.20
7,383 | -0.39 | -0.19
7,384 | -0.39 | -0.19
7,385 | -0.39 | -0.18
7,386 | -0.39 | -0.18
7,387 | -0.39 | -0.17
7,388 | -0.39 | -0.17
7,389 | -0.39 | -0.16
7,390 | -0.16 | -0.15
7,391 | -0.16 | -0.14
7,392 | -0.16 | -0.12
7,393 | -0.16 | -0.11
7,394 | -0.16 | -0.10
7,395 | -0.16 | -0.09
7,396 | -0.16 | -0.08
7,397 | -0.16 | -0.06
7,398 | -0.16 | -0.05
7,399 | -0.16 | -0.04
7,400 | -0.16 | -0.03
7,401 | -0.16 | -0.02
7,402 | -0.16 | 0.00
7,403 | -0.16 | 0.01
7,404 | -0.16 | 0.02
7,405 | -0.16 | 0.03
7,406 | -0.16 | 0.04
7,407 | -0.16 | 0.06
7,408 | -0.16 | 0.07
7,409 | -0.16 | 0.08

7,410 | -0.16 | 0.09
7,411 | -0.16 | 0.10
7,412 | -0.16 | 0.12
7,413 | -0.16 | 0.13
7,414 | -0.16 | 0.14
7,415 | -0.16 | 0.15
7,416 | -0.16 | 0.16
7,417 | -0.16 | 0.18
7,418 | -0.16 | 0.19
7,419 | -0.16 | 0.20
7,420 | -0.16 | 0.21
7,421 | -0.16 | 0.22
7,422 | -0.16 | 0.24
7,423 | -0.16 | 0.25
7,424 | -0.16 | 0.26
7,425 | -0.16 | 0.27
7,426 | -0.16 | 0.28
7,427 | -0.16 | 0.30
7,428 | -0.16 | 0.31
7,429 | -0.16 | 0.32
7,430 | -0.16 | 0.33
7,431 | -0.16 | 0.34
7,432 | -0.16 | 0.36
7,433 | -0.16 | 0.37
7,434 | -0.16 | 0.38
7,435 | -0.16 | 0.39
7,436 | -0.16 | 0.40
7,437 | -0.16 | 0.42
7,438 | -0.16 | 0.43
7,439 | 0.44 | 0.44

7,440	0.44	0.44	7,470	0.44	0.41
7,441	0.44	0.44	7,471	0.44	0.41
7,442	0.44	0.44	7,472	0.44	0.41
7,443	0.44	0.43	7,473	0.44	0.40
7,444	0.44	0.43	7,474	0.44	0.40
7,445	0.44	0.43	7,475	0.44	0.40
7,446	0.44	0.43	7,476	0.44	0.40
7,447	0.44	0.43	7,477	0.44	0.40
7,448	0.44	0.43	7,478	0.44	0.40
7,449	0.44	0.43	7,479	0.44	0.40
7,450	0.44	0.43	7,480	0.44	0.40
7,451	0.44	0.43	7,481	0.44	0.40
7,452	0.44	0.43	7,482	0.44	0.40
7,453	0.44	0.42	7,483	0.44	0.39
7,454	0.44	0.42	7,484	0.44	0.40
7,455	0.44	0.42	7,485	0.44	0.41
7,456	0.44	0.42	7,486	0.44	0.42
7,457	0.44	0.42	7,487	0.39	0.43
7,458	0.44	0.42	7,488	0.39	0.44
7,459	0.44	0.42	7,489	0.39	0.45
7,460	0.44	0.42	7,490	0.39	0.46
7,461	0.44	0.42	7,491	0.39	0.47
7,462	0.44	0.42	7,492	0.39	0.48
7,463	0.44	0.41	7,493	0.39	0.49
7,464	0.44	0.41	7,494	0.39	0.50
7,465	0.44	0.41	7,495	0.39	0.51
7,466	0.44	0.41	7,496	0.39	0.52
7,467	0.44	0.41	7,497	0.39	0.53
7,468	0.44	0.41	7,498	0.39	0.54
7,469	0.44	0.41	7,499	0.39	0.55

7,500 | 0.39 | 0.56
7,501 | 0.39 | 0.57
7,502 | 0.39 | 0.58
7,503 | 0.39 | 0.59
7,504 | 0.39 | 0.60
7,505 | 0.39 | 0.61
7,506 | 0.39 | 0.62
7,507 | 0.39 | 0.63
7,508 | 0.39 | 0.65
7,509 | 0.39 | 0.66
7,510 | 0.39 | 0.67
7,511 | 0.39 | 0.68
7,512 | 0.39 | 0.69
7,513 | 0.39 | 0.70
7,514 | 0.39 | 0.71
7,515 | 0.39 | 0.72
7,516 | 0.39 | 0.73
7,517 | 0.39 | 0.74
7,518 | 0.39 | 0.75
7,519 | 0.39 | 0.76
7,520 | 0.39 | 0.77
7,521 | 0.39 | 0.78
7,522 | 0.39 | 0.79
7,523 | 0.39 | 0.80
7,524 | 0.39 | 0.81
7,525 | 0.39 | 0.82
7,526 | 0.39 | 0.83
7,527 | 0.39 | 0.84
7,528 | 0.39 | 0.85
7,529 | 0.39 | 0.86

7,530 | 0.39 | 0.87
7,531 | 0.39 | 0.87
7,532 | 0.39 | 0.87
7,533 | 0.90 | 0.87
7,534 | 0.90 | 0.86
7,535 | 0.90 | 0.85
7,536 | 0.90 | 0.84
7,537 | 0.90 | 0.83
7,538 | 0.90 | 0.82
7,539 | 0.90 | 0.81
7,540 | 0.90 | 0.80
7,541 | 0.90 | 0.79
7,542 | 0.90 | 0.78
7,543 | 0.90 | 0.77
7,544 | 0.90 | 0.76
7,545 | 0.90 | 0.75
7,546 | 0.90 | 0.74
7,547 | 0.90 | 0.73
7,548 | 0.90 | 0.72
7,549 | 0.90 | 0.71
7,550 | 0.90 | 0.70
7,551 | 0.90 | 0.69
7,552 | 0.90 | 0.68
7,553 | 0.90 | 0.67
7,554 | 0.90 | 0.66
7,555 | 0.90 | 0.65
7,556 | 0.90 | 0.64
7,557 | 0.90 | 0.63
7,558 | 0.90 | 0.62
7,559 | 0.90 | 0.61

7,560 | 0.90 | 0.60   7,590 | 0.40 | 0.41
7,561 | 0.90 | 0.59   7,591 | 0.40 | 0.41
7,562 | 0.90 | 0.58   7,592 | 0.40 | 0.41
7,563 | 0.90 | 0.57   7,593 | 0.40 | 0.41
7,564 | 0.90 | 0.56   7,594 | 0.40 | 0.41
7,565 | 0.90 | 0.55   7,595 | 0.40 | 0.41
7,566 | 0.90 | 0.54   7,596 | 0.40 | 0.42
7,567 | 0.90 | 0.53   7,597 | 0.40 | 0.42
7,568 | 0.90 | 0.52   7,598 | 0.40 | 0.42
7,569 | 0.90 | 0.51   7,599 | 0.40 | 0.42
7,570 | 0.90 | 0.50   7,600 | 0.40 | 0.42
7,571 | 0.90 | 0.49   7,601 | 0.40 | 0.42
7,572 | 0.90 | 0.48   7,602 | 0.40 | 0.42
7,573 | 0.90 | 0.47   7,603 | 0.40 | 0.42
7,574 | 0.90 | 0.46   7,604 | 0.40 | 0.42
7,575 | 0.90 | 0.45   7,605 | 0.40 | 0.42
7,576 | 0.90 | 0.44   7,606 | 0.40 | 0.42
7,577 | 0.90 | 0.43   7,607 | 0.40 | 0.42
7,578 | 0.90 | 0.42   7,608 | 0.40 | 0.42
7,579 | 0.90 | 0.41   7,609 | 0.40 | 0.43
7,580 | 0.40 | 0.40   7,610 | 0.40 | 0.43
7,581 | 0.40 | 0.40   7,611 | 0.40 | 0.43
7,582 | 0.40 | 0.40   7,612 | 0.40 | 0.43
7,583 | 0.40 | 0.40   7,613 | 0.40 | 0.43
7,584 | 0.40 | 0.41   7,614 | 0.40 | 0.43
7,585 | 0.40 | 0.41   7,615 | 0.40 | 0.43
7,586 | 0.40 | 0.41   7,616 | 0.40 | 0.43
7,587 | 0.40 | 0.41   7,617 | 0.40 | 0.43
7,588 | 0.40 | 0.41   7,618 | 0.40 | 0.43
7,589 | 0.40 | 0.41   7,619 | 0.40 | 0.43

7,620 | 0.40 | 0.43
7,621 | 0.40 | 0.44
7,622 | 0.40 | 0.44
7,623 | 0.40 | 0.44
7,624 | 0.40 | 0.44
7,625 | 0.40 | 0.44
7,626 | 0.40 | 0.44
7,627 | 0.44 | 0.44
7,628 | 0.44 | 0.44
7,629 | 0.44 | 0.43
7,630 | 0.44 | 0.43
7,631 | 0.44 | 0.43
7,632 | 0.44 | 0.43
7,633 | 0.44 | 0.43
7,634 | 0.44 | 0.43
7,635 | 0.44 | 0.43
7,636 | 0.44 | 0.43
7,637 | 0.44 | 0.42
7,638 | 0.44 | 0.42
7,639 | 0.44 | 0.42
7,640 | 0.44 | 0.42
7,641 | 0.44 | 0.42
7,642 | 0.44 | 0.42
7,643 | 0.44 | 0.42
7,644 | 0.44 | 0.42
7,645 | 0.44 | 0.41
7,646 | 0.44 | 0.41
7,647 | 0.44 | 0.41
7,648 | 0.44 | 0.41
7,649 | 0.44 | 0.41

7,650 | 0.44 | 0.41
7,651 | 0.44 | 0.41
7,652 | 0.44 | 0.41
7,653 | 0.44 | 0.41
7,654 | 0.44 | 0.40
7,655 | 0.44 | 0.40
7,656 | 0.44 | 0.40
7,657 | 0.44 | 0.40
7,658 | 0.44 | 0.40
7,659 | 0.44 | 0.40
7,660 | 0.44 | 0.40
7,661 | 0.44 | 0.40
7,662 | 0.44 | 0.39
7,663 | 0.44 | 0.39
7,664 | 0.44 | 0.39
7,665 | 0.44 | 0.39
7,666 | 0.44 | 0.39
7,667 | 0.44 | 0.39
7,668 | 0.44 | 0.39
7,669 | 0.44 | 0.39
7,670 | 0.44 | 0.38
7,671 | 0.44 | 0.38
7,672 | 0.44 | 0.38
7,673 | 0.44 | 0.37
7,674 | 0.38 | 0.35
7,675 | 0.38 | 0.34
7,676 | 0.38 | 0.33
7,677 | 0.38 | 0.31
7,678 | 0.38 | 0.30
7,679 | 0.38 | 0.28

7,680 | 0.38 | 0.27

7,681 | 0.38 | 0.26

7,682 | 0.38 | 0.24

7,683 | 0.38 | 0.23

7,684 | 0.38 | 0.22

7,685 | 0.38 | 0.20

7,686 | 0.38 | 0.19

7,687 | 0.38 | 0.18

7,688 | 0.38 | 0.16

7,689 | 0.38 | 0.15

7,690 | 0.38 | 0.14

7,691 | 0.38 | 0.12

7,692 | 0.38 | 0.11

7,693 | 0.38 | 0.09

7,694 | 0.38 | 0.08

7,695 | 0.38 | 0.07

7,696 | 0.38 | 0.05

7,697 | 0.38 | 0.04

7,698 | 0.38 | 0.03

7,699 | 0.38 | 0.01

7,700 | 0.38 | 0.00

7,701 | 0.38 | -0.01

7,702 | 0.38 | -0.03

7,703 | 0.38 | -0.04

7,704 | 0.38 | -0.06

7,705 | 0.38 | -0.07

7,706 | 0.38 | -0.08

7,707 | 0.38 | -0.10

7,708 | 0.38 | -0.11

7,709 | 0.38 | -0.12

7,710 | 0.38 | -0.14

7,711 | 0.38 | -0.15

7,712 | 0.38 | -0.16

7,713 | 0.38 | -0.18

7,714 | 0.38 | -0.19

7,715 | 0.38 | -0.20

7,716 | 0.38 | -0.22

7,717 | 0.38 | -0.23

7,718 | 0.38 | -0.25

7,719 | 0.38 | -0.26

7,720 | 0.38 | -0.27

7,721 | 0.38 | -0.29

7,722 | -0.30 | -0.30

7,723 | -0.30 | -0.29

7,724 | -0.30 | -0.29

7,725 | -0.30 | -0.29

7,726 | -0.30 | -0.29

7,727 | -0.30 | -0.28

7,728 | -0.30 | -0.28

7,729 | -0.30 | -0.28

7,730 | -0.30 | -0.27

7,731 | -0.30 | -0.27

7,732 | -0.30 | -0.27

7,733 | -0.30 | -0.26

7,734 | -0.30 | -0.26

7,735 | -0.30 | -0.26

7,736 | -0.30 | -0.26

7,737 | -0.30 | -0.25

7,738 | -0.30 | -0.25

7,739 | -0.30 | -0.25

7,740	-0.30	-0.24	7,770	-0.30	-0.15
7,741	-0.30	-0.24	7,771	-0.15	-0.16
7,742	-0.30	-0.24	7,772	-0.15	-0.16
7,743	-0.30	-0.23	7,773	-0.15	-0.17
7,744	-0.30	-0.23	7,774	-0.15	-0.18
7,745	-0.30	-0.23	7,775	-0.15	-0.18
7,746	-0.30	-0.23	7,776	-0.15	-0.19
7,747	-0.30	-0.22	7,777	-0.15	-0.20
7,748	-0.30	-0.22	7,778	-0.15	-0.20
7,749	-0.30	-0.22	7,779	-0.15	-0.21
7,750	-0.30	-0.21	7,780	-0.15	-0.22
7,751	-0.30	-0.21	7,781	-0.15	-0.22
7,752	-0.30	-0.21	7,782	-0.15	-0.23
7,753	-0.30	-0.20	7,783	-0.15	-0.24
7,754	-0.30	-0.20	7,784	-0.15	-0.24
7,755	-0.30	-0.20	7,785	-0.15	-0.25
7,756	-0.30	-0.20	7,786	-0.15	-0.26
7,757	-0.30	-0.19	7,787	-0.15	-0.26
7,758	-0.30	-0.19	7,788	-0.15	-0.27
7,759	-0.30	-0.19	7,789	-0.15	-0.28
7,760	-0.30	-0.18	7,790	-0.15	-0.28
7,761	-0.30	-0.18	7,791	-0.15	-0.29
7,762	-0.30	-0.18	7,792	-0.15	-0.30
7,763	-0.30	-0.17	7,793	-0.15	-0.30
7,764	-0.30	-0.17	7,794	-0.15	-0.31
7,765	-0.30	-0.17	7,795	-0.15	-0.32
7,766	-0.30	-0.17	7,796	-0.15	-0.32
7,767	-0.30	-0.16	7,797	-0.15	-0.33
7,768	-0.30	-0.16	7,798	-0.15	-0.33
7,769	-0.30	-0.16	7,799	-0.15	-0.34

7,800	-0.15	-0.35	7,830	-0.48	-0.49
7,801	-0.15	-0.35	7,831	-0.48	-0.49
7,802	-0.15	-0.36	7,832	-0.48	-0.49
7,803	-0.15	-0.37	7,833	-0.48	-0.49
7,804	-0.15	-0.37	7,834	-0.48	-0.49
7,805	-0.15	-0.38	7,835	-0.48	-0.49
7,806	-0.15	-0.39	7,836	-0.48	-0.49
7,807	-0.15	-0.39	7,837	-0.48	-0.49
7,808	-0.15	-0.40	7,838	-0.48	-0.49
7,809	-0.15	-0.41	7,839	-0.48	-0.49
7,810	-0.15	-0.41	7,840	-0.48	-0.49
7,811	-0.15	-0.42	7,841	-0.48	-0.49
7,812	-0.15	-0.43	7,842	-0.48	-0.49
7,813	-0.15	-0.43	7,843	-0.48	-0.49
7,814	-0.15	-0.44	7,844	-0.48	-0.49
7,815	-0.15	-0.45	7,845	-0.48	-0.50
7,816	-0.15	-0.45	7,846	-0.48	-0.50
7,817	-0.15	-0.46	7,847	-0.48	-0.50
7,818	-0.15	-0.47	7,848	-0.48	-0.50
7,819	-0.15	-0.47	7,849	-0.48	-0.50
7,820	-0.48	-0.48	7,850	-0.48	-0.50
7,821	-0.48	-0.48	7,851	-0.48	-0.50
7,822	-0.48	-0.48	7,852	-0.48	-0.50
7,823	-0.48	-0.48	7,853	-0.48	-0.50
7,824	-0.48	-0.48	7,854	-0.48	-0.50
7,825	-0.48	-0.48	7,855	-0.48	-0.50
7,826	-0.48	-0.48	7,856	-0.48	-0.50
7,827	-0.48	-0.48	7,857	-0.48	-0.50
7,828	-0.48	-0.48	7,858	-0.48	-0.50
7,829	-0.48	-0.49	7,859	-0.48	-0.50

7,860	-0.48	-0.50
7,861	-0.48	-0.50
7,862	-0.48	-0.51
7,863	-0.48	-0.51
7,864	-0.48	-0.51
7,865	-0.48	-0.51
7,866	-0.48	-0.51
7,867	-0.48	-0.51
7,868	-0.48	-0.51
7,869	-0.48	-0.51
7,870	-0.51	-0.51
7,871	-0.51	-0.51
7,872	-0.51	-0.50
7,873	-0.51	-0.50
7,874	-0.51	-0.49
7,875	-0.51	-0.49
7,876	-0.51	-0.48
7,877	-0.51	-0.48
7,878	-0.51	-0.47
7,879	-0.51	-0.47
7,880	-0.51	-0.46
7,881	-0.51	-0.46
7,882	-0.51	-0.45
7,883	-0.51	-0.45
7,884	-0.51	-0.44
7,885	-0.51	-0.44
7,886	-0.51	-0.43
7,887	-0.51	-0.43
7,888	-0.51	-0.42
7,889	-0.51	-0.42
7,890	-0.51	-0.41
7,891	-0.51	-0.41
7,892	-0.51	-0.40
7,893	-0.51	-0.40
7,894	-0.51	-0.39
7,895	-0.51	-0.39
7,896	-0.51	-0.39
7,897	-0.51	-0.38
7,898	-0.51	-0.38
7,899	-0.51	-0.37
7,900	-0.51	-0.37
7,901	-0.51	-0.36
7,902	-0.51	-0.36
7,903	-0.51	-0.35
7,904	-0.51	-0.35
7,905	-0.51	-0.34
7,906	-0.51	-0.34
7,907	-0.51	-0.33
7,908	-0.51	-0.33
7,909	-0.51	-0.32
7,910	-0.51	-0.32
7,911	-0.51	-0.31
7,912	-0.51	-0.31
7,913	-0.51	-0.30
7,914	-0.51	-0.30
7,915	-0.51	-0.29
7,916	-0.51	-0.29
7,917	-0.51	-0.28
7,918	-0.51	-0.28
7,919	-0.51	-0.27

7,920 | -0.27 | -0.27
7,921 | -0.27 | -0.27
7,922 | -0.27 | -0.26
7,923 | -0.27 | -0.26
7,924 | -0.27 | -0.26
7,925 | -0.27 | -0.26
7,926 | -0.27 | -0.25
7,927 | -0.27 | -0.25
7,928 | -0.27 | -0.25
7,929 | -0.27 | -0.24
7,930 | -0.27 | -0.24
7,931 | -0.27 | -0.24
7,932 | -0.27 | -0.23
7,933 | -0.27 | -0.23
7,934 | -0.27 | -0.23
7,935 | -0.27 | -0.23
7,936 | -0.27 | -0.22
7,937 | -0.27 | -0.22
7,938 | -0.27 | -0.22
7,939 | -0.27 | -0.21
7,940 | -0.27 | -0.21
7,941 | -0.27 | -0.21
7,942 | -0.27 | -0.20
7,943 | -0.27 | -0.20
7,944 | -0.27 | -0.20
7,945 | -0.27 | -0.20
7,946 | -0.27 | -0.19
7,947 | -0.27 | -0.19
7,948 | -0.27 | -0.19
7,949 | -0.27 | -0.18

7,950 | -0.27 | -0.18
7,951 | -0.27 | -0.18
7,952 | -0.27 | -0.17
7,953 | -0.27 | -0.17
7,954 | -0.27 | -0.17
7,955 | -0.27 | -0.17
7,956 | -0.27 | -0.16
7,957 | -0.27 | -0.16
7,958 | -0.27 | -0.16
7,959 | -0.27 | -0.15
7,960 | -0.27 | -0.15
7,961 | -0.27 | -0.15
7,962 | -0.27 | -0.14
7,963 | -0.27 | -0.14
7,964 | -0.27 | -0.14
7,965 | -0.27 | -0.14
7,966 | -0.27 | -0.13
7,967 | -0.27 | -0.13
7,968 | -0.27 | -0.13
7,969 | -0.27 | -0.12
7,970 | -0.12 | -0.12
7,971 | -0.12 | -0.12
7,972 | -0.12 | -0.12
7,973 | -0.12 | -0.13
7,974 | -0.12 | -0.13
7,975 | -0.12 | -0.13
7,976 | -0.12 | -0.13
7,977 | -0.12 | -0.13
7,978 | -0.12 | -0.14
7,979 | -0.12 | -0.14

7,980	-0.12	-0.14	8,010	-0.12	-0.20
7,981	-0.12	-0.14	8,011	-0.12	-0.20
7,982	-0.12	-0.14	8,012	-0.12	-0.20
7,983	-0.12	-0.15	8,013	-0.12	-0.21
7,984	-0.12	-0.15	8,014	-0.12	-0.21
7,985	-0.12	-0.15	8,015	-0.12	-0.21
7,986	-0.12	-0.15	8,016	-0.12	-0.21
7,987	-0.12	-0.15	8,017	-0.12	-0.21
7,988	-0.12	-0.16	8,018	-0.12	-0.22
7,989	-0.12	-0.16	8,019	-0.12	-0.20
7,990	-0.12	-0.16	8,020	-0.22	-0.18
7,991	-0.12	-0.16	8,021	-0.22	-0.16
7,992	-0.12	-0.16	8,022	-0.22	-0.14
7,993	-0.12	-0.17	8,023	-0.22	-0.13
7,994	-0.12	-0.17	8,024	-0.22	-0.11
7,995	-0.12	-0.17	8,025	-0.22	-0.09
7,996	-0.12	-0.17	8,026	-0.22	-0.07
7,997	-0.12	-0.17	8,027	-0.22	-0.05
7,998	-0.12	-0.18	8,028	-0.22	-0.03
7,999	-0.12	-0.18	8,029	-0.22	-0.01
8,000	-0.12	-0.18	8,030	-0.22	0.01
8,001	-0.12	-0.18	8,031	-0.22	0.02
8,002	-0.12	-0.18	8,032	-0.22	0.04
8,003	-0.12	-0.19	8,033	-0.22	0.06
8,004	-0.12	-0.19	8,034	-0.22	0.08
8,005	-0.12	-0.19	8,035	-0.22	0.10
8,006	-0.12	-0.19	8,036	-0.22	0.12
8,007	-0.12	-0.19	8,037	-0.22	0.14
8,008	-0.12	-0.20	8,038	-0.22	0.16
8,009	-0.12	-0.20	8,039	-0.22	0.17

8,040 | -0.22 | 0.19

8,041 | -0.22 | 0.21

8,042 | -0.22 | 0.23

8,043 | -0.22 | 0.25

8,044 | -0.22 | 0.27

8,045 | -0.22 | 0.29

8,046 | -0.22 | 0.31

8,047 | -0.22 | 0.33

8,048 | -0.22 | 0.34

8,049 | -0.22 | 0.36

8,050 | -0.22 | 0.38

8,051 | -0.22 | 0.40

8,052 | -0.22 | 0.42

8,053 | -0.22 | 0.44

8,054 | -0.22 | 0.46

8,055 | -0.22 | 0.48

8,056 | -0.22 | 0.49

8,057 | -0.22 | 0.51

8,058 | -0.22 | 0.53

8,059 | -0.22 | 0.55

8,060 | -0.22 | 0.57

8,061 | -0.22 | 0.59

8,062 | -0.22 | 0.61

8,063 | -0.22 | 0.63

8,064 | -0.22 | 0.64

8,065 | -0.22 | 0.70

8,066 | -0.22 | 0.75

8,067 | -0.22 | 0.81

8,068 | 0.72 | 0.86

8,069 | 0.72 | 0.90

8,070 | 0.72 | 0.93

8,071 | 0.72 | 0.97

8,072 | 0.72 | 1.00

8,073 | 0.72 | 1.04

8,074 | 0.72 | 1.07

8,075 | 0.72 | 1.11

8,076 | 0.72 | 1.14

8,077 | 0.72 | 1.18

8,078 | 0.72 | 1.21

8,079 | 0.72 | 1.25

8,080 | 0.72 | 1.28

8,081 | 0.72 | 1.32

8,082 | 0.72 | 1.35

8,083 | 0.72 | 1.39

8,084 | 0.72 | 1.42

8,085 | 0.72 | 1.46

8,086 | 0.72 | 1.49

8,087 | 0.72 | 1.53

8,088 | 0.72 | 1.56

8,089 | 0.72 | 1.60

8,090 | 0.72 | 1.64

8,091 | 0.72 | 1.67

8,092 | 0.72 | 1.71

8,093 | 0.72 | 1.74

8,094 | 0.72 | 1.78

8,095 | 0.72 | 1.81

8,096 | 0.72 | 1.85

8,097 | 0.72 | 1.88

8,098 | 0.72 | 1.92

8,099 | 0.72 | 1.95

8,100	0.72	1.99	8,130	2.48	1.83
8,101	0.72	2.02	8,131	2.48	1.80
8,102	0.72	2.06	8,132	2.48	1.77
8,103	0.72	2.09	8,133	2.48	1.75
8,104	0.72	2.13	8,134	2.48	1.72
8,105	0.72	2.16	8,135	2.48	1.69
8,106	0.72	2.20	8,136	2.48	1.66
8,107	0.72	2.23	8,137	2.48	1.63
8,108	0.72	2.27	8,138	2.48	1.60
8,109	0.72	2.27	8,139	2.48	1.57
8,110	0.72	2.28	8,140	2.48	1.54
8,111	0.72	2.29	8,141	2.48	1.51
8,112	0.72	2.29	8,142	2.48	1.48
8,113	0.72	2.30	8,143	2.48	1.45
8,114	2.48	2.30	8,144	2.48	1.42
8,115	2.48	2.27	8,145	2.48	1.39
8,116	2.48	2.24	8,146	2.48	1.36
8,117	2.48	2.22	8,147	2.48	1.33
8,118	2.48	2.19	8,148	2.48	1.30
8,119	2.48	2.16	8,149	2.48	1.27
8,120	2.48	2.13	8,150	2.48	1.25
8,121	2.48	2.10	8,151	2.48	1.22
8,122	2.48	2.07	8,152	2.48	1.19
8,123	2.48	2.04	8,153	2.48	1.16
8,124	2.48	2.01	8,154	2.48	1.10
8,125	2.48	1.98	8,155	2.48	1.04
8,126	2.48	1.95	8,156	2.48	0.98
8,127	2.48	1.92	8,157	2.48	0.92
8,128	2.48	1.89	8,158	1.01	0.86
8,129	2.48	1.86	8,159	1.01	0.83

8,160 | 1.01 | 0.81
8,161 | 1.01 | 0.78
8,162 | 1.01 | 0.75
8,163 | 1.01 | 0.72
8,164 | 1.01 | 0.69
8,165 | 1.01 | 0.66
8,166 | 1.01 | 0.63
8,167 | 1.01 | 0.60
8,168 | 1.01 | 0.57
8,169 | 1.01 | 0.54
8,170 | 1.01 | 0.51
8,171 | 1.01 | 0.48
8,172 | 1.01 | 0.46
8,173 | 1.01 | 0.43
8,174 | 1.01 | 0.40
8,175 | 1.01 | 0.37
8,176 | 1.01 | 0.34
8,177 | 1.01 | 0.31
8,178 | 1.01 | 0.28
8,179 | 1.01 | 0.25
8,180 | 1.01 | 0.22
8,181 | 1.01 | 0.19
8,182 | 1.01 | 0.16
8,183 | 1.01 | 0.13
8,184 | 1.01 | 0.10
8,185 | 1.01 | 0.08
8,186 | 1.01 | 0.05
8,187 | 1.01 | 0.02
8,188 | 1.01 | -0.01
8,189 | 1.01 | -0.04

8,190 | 1.01 | -0.07
8,191 | 1.01 | -0.10
8,192 | 1.01 | -0.13
8,193 | 1.01 | -0.16
8,194 | 1.01 | -0.19
8,195 | 1.01 | -0.22
8,196 | 1.01 | -0.25
8,197 | 1.01 | -0.27
8,198 | 1.01 | -0.30
8,199 | 1.01 | -0.33
8,200 | 1.01 | -0.36
8,201 | 1.01 | -0.39
8,202 | 1.01 | -0.42
8,203 | -0.45 | -0.45
8,204 | -0.45 | -0.45
8,205 | -0.45 | -0.44
8,206 | -0.45 | -0.44
8,207 | -0.45 | -0.44
8,208 | -0.45 | -0.44
8,209 | -0.45 | -0.44
8,210 | -0.45 | -0.44
8,211 | -0.45 | -0.43
8,212 | -0.45 | -0.43
8,213 | -0.45 | -0.43
8,214 | -0.45 | -0.43
8,215 | -0.45 | -0.43
8,216 | -0.45 | -0.42
8,217 | -0.45 | -0.42
8,218 | -0.45 | -0.42
8,219 | -0.45 | -0.42

8,220	-0.45	-0.42	8,250	-0.45	-0.36
8,221	-0.45	-0.42	8,251	-0.45	-0.36
8,222	-0.45	-0.41	8,252	-0.36	-0.36
8,223	-0.45	-0.41	8,253	-0.36	-0.36
8,224	-0.45	-0.41	8,254	-0.36	-0.35
8,225	-0.45	-0.41	8,255	-0.36	-0.35
8,226	-0.45	-0.41	8,256	-0.36	-0.34
8,227	-0.45	-0.41	8,257	-0.36	-0.34
8,228	-0.45	-0.40	8,258	-0.36	-0.33
8,229	-0.45	-0.40	8,259	-0.36	-0.33
8,230	-0.45	-0.40	8,260	-0.36	-0.32
8,231	-0.45	-0.40	8,261	-0.36	-0.32
8,232	-0.45	-0.40	8,262	-0.36	-0.31
8,233	-0.45	-0.39	8,263	-0.36	-0.31
8,234	-0.45	-0.39	8,264	-0.36	-0.30
8,235	-0.45	-0.39	8,265	-0.36	-0.30
8,236	-0.45	-0.39	8,266	-0.36	-0.30
8,237	-0.45	-0.39	8,267	-0.36	-0.29
8,238	-0.45	-0.39	8,268	-0.36	-0.29
8,239	-0.45	-0.38	8,269	-0.36	-0.28
8,240	-0.45	-0.38	8,270	-0.36	-0.28
8,241	-0.45	-0.38	8,271	-0.36	-0.27
8,242	-0.45	-0.38	8,272	-0.36	-0.27
8,243	-0.45	-0.38	8,273	-0.36	-0.26
8,244	-0.45	-0.37	8,274	-0.36	-0.26
8,245	-0.45	-0.37	8,275	-0.36	-0.25
8,246	-0.45	-0.37	8,276	-0.36	-0.25
8,247	-0.45	-0.37	8,277	-0.36	-0.25
8,248	-0.45	-0.37	8,278	-0.36	-0.24
8,249	-0.45	-0.37	8,279	-0.36	-0.24

8,280	-0.36	-0.23
8,281	-0.36	-0.23
8,282	-0.36	-0.22
8,283	-0.36	-0.22
8,284	-0.36	-0.21
8,285	-0.36	-0.21
8,286	-0.36	-0.20
8,287	-0.36	-0.20
8,288	-0.36	-0.19
8,289	-0.36	-0.19
8,290	-0.36	-0.19
8,291	-0.36	-0.18
8,292	-0.36	-0.18
8,293	-0.36	-0.17
8,294	-0.36	-0.17
8,295	-0.36	-0.16
8,296	-0.36	-0.16
8,297	-0.36	-0.15
8,298	-0.36	-0.15
8,299	-0.36	-0.14
8,300	-0.36	-0.14
8,301	-0.36	-0.13
8,302	-0.13	-0.12
8,303	-0.13	-0.12
8,304	-0.13	-0.11
8,305	-0.13	-0.10
8,306	-0.13	-0.10
8,307	-0.13	-0.09
8,308	-0.13	-0.08
8,309	-0.13	-0.08
8,310	-0.13	-0.07
8,311	-0.13	-0.06
8,312	-0.13	-0.06
8,313	-0.13	-0.05
8,314	-0.13	-0.04
8,315	-0.13	-0.03
8,316	-0.13	-0.03
8,317	-0.13	-0.02
8,318	-0.13	-0.01
8,319	-0.13	-0.01
8,320	-0.13	0.00
8,321	-0.13	0.01
8,322	-0.13	0.01
8,323	-0.13	0.02
8,324	-0.13	0.03
8,325	-0.13	0.03
8,326	-0.13	0.04
8,327	-0.13	0.05
8,328	-0.13	0.05
8,329	-0.13	0.06
8,330	-0.13	0.07
8,331	-0.13	0.07
8,332	-0.13	0.08
8,333	-0.13	0.09
8,334	-0.13	0.09
8,335	-0.13	0.10
8,336	-0.13	0.11
8,337	-0.13	0.11
8,338	-0.13	0.12
8,339	-0.13	0.13

8,340 | -0.13 | 0.14
8,341 | -0.13 | 0.14
8,342 | -0.13 | 0.15
8,343 | -0.13 | 0.16
8,344 | -0.13 | 0.16
8,345 | -0.13 | 0.17
8,346 | -0.13 | 0.18
8,347 | -0.13 | 0.18
8,348 | -0.13 | 0.19
8,349 | -0.13 | 0.20
8,350 | -0.13 | 0.20
8,351 | 0.21 | 0.21
8,352 | 0.21 | 0.21
8,353 | 0.21 | 0.21
8,354 | 0.21 | 0.21
8,355 | 0.21 | 0.20
8,356 | 0.21 | 0.20
8,357 | 0.21 | 0.20
8,358 | 0.21 | 0.20
8,359 | 0.21 | 0.20
8,360 | 0.21 | 0.20
8,361 | 0.21 | 0.20
8,362 | 0.21 | 0.20
8,363 | 0.21 | 0.20
8,364 | 0.21 | 0.20
8,365 | 0.21 | 0.19
8,366 | 0.21 | 0.19
8,367 | 0.21 | 0.19
8,368 | 0.21 | 0.19
8,369 | 0.21 | 0.19

8,370 | 0.21 | 0.19
8,371 | 0.21 | 0.19
8,372 | 0.21 | 0.19
8,373 | 0.21 | 0.19
8,374 | 0.21 | 0.19
8,375 | 0.21 | 0.18
8,376 | 0.21 | 0.18
8,377 | 0.21 | 0.18
8,378 | 0.21 | 0.18
8,379 | 0.21 | 0.18
8,380 | 0.21 | 0.18
8,381 | 0.21 | 0.18
8,382 | 0.21 | 0.18
8,383 | 0.21 | 0.18
8,384 | 0.21 | 0.18
8,385 | 0.21 | 0.17
8,386 | 0.21 | 0.17
8,387 | 0.21 | 0.17
8,388 | 0.21 | 0.17
8,389 | 0.21 | 0.17
8,390 | 0.21 | 0.17
8,391 | 0.21 | 0.17
8,392 | 0.21 | 0.17
8,393 | 0.21 | 0.17
8,394 | 0.21 | 0.17
8,395 | 0.21 | 0.16
8,396 | 0.21 | 0.16
8,397 | 0.21 | 0.16
8,398 | 0.21 | 0.16
8,399 | 0.16 | 0.15

8,400 | 0.16 | 0.15
8,401 | 0.16 | 0.14
8,402 | 0.16 | 0.14
8,403 | 0.16 | 0.13
8,404 | 0.16 | 0.13
8,405 | 0.16 | 0.12
8,406 | 0.16 | 0.12
8,407 | 0.16 | 0.11
8,408 | 0.16 | 0.11
8,409 | 0.16 | 0.10
8,410 | 0.16 | 0.10
8,411 | 0.16 | 0.09
8,412 | 0.16 | 0.09
8,413 | 0.16 | 0.08
8,414 | 0.16 | 0.08
8,415 | 0.16 | 0.07
8,416 | 0.16 | 0.07
8,417 | 0.16 | 0.06
8,418 | 0.16 | 0.06
8,419 | 0.16 | 0.05
8,420 | 0.16 | 0.05
8,421 | 0.16 | 0.04
8,422 | 0.16 | 0.04
8,423 | 0.16 | 0.03
8,424 | 0.16 | 0.02
8,425 | 0.16 | 0.02
8,426 | 0.16 | 0.01
8,427 | 0.16 | 0.01
8,428 | 0.16 | 0.00
8,429 | 0.16 | 0.00

8,430 | 0.16 | -0.01
8,431 | 0.16 | -0.01
8,432 | 0.16 | -0.02
8,433 | 0.16 | -0.02
8,434 | 0.16 | -0.03
8,435 | 0.16 | -0.03
8,436 | 0.16 | -0.04
8,437 | 0.16 | -0.04
8,438 | 0.16 | -0.05
8,439 | 0.16 | -0.05
8,440 | 0.16 | -0.06
8,441 | 0.16 | -0.06
8,442 | 0.16 | -0.07
8,443 | 0.16 | -0.07
8,444 | 0.16 | -0.08
8,445 | 0.16 | -0.08
8,446 | 0.16 | -0.09
8,447 | 0.16 | -0.09
8,448 | -0.10 | -0.10
8,449 | -0.10 | -0.11
8,450 | -0.10 | -0.11
8,451 | -0.10 | -0.11
8,452 | -0.10 | -0.11
8,453 | -0.10 | -0.12
8,454 | -0.10 | -0.12
8,455 | -0.10 | -0.12
8,456 | -0.10 | -0.12
8,457 | -0.10 | -0.13
8,458 | -0.10 | -0.13
8,459 | -0.10 | -0.13

8,460	-0.10	-0.13	8,490	-0.10	-0.21
8,461	-0.10	-0.14	8,491	-0.10	-0.21
8,462	-0.10	-0.14	8,492	-0.10	-0.22
8,463	-0.10	-0.14	8,493	-0.10	-0.22
8,464	-0.10	-0.14	8,494	-0.10	-0.22
8,465	-0.10	-0.15	8,495	-0.10	-0.22
8,466	-0.10	-0.15	8,496	-0.10	-0.23
8,467	-0.10	-0.15	8,497	-0.23	-0.23
8,468	-0.10	-0.15	8,498	-0.23	-0.23
8,469	-0.10	-0.16	8,499	-0.23	-0.23
8,470	-0.10	-0.16	8,500	-0.23	-0.23
8,471	-0.10	-0.16	8,501	-0.23	-0.23
8,472	-0.10	-0.17	8,502	-0.23	-0.23
8,473	-0.10	-0.17	8,503	-0.23	-0.23
8,474	-0.10	-0.17	8,504	-0.23	-0.23
8,475	-0.10	-0.17	8,505	-0.23	-0.24
8,476	-0.10	-0.18	8,506	-0.23	-0.24
8,477	-0.10	-0.18	8,507	-0.23	-0.24
8,478	-0.10	-0.18	8,508	-0.23	-0.24
8,479	-0.10	-0.18	8,509	-0.23	-0.24
8,480	-0.10	-0.19	8,510	-0.23	-0.24
8,481	-0.10	-0.19	8,511	-0.23	-0.24
8,482	-0.10	-0.19	8,512	-0.23	-0.24
8,483	-0.10	-0.19	8,513	-0.23	-0.24
8,484	-0.10	-0.20	8,514	-0.23	-0.24
8,485	-0.10	-0.20	8,515	-0.23	-0.24
8,486	-0.10	-0.20	8,516	-0.23	-0.24
8,487	-0.10	-0.20	8,517	-0.23	-0.24
8,488	-0.10	-0.21	8,518	-0.23	-0.24
8,489	-0.10	-0.21	8,519	-0.23	-0.24

8,520	-0.23	-0.24	8,550	-0.26	-0.27
8,521	-0.23	-0.25	8,551	-0.26	-0.27
8,522	-0.23	-0.25	8,552	-0.26	-0.27
8,523	-0.23	-0.25	8,553	-0.26	-0.27
8,524	-0.23	-0.25	8,554	-0.26	-0.27
8,525	-0.23	-0.25	8,555	-0.26	-0.27
8,526	-0.23	-0.25	8,556	-0.26	-0.27
8,527	-0.23	-0.25	8,557	-0.26	-0.27
8,528	-0.23	-0.25	8,558	-0.26	-0.27
8,529	-0.23	-0.25	8,559	-0.26	-0.27
8,530	-0.23	-0.25	8,560	-0.26	-0.28
8,531	-0.23	-0.25	8,561	-0.26	-0.28
8,532	-0.23	-0.25	8,562	-0.26	-0.28
8,533	-0.23	-0.25	8,563	-0.26	-0.28
8,534	-0.23	-0.25	8,564	-0.26	-0.28
8,535	-0.23	-0.25	8,565	-0.26	-0.28
8,536	-0.23	-0.25	8,566	-0.26	-0.28
8,537	-0.23	-0.25	8,567	-0.26	-0.28
8,538	-0.23	-0.26	8,568	-0.26	-0.28
8,539	-0.23	-0.26	8,569	-0.26	-0.28
8,540	-0.23	-0.26	8,570	-0.26	-0.29
8,541	-0.23	-0.26	8,571	-0.26	-0.29
8,542	-0.23	-0.26	8,572	-0.26	-0.29
8,543	-0.23	-0.26	8,573	-0.26	-0.29
8,544	-0.23	-0.26	8,574	-0.26	-0.29
8,545	-0.23	-0.26	8,575	-0.26	-0.29
8,546	-0.26	-0.26	8,576	-0.26	-0.29
8,547	-0.26	-0.26	8,577	-0.26	-0.29
8,548	-0.26	-0.26	8,578	-0.26	-0.29
8,549	-0.26	-0.26	8,579	-0.26	-0.29

8,580 | -0.26 | -0.30
8,581 | -0.26 | -0.30
8,582 | -0.26 | -0.30
8,583 | -0.26 | -0.30
8,584 | -0.26 | -0.30
8,585 | -0.26 | -0.30
8,586 | -0.26 | -0.30
8,587 | -0.26 | -0.30
8,588 | -0.26 | -0.30
8,589 | -0.26 | -0.30
8,590 | -0.26 | -0.31
8,591 | -0.26 | -0.31
8,592 | -0.26 | -0.31
8,593 | -0.26 | -0.31
8,594 | -0.26 | -0.31
8,595 | -0.31 | -0.31
8,596 | -0.31 | -0.31
8,597 | -0.31 | -0.30
8,598 | -0.31 | -0.30
8,599 | -0.31 | -0.30
8,600 | -0.31 | -0.29
8,601 | -0.31 | -0.29
8,602 | -0.31 | -0.29
8,603 | -0.31 | -0.28
8,604 | -0.31 | -0.28
8,605 | -0.31 | -0.28
8,606 | -0.31 | -0.27
8,607 | -0.31 | -0.27
8,608 | -0.31 | -0.27
8,609 | -0.31 | -0.26

8,610 | -0.31 | -0.26
8,611 | -0.31 | -0.26
8,612 | -0.31 | -0.25
8,613 | -0.31 | -0.25
8,614 | -0.31 | -0.25
8,615 | -0.31 | -0.24
8,616 | -0.31 | -0.24
8,617 | -0.31 | -0.24
8,618 | -0.31 | -0.23
8,619 | -0.31 | -0.23
8,620 | -0.31 | -0.23
8,621 | -0.31 | -0.22
8,622 | -0.31 | -0.22
8,623 | -0.31 | -0.21
8,624 | -0.31 | -0.21
8,625 | -0.31 | -0.21
8,626 | -0.31 | -0.20
8,627 | -0.31 | -0.20
8,628 | -0.31 | -0.20
8,629 | -0.31 | -0.19
8,630 | -0.31 | -0.19
8,631 | -0.31 | -0.19
8,632 | -0.31 | -0.18
8,633 | -0.31 | -0.18
8,634 | -0.31 | -0.18
8,635 | -0.31 | -0.17
8,636 | -0.31 | -0.17
8,637 | -0.31 | -0.17
8,638 | -0.31 | -0.16
8,639 | -0.31 | -0.16

```
8,640 | -0.31 | -0.16        8,670 | -0.14 | 0.35
8,641 | -0.31 | -0.15        8,671 | -0.14 | 0.36
8,642 | -0.31 | -0.15        8,672 | -0.14 | 0.38
8,643 | -0.31 | -0.15        8,673 | -0.14 | 0.40
8,644 | -0.31 | -0.13        8,674 | -0.14 | 0.42
8,645 | -0.14 | -0.10        8,675 | -0.14 | 0.44
8,646 | -0.14 | -0.09        8,676 | -0.14 | 0.45
8,647 | -0.14 | -0.07        8,677 | -0.14 | 0.47
8,648 | -0.14 | -0.05        8,678 | -0.14 | 0.49
8,649 | -0.14 | -0.03        8,679 | -0.14 | 0.51
8,650 | -0.14 | -0.01        8,680 | -0.14 | 0.53
8,651 | -0.14 | 0.00         8,681 | -0.14 | 0.54
8,652 | -0.14 | 0.02         8,682 | -0.14 | 0.56
8,653 | -0.14 | 0.04         8,683 | -0.14 | 0.58
8,654 | -0.14 | 0.06         8,684 | -0.14 | 0.60
8,655 | -0.14 | 0.08         8,685 | -0.14 | 0.62
8,656 | -0.14 | 0.09         8,686 | -0.14 | 0.63
8,657 | -0.14 | 0.11         8,687 | -0.14 | 0.65
8,658 | -0.14 | 0.13         8,688 | -0.14 | 0.67
8,659 | -0.14 | 0.15         8,689 | -0.14 | 0.69
8,660 | -0.14 | 0.17         8,690 | -0.14 | 0.71
8,661 | -0.14 | 0.18         8,691 | -0.14 | 0.72
8,662 | -0.14 | 0.20         8,692 | -0.14 | 0.74
8,663 | -0.14 | 0.22         8,693 | 0.76 | 0.75
8,664 | -0.14 | 0.24         8,694 | 0.76 | 0.75
8,665 | -0.14 | 0.26         8,695 | 0.76 | 0.74
8,666 | -0.14 | 0.27         8,696 | 0.76 | 0.74
8,667 | -0.14 | 0.29         8,697 | 0.76 | 0.73
8,668 | -0.14 | 0.31         8,698 | 0.76 | 0.73
8,669 | -0.14 | 0.33         8,699 | 0.76 | 0.73
```

8,700 | 0.76 | 0.72
8,701 | 0.76 | 0.72
8,702 | 0.76 | 0.71
8,703 | 0.76 | 0.71
8,704 | 0.76 | 0.71
8,705 | 0.76 | 0.70
8,706 | 0.76 | 0.70
8,707 | 0.76 | 0.69
8,708 | 0.76 | 0.69
8,709 | 0.76 | 0.68
8,710 | 0.76 | 0.68
8,711 | 0.76 | 0.68
8,712 | 0.76 | 0.67
8,713 | 0.76 | 0.67
8,714 | 0.76 | 0.66
8,715 | 0.76 | 0.66
8,716 | 0.76 | 0.66
8,717 | 0.76 | 0.65
8,718 | 0.76 | 0.65
8,719 | 0.76 | 0.64
8,720 | 0.76 | 0.64
8,721 | 0.76 | 0.63
8,722 | 0.76 | 0.63
8,723 | 0.76 | 0.63
8,724 | 0.76 | 0.62
8,725 | 0.76 | 0.62
8,726 | 0.76 | 0.61
8,727 | 0.76 | 0.61
8,728 | 0.76 | 0.60
8,729 | 0.76 | 0.60

8,730 | 0.76 | 0.60
8,731 | 0.76 | 0.59
8,732 | 0.76 | 0.59
8,733 | 0.76 | 0.58
8,734 | 0.76 | 0.58
8,735 | 0.76 | 0.58
8,736 | 0.76 | 0.57
8,737 | 0.76 | 0.57
8,738 | 0.76 | 0.56
8,739 | 0.76 | 0.54
8,740 | 0.76 | 0.52
8,741 | 0.55 | 0.49
8,742 | 0.55 | 0.47
8,743 | 0.55 | 0.46
8,744 | 0.55 | 0.44
8,745 | 0.55 | 0.42
8,746 | 0.55 | 0.40
8,747 | 0.55 | 0.38
8,748 | 0.55 | 0.36
8,749 | 0.55 | 0.34
8,750 | 0.55 | 0.32
8,751 | 0.55 | 0.31
8,752 | 0.55 | 0.29
8,753 | 0.55 | 0.27
8,754 | 0.55 | 0.25
8,755 | 0.55 | 0.23
8,756 | 0.55 | 0.21
8,757 | 0.55 | 0.19
8,758 | 0.55 | 0.17
8,759 | 0.55 | 0.16

8,760	0.55	0.14	8,790	-0.39	-0.37
8,761	0.55	0.12	8,791	-0.39	-0.37
8,762	0.55	0.10	8,792	-0.39	-0.36
8,763	0.55	0.08	8,793	-0.39	-0.36
8,764	0.55	0.06	8,794	-0.39	-0.35
8,765	0.55	0.04	8,795	-0.39	-0.35
8,766	0.55	0.02	8,796	-0.39	-0.34
8,767	0.55	0.00	8,797	-0.39	-0.33
8,768	0.55	-0.01	8,798	-0.39	-0.33
8,769	0.55	-0.03	8,799	-0.39	-0.32
8,770	0.55	-0.05	8,800	-0.39	-0.32
8,771	0.55	-0.07	8,801	-0.39	-0.31
8,772	0.55	-0.09	8,802	-0.39	-0.31
8,773	0.55	-0.11	8,803	-0.39	-0.30
8,774	0.55	-0.13	8,804	-0.39	-0.29
8,775	0.55	-0.15	8,805	-0.39	-0.29
8,776	0.55	-0.16	8,806	-0.39	-0.28
8,777	0.55	-0.18	8,807	-0.39	-0.28
8,778	0.55	-0.20	8,808	-0.39	-0.27
8,779	0.55	-0.22	8,809	-0.39	-0.27
8,780	0.55	-0.24	8,810	-0.39	-0.26
8,781	0.55	-0.26	8,811	-0.39	-0.26
8,782	0.55	-0.28	8,812	-0.39	-0.25
8,783	0.55	-0.30	8,813	-0.39	-0.24
8,784	0.55	-0.31	8,814	-0.39	-0.24
8,785	0.55	-0.33	8,815	-0.39	-0.23
8,786	0.55	-0.35	8,816	-0.39	-0.23
8,787	0.55	-0.37	8,817	-0.39	-0.22
8,788	-0.39	-0.38	8,818	-0.39	-0.22
8,789	-0.39	-0.38	8,819	-0.39	-0.21

8,820	-0.39	-0.21	8,850	-0.11	-0.11
8,821	-0.39	-0.20	8,851	-0.11	-0.11
8,822	-0.39	-0.19	8,852	-0.11	-0.11
8,823	-0.39	-0.19	8,853	-0.11	-0.11
8,824	-0.39	-0.18	8,854	-0.11	-0.11
8,825	-0.39	-0.18	8,855	-0.11	-0.11
8,826	-0.39	-0.17	8,856	-0.11	-0.11
8,827	-0.39	-0.17	8,857	-0.11	-0.11
8,828	-0.39	-0.16	8,858	-0.11	-0.11
8,829	-0.39	-0.15	8,859	-0.11	-0.11
8,830	-0.39	-0.15	8,860	-0.11	-0.11
8,831	-0.39	-0.14	8,861	-0.11	-0.11
8,832	-0.39	-0.14	8,862	-0.11	-0.12
8,833	-0.39	-0.13	8,863	-0.11	-0.12
8,834	-0.39	-0.13	8,864	-0.11	-0.12
8,835	-0.39	-0.12	8,865	-0.11	-0.12
8,836	-0.39	-0.12	8,866	-0.11	-0.12
8,837	-0.11	-0.11	8,867	-0.11	-0.12
8,838	-0.11	-0.11	8,868	-0.11	-0.12
8,839	-0.11	-0.11	8,869	-0.11	-0.12
8,840	-0.11	-0.11	8,870	-0.11	-0.12
8,841	-0.11	-0.11	8,871	-0.11	-0.12
8,842	-0.11	-0.11	8,872	-0.11	-0.12
8,843	-0.11	-0.11	8,873	-0.11	-0.12
8,844	-0.11	-0.11	8,874	-0.11	-0.12
8,845	-0.11	-0.11	8,875	-0.11	-0.12
8,846	-0.11	-0.11	8,876	-0.11	-0.12
8,847	-0.11	-0.11	8,877	-0.11	-0.12
8,848	-0.11	-0.11	8,878	-0.11	-0.12
8,849	-0.11	-0.11	8,879	-0.11	-0.12

8,880	-0.11	-0.12	8,910	-0.12	-0.28
8,881	-0.11	-0.12	8,911	-0.12	-0.29
8,882	-0.11	-0.12	8,912	-0.12	-0.30
8,883	-0.11	-0.12	8,913	-0.12	-0.30
8,884	-0.11	-0.12	8,914	-0.12	-0.31
8,885	-0.11	-0.12	8,915	-0.12	-0.32
8,886	-0.11	-0.12	8,916	-0.12	-0.32
8,887	-0.12	-0.13	8,917	-0.12	-0.33
8,888	-0.12	-0.13	8,918	-0.12	-0.34
8,889	-0.12	-0.14	8,919	-0.12	-0.34
8,890	-0.12	-0.15	8,920	-0.12	-0.35
8,891	-0.12	-0.15	8,921	-0.12	-0.36
8,892	-0.12	-0.16	8,922	-0.12	-0.36
8,893	-0.12	-0.17	8,923	-0.12	-0.37
8,894	-0.12	-0.17	8,924	-0.12	-0.38
8,895	-0.12	-0.18	8,925	-0.12	-0.39
8,896	-0.12	-0.19	8,926	-0.12	-0.39
8,897	-0.12	-0.19	8,927	-0.12	-0.40
8,898	-0.12	-0.20	8,928	-0.12	-0.41
8,899	-0.12	-0.21	8,929	-0.12	-0.41
8,900	-0.12	-0.22	8,930	-0.12	-0.42
8,901	-0.12	-0.22	8,931	-0.12	-0.43
8,902	-0.12	-0.23	8,932	-0.12	-0.43
8,903	-0.12	-0.24	8,933	-0.12	-0.44
8,904	-0.12	-0.24	8,934	-0.12	-0.45
8,905	-0.12	-0.25	8,935	-0.12	-0.45
8,906	-0.12	-0.26	8,936	-0.46	-0.46
8,907	-0.12	-0.26	8,937	-0.46	-0.45
8,908	-0.12	-0.27	8,938	-0.46	-0.44
8,909	-0.12	-0.28	8,939	-0.46	-0.44

8,940	-0.46	-0.43	8,970	-0.46	-0.19
8,941	-0.46	-0.42	8,971	-0.46	-0.18
8,942	-0.46	-0.41	8,972	-0.46	-0.17
8,943	-0.46	-0.40	8,973	-0.46	-0.16
8,944	-0.46	-0.40	8,974	-0.46	-0.16
8,945	-0.46	-0.39	8,975	-0.46	-0.15
8,946	-0.46	-0.38	8,976	-0.46	-0.14
8,947	-0.46	-0.37	8,977	-0.46	-0.13
8,948	-0.46	-0.36	8,978	-0.46	-0.12
8,949	-0.46	-0.36	8,979	-0.46	-0.12
8,950	-0.46	-0.35	8,980	-0.46	-0.11
8,951	-0.46	-0.34	8,981	-0.46	-0.10
8,952	-0.46	-0.33	8,982	-0.46	-0.09
8,953	-0.46	-0.32	8,983	-0.46	-0.08
8,954	-0.46	-0.32	8,984	-0.46	-0.08
8,955	-0.46	-0.31	8,985	-0.46	-0.07
8,956	-0.46	-0.30	8,986	-0.06	-0.06
8,957	-0.46	-0.29	8,987	-0.06	-0.06
8,958	-0.46	-0.28	8,988	-0.06	-0.06
8,959	-0.46	-0.28	8,989	-0.06	-0.05
8,960	-0.46	-0.27	8,990	-0.06	-0.05
8,961	-0.46	-0.26	8,991	-0.06	-0.05
8,962	-0.46	-0.25	8,992	-0.06	-0.05
8,963	-0.46	-0.24	8,993	-0.06	-0.05
8,964	-0.46	-0.24	8,994	-0.06	-0.05
8,965	-0.46	-0.23	8,995	-0.06	-0.05
8,966	-0.46	-0.22	8,996	-0.06	-0.04
8,967	-0.46	-0.21	8,997	-0.06	-0.04
8,968	-0.46	-0.20	8,998	-0.06	-0.04
8,969	-0.46	-0.20	8,999	-0.06	-0.04

9,000	-0.06	-0.04		9,030	-0.06	0.00
9,001	-0.06	-0.04		9,031	-0.06	0.00
9,002	-0.06	-0.04		9,032	-0.06	0.01
9,003	-0.06	-0.03		9,033	-0.06	0.01
9,004	-0.06	-0.03		9,034	-0.06	0.01
9,005	-0.06	-0.03		9,035	0.01	0.01
9,006	-0.06	-0.03		9,036	0.01	0.00
9,007	-0.06	-0.03		9,037	0.01	0.00
9,008	-0.06	-0.03		9,038	0.01	-0.01
9,009	-0.06	-0.03		9,039	0.01	-0.01
9,010	-0.06	-0.03		9,040	0.01	-0.01
9,011	-0.06	-0.02		9,041	0.01	-0.02
9,012	-0.06	-0.02		9,042	0.01	-0.02
9,013	-0.06	-0.02		9,043	0.01	-0.02
9,014	-0.06	-0.02		9,044	0.01	-0.03
9,015	-0.06	-0.02		9,045	0.01	-0.03
9,016	-0.06	-0.02		9,046	0.01	-0.04
9,017	-0.06	-0.02		9,047	0.01	-0.04
9,018	-0.06	-0.01		9,048	0.01	-0.04
9,019	-0.06	-0.01		9,049	0.01	-0.05
9,020	-0.06	-0.01		9,050	0.01	-0.05
9,021	-0.06	-0.01		9,051	0.01	-0.05
9,022	-0.06	-0.01		9,052	0.01	-0.06
9,023	-0.06	-0.01		9,053	0.01	-0.06
9,024	-0.06	-0.01		9,054	0.01	-0.07
9,025	-0.06	0.00		9,055	0.01	-0.07
9,026	-0.06	0.00		9,056	0.01	-0.07
9,027	-0.06	0.00		9,057	0.01	-0.08
9,028	-0.06	0.00		9,058	0.01	-0.08
9,029	-0.06	0.00		9,059	0.01	-0.09

9,060	0.01	-0.09	9,090	-0.18	-0.17
9,061	0.01	-0.09	9,091	-0.18	-0.17
9,062	0.01	-0.10	9,092	-0.18	-0.16
9,063	0.01	-0.10	9,093	-0.18	-0.16
9,064	0.01	-0.10	9,094	-0.18	-0.16
9,065	0.01	-0.11	9,095	-0.18	-0.16
9,066	0.01	-0.11	9,096	-0.18	-0.16
9,067	0.01	-0.12	9,097	-0.18	-0.15
9,068	0.01	-0.12	9,098	-0.18	-0.15
9,069	0.01	-0.12	9,099	-0.18	-0.15
9,070	0.01	-0.13	9,100	-0.18	-0.15
9,071	0.01	-0.13	9,101	-0.18	-0.15
9,072	0.01	-0.13	9,102	-0.18	-0.15
9,073	0.01	-0.14	9,103	-0.18	-0.14
9,074	0.01	-0.14	9,104	-0.18	-0.14
9,075	0.01	-0.15	9,105	-0.18	-0.14
9,076	0.01	-0.15	9,106	-0.18	-0.14
9,077	0.01	-0.15	9,107	-0.18	-0.14
9,078	0.01	-0.16	9,108	-0.18	-0.14
9,079	0.01	-0.16	9,109	-0.18	-0.13
9,080	0.01	-0.16	9,110	-0.18	-0.13
9,081	0.01	-0.17	9,111	-0.18	-0.13
9,082	0.01	-0.17	9,112	-0.18	-0.13
9,083	0.01	-0.18	9,113	-0.18	-0.13
9,084	-0.18	-0.18	9,114	-0.18	-0.12
9,085	-0.18	-0.18	9,115	-0.18	-0.12
9,086	-0.18	-0.17	9,116	-0.18	-0.12
9,087	-0.18	-0.17	9,117	-0.18	-0.12
9,088	-0.18	-0.17	9,118	-0.18	-0.12
9,089	-0.18	-0.17	9,119	-0.18	-0.12

9,120	-0.18	-0.11
9,121	-0.18	-0.11
9,122	-0.18	-0.11
9,123	-0.18	-0.11
9,124	-0.18	-0.11
9,125	-0.18	-0.10
9,126	-0.18	-0.10
9,127	-0.18	-0.10
9,128	-0.18	-0.10
9,129	-0.18	-0.10
9,130	-0.18	-0.10
9,131	-0.18	-0.09
9,132	-0.18	-0.08
9,133	-0.09	-0.07
9,134	-0.09	-0.07
9,135	-0.09	-0.06
9,136	-0.09	-0.05
9,137	-0.09	-0.04
9,138	-0.09	-0.04
9,139	-0.09	-0.03
9,140	-0.09	-0.02
9,141	-0.09	-0.01
9,142	-0.09	-0.01
9,143	-0.09	0.00
9,144	-0.09	0.01
9,145	-0.09	0.02
9,146	-0.09	0.02
9,147	-0.09	0.03
9,148	-0.09	0.04
9,149	-0.09	0.05
9,150	-0.09	0.05
9,151	-0.09	0.06
9,152	-0.09	0.07
9,153	-0.09	0.08
9,154	-0.09	0.08
9,155	-0.09	0.09
9,156	-0.09	0.10
9,157	-0.09	0.11
9,158	-0.09	0.12
9,159	-0.09	0.12
9,160	-0.09	0.13
9,161	-0.09	0.14
9,162	-0.09	0.15
9,163	-0.09	0.15
9,164	-0.09	0.16
9,165	-0.09	0.17
9,166	-0.09	0.18
9,167	-0.09	0.18
9,168	-0.09	0.19
9,169	-0.09	0.20
9,170	-0.09	0.21
9,171	-0.09	0.21
9,172	-0.09	0.22
9,173	-0.09	0.23
9,174	-0.09	0.24
9,175	-0.09	0.24
9,176	-0.09	0.25
9,177	-0.09	0.26
9,178	-0.09	0.27
9,179	-0.09	0.27

9,180 | -0.09 | 0.29
9,181 | 0.29 | 0.31
9,182 | 0.29 | 0.31
9,183 | 0.29 | 0.32
9,184 | 0.29 | 0.33
9,185 | 0.29 | 0.34
9,186 | 0.29 | 0.34
9,187 | 0.29 | 0.35
9,188 | 0.29 | 0.36
9,189 | 0.29 | 0.37
9,190 | 0.29 | 0.38
9,191 | 0.29 | 0.38
9,192 | 0.29 | 0.39
9,193 | 0.29 | 0.40
9,194 | 0.29 | 0.41
9,195 | 0.29 | 0.41
9,196 | 0.29 | 0.42
9,197 | 0.29 | 0.43
9,198 | 0.29 | 0.44
9,199 | 0.29 | 0.45
9,200 | 0.29 | 0.45
9,201 | 0.29 | 0.46
9,202 | 0.29 | 0.47
9,203 | 0.29 | 0.48
9,204 | 0.29 | 0.49
9,205 | 0.29 | 0.49
9,206 | 0.29 | 0.50
9,207 | 0.29 | 0.51
9,208 | 0.29 | 0.52
9,209 | 0.29 | 0.52

9,210 | 0.29 | 0.53
9,211 | 0.29 | 0.54
9,212 | 0.29 | 0.55
9,213 | 0.29 | 0.56
9,214 | 0.29 | 0.56
9,215 | 0.29 | 0.57
9,216 | 0.29 | 0.58
9,217 | 0.29 | 0.59
9,218 | 0.29 | 0.59
9,219 | 0.29 | 0.60
9,220 | 0.29 | 0.61
9,221 | 0.29 | 0.62
9,222 | 0.29 | 0.63
9,223 | 0.29 | 0.63
9,224 | 0.29 | 0.64
9,225 | 0.29 | 0.65
9,226 | 0.29 | 0.66
9,227 | 0.29 | 0.66
9,228 | 0.29 | 0.66
9,229 | 0.68 | 0.67
9,230 | 0.68 | 0.66
9,231 | 0.68 | 0.66
9,232 | 0.68 | 0.65
9,233 | 0.68 | 0.65
9,234 | 0.68 | 0.64
9,235 | 0.68 | 0.64
9,236 | 0.68 | 0.63
9,237 | 0.68 | 0.63
9,238 | 0.68 | 0.62
9,239 | 0.68 | 0.62

9,240	0.68	0.62	9,270	0.68	0.47
9,241	0.68	0.61	9,271	0.68	0.47
9,242	0.68	0.61	9,272	0.68	0.47
9,243	0.68	0.60	9,273	0.68	0.46
9,244	0.68	0.60	9,274	0.68	0.45
9,245	0.68	0.59	9,275	0.68	0.43
9,246	0.68	0.59	9,276	0.45	0.42
9,247	0.68	0.58	9,277	0.45	0.41
9,248	0.68	0.58	9,278	0.45	0.40
9,249	0.68	0.57	9,279	0.45	0.39
9,250	0.68	0.57	9,280	0.45	0.39
9,251	0.68	0.57	9,281	0.45	0.38
9,252	0.68	0.56	9,282	0.45	0.37
9,253	0.68	0.56	9,283	0.45	0.36
9,254	0.68	0.55	9,284	0.45	0.35
9,255	0.68	0.55	9,285	0.45	0.34
9,256	0.68	0.54	9,286	0.45	0.33
9,257	0.68	0.54	9,287	0.45	0.33
9,258	0.68	0.53	9,288	0.45	0.32
9,259	0.68	0.53	9,289	0.45	0.31
9,260	0.68	0.52	9,290	0.45	0.30
9,261	0.68	0.52	9,291	0.45	0.29
9,262	0.68	0.51	9,292	0.45	0.28
9,263	0.68	0.51	9,293	0.45	0.27
9,264	0.68	0.50	9,294	0.45	0.27
9,265	0.68	0.50	9,295	0.45	0.26
9,266	0.68	0.49	9,296	0.45	0.25
9,267	0.68	0.49	9,297	0.45	0.24
9,268	0.68	0.48	9,298	0.45	0.23
9,269	0.68	0.48	9,299	0.45	0.22

9,300 | 0.45 | 0.21
9,301 | 0.45 | 0.21
9,302 | 0.45 | 0.20
9,303 | 0.45 | 0.19
9,304 | 0.45 | 0.18
9,305 | 0.44 | 0.17
9,306 | 0.44 | 0.16
9,307 | 0.44 | 0.15
9,308 | 0.44 | 0.15
9,309 | 0.44 | 0.14
9,310 | 0.44 | 0.13
9,311 | 0.44 | 0.12
9,312 | 0.44 | 0.11
9,313 | 0.44 | 0.10
9,314 | 0.44 | 0.10
9,315 | 0.44 | 0.09
9,316 | 0.44 | 0.08
9,317 | 0.44 | 0.07
9,318 | 0.44 | 0.06
9,319 | 0.44 | 0.05
9,320 | 0.44 | 0.05
9,321 | 0.44 | 0.04
9,322 | 0.44 | 0.03
9,323 | 0.02 | 0.01
9,324 | 0.02 | 0.00
9,325 | 0.02 | -0.01
9,326 | 0.02 | -0.01
9,327 | 0.02 | -0.02
9,328 | 0.02 | -0.03
9,329 | 0.02 | -0.04

9,330 | 0.02 | -0.05
9,331 | 0.02 | -0.06
9,332 | 0.02 | -0.07
9,333 | 0.02 | -0.07
9,334 | 0.02 | -0.08
9,335 | 0.02 | -0.09
9,336 | 0.02 | -0.10
9,337 | 0.02 | -0.11
9,338 | 0.02 | -0.12
9,339 | 0.02 | -0.13
9,340 | 0.02 | -0.13
9,341 | 0.02 | -0.14
9,342 | 0.02 | -0.15
9,343 | 0.02 | -0.16
9,344 | 0.02 | -0.17
9,345 | 0.02 | -0.18
9,346 | 0.02 | -0.19
9,347 | 0.02 | -0.20
9,348 | 0.02 | -0.20
9,349 | 0.02 | -0.21
9,350 | 0.02 | -0.22
9,351 | 0.02 | -0.23
9,352 | 0.02 | -0.24
9,353 | 0.02 | -0.25
9,354 | 0.02 | -0.26
9,355 | 0.02 | -0.26
9,356 | 0.02 | -0.27
9,357 | 0.02 | -0.28
9,358 | 0.02 | -0.29
9,359 | 0.02 | -0.30

9,360	0.02	-0.31	9,390	-0.41	-0.37
9,361	0.02	-0.32	9,391	-0.41	-0.36
9,362	0.02	-0.32	9,392	-0.41	-0.36
9,363	0.02	-0.33	9,393	-0.41	-0.36
9,364	0.02	-0.34	9,394	-0.41	-0.36
9,365	0.02	-0.35	9,395	-0.41	-0.35
9,366	0.02	-0.36	9,396	-0.41	-0.35
9,367	0.02	-0.37	9,397	-0.41	-0.35
9,368	0.02	-0.38	9,398	-0.41	-0.35
9,369	0.02	-0.38	9,399	-0.41	-0.35
9,370	0.02	-0.39	9,400	-0.41	-0.34
9,371	0.02	-0.40	9,401	-0.41	-0.34
9,372	-0.41	-0.41	9,402	-0.41	-0.34
9,373	-0.41	-0.41	9,403	-0.41	-0.34
9,374	-0.41	-0.41	9,404	-0.41	-0.33
9,375	-0.41	-0.40	9,405	-0.41	-0.33
9,376	-0.41	-0.40	9,406	-0.41	-0.33
9,377	-0.41	-0.40	9,407	-0.41	-0.33
9,378	-0.41	-0.40	9,408	-0.41	-0.32
9,379	-0.41	-0.39	9,409	-0.41	-0.32
9,380	-0.41	-0.39	9,410	-0.41	-0.32
9,381	-0.41	-0.39	9,411	-0.41	-0.32
9,382	-0.41	-0.39	9,412	-0.41	-0.31
9,383	-0.41	-0.38	9,413	-0.41	-0.31
9,384	-0.41	-0.38	9,414	-0.41	-0.31
9,385	-0.41	-0.38	9,415	-0.41	-0.31
9,386	-0.41	-0.38	9,416	-0.41	-0.30
9,387	-0.41	-0.37	9,417	-0.41	-0.30
9,388	-0.41	-0.37	9,418	-0.41	-0.30
9,389	-0.41	-0.37	9,419	-0.41	-0.30

9,420 | -0.41 | -0.29
9,421 | -0.41 | -0.29
9,422 | -0.29 | -0.28
9,423 | -0.29 | -0.27
9,424 | -0.29 | -0.26
9,425 | -0.29 | -0.25
9,426 | -0.29 | -0.24
9,427 | -0.29 | -0.23
9,428 | -0.29 | -0.21
9,429 | -0.29 | -0.20
9,430 | -0.29 | -0.19
9,431 | -0.29 | -0.18
9,432 | -0.29 | -0.17
9,433 | -0.29 | -0.16
9,434 | -0.29 | -0.15
9,435 | -0.29 | -0.14
9,436 | -0.29 | -0.13
9,437 | -0.29 | -0.12
9,438 | -0.29 | -0.11
9,439 | -0.29 | -0.10
9,440 | -0.29 | -0.08
9,441 | -0.29 | -0.07
9,442 | -0.29 | -0.06
9,443 | -0.29 | -0.05
9,444 | -0.29 | -0.04
9,445 | -0.29 | -0.03
9,446 | -0.29 | -0.02
9,447 | -0.29 | -0.01
9,448 | -0.29 | 0.00
9,449 | -0.29 | 0.01

9,450 | -0.29 | 0.02
9,451 | -0.29 | 0.03
9,452 | -0.29 | 0.04
9,453 | -0.29 | 0.06
9,454 | -0.29 | 0.07
9,455 | -0.29 | 0.08
9,456 | -0.29 | 0.09
9,457 | -0.29 | 0.10
9,458 | -0.29 | 0.11
9,459 | -0.29 | 0.12
9,460 | -0.29 | 0.13
9,461 | -0.29 | 0.14
9,462 | -0.29 | 0.15
9,463 | -0.29 | 0.16
9,464 | -0.29 | 0.17
9,465 | -0.29 | 0.19
9,466 | -0.29 | 0.20
9,467 | -0.29 | 0.21
9,468 | -0.29 | 0.22
9,469 | -0.29 | 0.23
9,470 | -0.29 | 0.24
9,471 | 0.25 | 0.25
9,472 | 0.25 | 0.25
9,473 | 0.25 | 0.25
9,474 | 0.25 | 0.25
9,475 | 0.25 | 0.25
9,476 | 0.25 | 0.25
9,477 | 0.25 | 0.25
9,478 | 0.25 | 0.25
9,479 | 0.25 | 0.25

9,480	0.25	0.25
9,481	0.25	0.25
9,482	0.25	0.25
9,483	0.25	0.25
9,484	0.25	0.25
9,485	0.25	0.25
9,486	0.25	0.25
9,487	0.25	0.25
9,488	0.25	0.25
9,489	0.25	0.25
9,490	0.25	0.25
9,491	0.25	0.25
9,492	0.25	0.25
9,493	0.25	0.25
9,494	0.25	0.25
9,495	0.25	0.25
9,496	0.25	0.25
9,497	0.25	0.25
9,498	0.25	0.25
9,499	0.25	0.25
9,500	0.25	0.25
9,501	0.25	0.25
9,502	0.25	0.25
9,503	0.25	0.25
9,504	0.25	0.25
9,505	0.25	0.25
9,506	0.25	0.25
9,507	0.25	0.25
9,508	0.25	0.25
9,509	0.25	0.25
9,510	0.25	0.25
9,511	0.25	0.25
9,512	0.25	0.25
9,513	0.25	0.25
9,514	0.25	0.25
9,515	0.25	0.25
9,516	0.25	0.25
9,517	0.25	0.25
9,518	0.25	0.26
9,519	0.25	0.27
9,520	0.25	0.29
9,521	0.25	0.30
9,522	0.25	0.31
9,523	0.25	0.32
9,524	0.25	0.33
9,525	0.25	0.34
9,526	0.25	0.36
9,527	0.25	0.37
9,528	0.25	0.38
9,529	0.25	0.39
9,530	0.25	0.40
9,531	0.25	0.42
9,532	0.25	0.43
9,533	0.25	0.44
9,534	0.25	0.45
9,535	0.25	0.46
9,536	0.25	0.47
9,537	0.25	0.49
9,538	0.25	0.50
9,539	0.25	0.51

9,540 | 0.25 | 0.52
9,541 | 0.25 | 0.53
9,542 | 0.25 | 0.55
9,543 | 0.25 | 0.56
9,544 | 0.25 | 0.57
9,545 | 0.25 | 0.58
9,546 | 0.25 | 0.59
9,547 | 0.25 | 0.60
9,548 | 0.25 | 0.62
9,549 | 0.25 | 0.63
9,550 | 0.25 | 0.64
9,551 | 0.25 | 0.65
9,552 | 0.25 | 0.66
9,553 | 0.25 | 0.67
9,554 | 0.25 | 0.69
9,555 | 0.25 | 0.70
9,556 | 0.25 | 0.71
9,557 | 0.25 | 0.72
9,558 | 0.25 | 0.73
9,559 | 0.25 | 0.75
9,560 | 0.25 | 0.76
9,561 | 0.25 | 0.77
9,562 | 0.25 | 0.78
9,563 | 0.25 | 0.79
9,564 | 0.25 | 0.81
9,565 | 0.25 | 0.82
9,566 | 0.25 | 0.83
9,567 | 0.84 | 0.85
9,568 | 0.84 | 0.85
9,569 | 0.84 | 0.85

9,570 | 0.84 | 0.85
9,571 | 0.84 | 0.85
9,572 | 0.84 | 0.85
9,573 | 0.84 | 0.86
9,574 | 0.84 | 0.86
9,575 | 0.84 | 0.86
9,576 | 0.84 | 0.86
9,577 | 0.84 | 0.86
9,578 | 0.84 | 0.86
9,579 | 0.84 | 0.87
9,580 | 0.84 | 0.87
9,581 | 0.84 | 0.87
9,582 | 0.84 | 0.87
9,583 | 0.84 | 0.87
9,584 | 0.84 | 0.87
9,585 | 0.84 | 0.88
9,586 | 0.84 | 0.88
9,587 | 0.84 | 0.88
9,588 | 0.84 | 0.88
9,589 | 0.84 | 0.88
9,590 | 0.84 | 0.88
9,591 | 0.84 | 0.88
9,592 | 0.84 | 0.89
9,593 | 0.84 | 0.89
9,594 | 0.84 | 0.89
9,595 | 0.84 | 0.89
9,596 | 0.84 | 0.89
9,597 | 0.84 | 0.89
9,598 | 0.84 | 0.90
9,599 | 0.84 | 0.90

9,600 | 0.84 | 0.90
9,601 | 0.84 | 0.90
9,602 | 0.84 | 0.90
9,603 | 0.84 | 0.90
9,604 | 0.84 | 0.91
9,605 | 0.84 | 0.91
9,606 | 0.84 | 0.91
9,607 | 0.84 | 0.91
9,608 | 0.84 | 0.91
9,609 | 0.84 | 0.91
9,610 | 0.84 | 0.92
9,611 | 0.84 | 0.90
9,612 | 0.84 | 0.88
9,613 | 0.92 | 0.86
9,614 | 0.92 | 0.84
9,615 | 0.92 | 0.82
9,616 | 0.92 | 0.80
9,617 | 0.92 | 0.78
9,618 | 0.92 | 0.76
9,619 | 0.92 | 0.75
9,620 | 0.92 | 0.73
9,621 | 0.92 | 0.71
9,622 | 0.92 | 0.69
9,623 | 0.92 | 0.67
9,624 | 0.92 | 0.65
9,625 | 0.92 | 0.63
9,626 | 0.92 | 0.61
9,627 | 0.92 | 0.59
9,628 | 0.92 | 0.57
9,629 | 0.92 | 0.55

9,630 | 0.92 | 0.53
9,631 | 0.92 | 0.51
9,632 | 0.92 | 0.49
9,633 | 0.92 | 0.47
9,634 | 0.92 | 0.45
9,635 | 0.92 | 0.44
9,636 | 0.92 | 0.42
9,637 | 0.92 | 0.40
9,638 | 0.92 | 0.38
9,639 | 0.92 | 0.36
9,640 | 0.92 | 0.34
9,641 | 0.92 | 0.32
9,642 | 0.92 | 0.30
9,643 | 0.92 | 0.28
9,644 | 0.92 | 0.26
9,645 | 0.92 | 0.24
9,646 | 0.92 | 0.22
9,647 | 0.92 | 0.20
9,648 | 0.92 | 0.18
9,649 | 0.92 | 0.16
9,650 | 0.92 | 0.14
9,651 | 0.92 | 0.12
9,652 | 0.92 | 0.11
9,653 | 0.92 | 0.09
9,654 | 0.92 | 0.07
9,655 | 0.92 | 0.05
9,656 | 0.92 | 0.03
9,657 | 0.92 | 0.01
9,658 | 0.92 | -0.01
9,659 | 0.92 | -0.04

9,660	-0.05	-0.06	9,690	-0.05	-0.27
9,661	-0.05	-0.07	9,691	-0.05	-0.27
9,662	-0.05	-0.08	9,692	-0.05	-0.28
9,663	-0.05	-0.08	9,693	-0.05	-0.29
9,664	-0.05	-0.09	9,694	-0.05	-0.29
9,665	-0.05	-0.10	9,695	-0.05	-0.30
9,666	-0.05	-0.10	9,696	-0.05	-0.31
9,667	-0.05	-0.11	9,697	-0.05	-0.32
9,668	-0.05	-0.12	9,698	-0.05	-0.32
9,669	-0.05	-0.12	9,699	-0.05	-0.33
9,670	-0.05	-0.13	9,700	-0.05	-0.34
9,671	-0.05	-0.14	9,701	-0.05	-0.34
9,672	-0.05	-0.15	9,702	-0.05	-0.35
9,673	-0.05	-0.15	9,703	-0.05	-0.36
9,674	-0.05	-0.16	9,704	-0.05	-0.36
9,675	-0.05	-0.17	9,705	-0.05	-0.37
9,676	-0.05	-0.17	9,706	-0.05	-0.38
9,677	-0.05	-0.18	9,707	-0.05	-0.38
9,678	-0.05	-0.19	9,708	-0.39	-0.39
9,679	-0.05	-0.19	9,709	-0.39	-0.39
9,680	-0.05	-0.20	9,710	-0.39	-0.39
9,681	-0.05	-0.21	9,711	-0.39	-0.39
9,682	-0.05	-0.21	9,712	-0.39	-0.39
9,683	-0.05	-0.22	9,713	-0.39	-0.40
9,684	-0.05	-0.23	9,714	-0.39	-0.40
9,685	-0.05	-0.23	9,715	-0.39	-0.40
9,686	-0.05	-0.24	9,716	-0.39	-0.40
9,687	-0.05	-0.25	9,717	-0.39	-0.40
9,688	-0.05	-0.25	9,718	-0.39	-0.40
9,689	-0.05	-0.26	9,719	-0.39	-0.40

9,720 | -0.39 | -0.40
9,721 | -0.39 | -0.40
9,722 | -0.39 | -0.40
9,723 | -0.39 | -0.41
9,724 | -0.39 | -0.41
9,725 | -0.39 | -0.41
9,726 | -0.39 | -0.41
9,727 | -0.39 | -0.41
9,728 | -0.39 | -0.41
9,729 | -0.39 | -0.41
9,730 | -0.39 | -0.41
9,731 | -0.39 | -0.41
9,732 | -0.39 | -0.41
9,733 | -0.39 | -0.42
9,734 | -0.39 | -0.42
9,735 | -0.39 | -0.42
9,736 | -0.39 | -0.42
9,737 | -0.39 | -0.42
9,738 | -0.39 | -0.42
9,739 | -0.39 | -0.42
9,740 | -0.39 | -0.42
9,741 | -0.39 | -0.42
9,742 | -0.39 | -0.42
9,743 | -0.39 | -0.43
9,744 | -0.39 | -0.43
9,745 | -0.39 | -0.43
9,746 | -0.39 | -0.43
9,747 | -0.39 | -0.43
9,748 | -0.39 | -0.43
9,749 | -0.39 | -0.43

9,750 | -0.39 | -0.43
9,751 | -0.39 | -0.43
9,752 | -0.39 | -0.43
9,753 | -0.39 | -0.44
9,754 | -0.39 | -0.44
9,755 | -0.39 | -0.44
9,756 | -0.39 | -0.44
9,757 | -0.39 | -0.44
9,758 | -0.44 | -0.44
9,759 | -0.44 | -0.44
9,760 | -0.44 | -0.44
9,761 | -0.44 | -0.43
9,762 | -0.44 | -0.43
9,763 | -0.44 | -0.43
9,764 | -0.44 | -0.43
9,765 | -0.44 | -0.42
9,766 | -0.44 | -0.42
9,767 | -0.44 | -0.42
9,768 | -0.44 | -0.42
9,769 | -0.44 | -0.42
9,770 | -0.44 | -0.41
9,771 | -0.44 | -0.41
9,772 | -0.44 | -0.41
9,773 | -0.44 | -0.41
9,774 | -0.44 | -0.40
9,775 | -0.44 | -0.40
9,776 | -0.44 | -0.40
9,777 | -0.44 | -0.40
9,778 | -0.44 | -0.40
9,779 | -0.44 | -0.39

9,780	-0.44	-0.39	9,810	-0.33	-0.31
9,781	-0.44	-0.39	9,811	-0.33	-0.30
9,782	-0.44	-0.39	9,812	-0.33	-0.29
9,783	-0.44	-0.39	9,813	-0.33	-0.28
9,784	-0.44	-0.38	9,814	-0.33	-0.28
9,785	-0.44	-0.38	9,815	-0.33	-0.27
9,786	-0.44	-0.38	9,816	-0.33	-0.26
9,787	-0.44	-0.38	9,817	-0.33	-0.25
9,788	-0.44	-0.37	9,818	-0.33	-0.25
9,789	-0.44	-0.37	9,819	-0.33	-0.24
9,790	-0.44	-0.37	9,820	-0.33	-0.23
9,791	-0.44	-0.37	9,821	-0.33	-0.22
9,792	-0.44	-0.37	9,822	-0.33	-0.22
9,793	-0.44	-0.36	9,823	-0.33	-0.21
9,794	-0.44	-0.36	9,824	-0.33	-0.20
9,795	-0.44	-0.36	9,825	-0.33	-0.19
9,796	-0.44	-0.36	9,826	-0.33	-0.19
9,797	-0.44	-0.35	9,827	-0.33	-0.18
9,798	-0.44	-0.35	9,828	-0.33	-0.17
9,799	-0.44	-0.35	9,829	-0.33	-0.16
9,800	-0.44	-0.35	9,830	-0.33	-0.16
9,801	-0.44	-0.35	9,831	-0.33	-0.15
9,802	-0.44	-0.34	9,832	-0.33	-0.14
9,803	-0.44	-0.34	9,833	-0.33	-0.13
9,804	-0.44	-0.34	9,834	-0.33	-0.12
9,805	-0.44	-0.34	9,835	-0.33	-0.12
9,806	-0.44	-0.33	9,836	-0.33	-0.11
9,807	-0.44	-0.33	9,837	-0.33	-0.10
9,808	-0.33	-0.32	9,838	-0.33	-0.09
9,809	-0.33	-0.31	9,839	-0.33	-0.09

9,840	-0.33	-0.08	9,870	0.05	0.05
9,841	-0.33	-0.07	9,871	0.05	0.05
9,842	-0.33	-0.06	9,872	0.05	0.05
9,843	-0.33	-0.06	9,873	0.05	0.05
9,844	-0.33	-0.05	9,874	0.05	0.05
9,845	-0.33	-0.04	9,875	0.05	0.05
9,846	-0.33	-0.03	9,876	0.05	0.05
9,847	-0.33	-0.03	9,877	0.05	0.05
9,848	-0.33	-0.02	9,878	0.05	0.05
9,849	-0.33	-0.01	9,879	0.05	0.05
9,850	-0.33	0.00	9,880	0.05	0.05
9,851	-0.33	0.00	9,881	0.05	0.05
9,852	-0.33	0.01	9,882	0.05	0.04
9,853	-0.33	0.02	9,883	0.05	0.04
9,854	-0.33	0.03	9,884	0.05	0.04
9,855	-0.33	0.03	9,885	0.05	0.04
9,856	-0.33	0.04	9,886	0.05	0.04
9,857	0.05	0.05	9,887	0.05	0.04
9,858	0.05	0.05	9,888	0.05	0.04
9,859	0.05	0.05	9,889	0.05	0.04
9,860	0.05	0.05	9,890	0.05	0.04
9,861	0.05	0.05	9,891	0.05	0.04
9,862	0.05	0.05	9,892	0.05	0.04
9,863	0.05	0.05	9,893	0.05	0.04
9,864	0.05	0.05	9,894	0.05	0.04
9,865	0.05	0.05	9,895	0.05	0.04
9,866	0.05	0.05	9,896	0.05	0.04
9,867	0.05	0.05	9,897	0.05	0.04
9,868	0.05	0.05	9,898	0.05	0.04
9,869	0.05	0.05	9,899	0.05	0.04

9,900	0.05	0.04	9,930	0.04	-0.10
9,901	0.05	0.04	9,931	0.04	-0.11
9,902	0.05	0.04	9,932	0.04	-0.11
9,903	0.05	0.04	9,933	0.04	-0.12
9,904	0.05	0.04	9,934	0.04	-0.12
9,905	0.05	0.04	9,935	0.04	-0.13
9,906	0.04	0.03	9,936	0.04	-0.13
9,907	0.04	0.03	9,937	0.04	-0.14
9,908	0.04	0.02	9,938	0.04	-0.14
9,909	0.04	0.02	9,939	0.04	-0.15
9,910	0.04	0.01	9,940	0.04	-0.16
9,911	0.04	0.01	9,941	0.04	-0.16
9,912	0.04	0.00	9,942	0.04	-0.17
9,913	0.04	0.00	9,943	0.04	-0.17
9,914	0.04	-0.01	9,944	0.04	-0.18
9,915	0.04	-0.02	9,945	0.04	-0.18
9,916	0.04	-0.02	9,946	0.04	-0.19
9,917	0.04	-0.03	9,947	0.04	-0.20
9,918	0.04	-0.03	9,948	0.04	-0.20
9,919	0.04	-0.04	9,949	0.04	-0.21
9,920	0.04	-0.04	9,950	0.04	-0.21
9,921	0.04	-0.05	9,951	0.04	-0.22
9,922	0.04	-0.06	9,952	0.04	----
9,923	0.04	-0.06	9,953	0.04	----
9,924	0.04	-0.07	9,954	0.04	----
9,925	0.04	-0.07	9,955	-0.24	----
9,926	0.04	-0.08	9,956	-0.24	----
9,927	0.04	-0.08	9,957	-0.24	----
9,928	0.04	-0.09	9,958	-0.24	----
9,929	0.04	-0.09	9,959	-0.24	----

9,960 | -0.24 | ----
9,961 | -0.24 | ----
9,962 | -0.24 | ----
9,963 | -0.24 | ----
9,964 | -0.24 | ----
9,965 | -0.24 | ----
9,966 | -0.24 | ----
9,967 | -0.24 | ----
9,968 | -0.24 | ----
9,969 | -0.24 | ----
9,970 | -0.24 | ----
9,971 | -0.24 | ----
9,972 | -0.24 | ----
9,973 | -0.24 | ----
9,974 | -0.24 | ----
9,975 | -0.24 | ----
9,976 | -0.24 | ----
9,977 | -0.24 | ----
9,978 | -0.24 | ----
9,979 | -0.24 | ----
9,980 | -0.24 | ----
9,981 | -0.24 | ----
9,982 | -0.24 | ----
9,983 | -0.24 | ----
9,984 | -0.24 | ----
9,985 | -0.24 | ----
9,986 | -0.24 | ----
9,987 | -0.24 | ----
9,988 | -0.24 | ----
9,989 | -0.24 | ----

9,990 | -0.24 | ----
9,991 | -0.24 | ----
9,992 | -0.24 | ----
9,993 | -0.24 | ----
9,994 | -0.24 | ----
9,995 | -0.24 | ----
9,996 | -0.24 | ----
9,997 | -0.24 | ----
9,998 | -0.24 | ----
9,999 | -0.24 | ----
10,000 | -0.24 | ----

DATASET 15

Dataset 15 (GlobalWindow°C). This dataset contains the following global variables: (Year)—calendar years from 1880 to 2014; (TDg)—global annual-scale atmospheric temperature deviations from calendar year 1880 to 2014; (TDgSW)—global window-scale atmospheric temperature deviations from calendar year 1929 to 2014, which are the mean temperature deviations obtained with a window of 50 years sliding in steps of 1 year. TDg and TDgSW are in degrees Celsius (°C) relative to the Vostok base Kyr1 = 0.00 °C.[375]

Year	TDg	TDgSW
2014	1.32	0.93
2013	1.24	0.91
2012	1.21	0.90
2011	1.19	0.89
2010	1.30	0.88
2009	1.23	0.87
2008	1.13	0.86
2007	1.27	0.85
2006	1.23	0.84
2005	1.30	0.82
2004	1.15	0.81
2003	1.24	0.79
2002	1.25	0.78
2001	1.17	0.77
2000	1.04	0.76
1999	1.04	0.75
1998	1.25	0.74
1997	1.10	0.72
1996	0.97	0.71
1995	1.07	0.71
1994	0.93	0.70
1993	0.85	0.69
1992	0.83	0.69
1991	1.02	0.69
1990	1.03	0.68
1989	0.88	0.68
1988	0.99	0.67
1987	0.92	0.66
1986	0.78	0.66
1985	0.72	0.65
1984	0.76	0.65
1983	0.91	0.65
1982	0.73	0.63
1981	0.92	0.63
1980	0.87	0.62
1979	0.76	0.62
1978	0.69	0.61
1977	0.79	0.60
1976	0.52	0.60
1975	0.63	0.60
1974	0.57	0.59
1973	0.80	0.59
1972	0.67	0.58
1971	0.58	0.57
1970	0.68	0.57
1969	0.70	0.56
1968	0.59	0.56
1967	0.63	0.55
1966	0.60	0.54
1965	0.54	0.54
1964	0.45	0.53
1963	0.72	0.53
1962	0.68	0.52
1961	0.70	0.52
1960	0.60	0.51
1959	0.67	0.50
1958	0.69	0.49
1957	0.67	0.48
1956	0.46	0.47

1955	0.52	0.47	1925	0.42	----
1954	0.52	0.46	1924	0.39	----
1953	0.73	0.46	1923	0.38	----
1952	0.66	0.45	1922	0.34	----
1951	0.58	0.44	1921	0.43	----
1950	0.45	0.44	1920	0.36	----
1949	0.53	0.44	1919	0.35	----
1948	0.54	0.44	1918	0.32	----
1947	0.59	0.43	1917	0.20	----
1946	0.56	0.43	1916	0.27	----
1945	0.64	0.43	1915	0.47	----
1944	0.77	0.42	1914	0.40	----
1943	0.69	0.41	1913	0.24	----
1942	0.69	0.40	1912	0.22	----
1941	0.71	0.40	1911	0.19	----
1940	0.70	0.39	1910	0.18	----
1939	0.64	0.38	1909	0.16	----
1938	0.69	0.38	1908	0.20	----
1937	0.67	0.37	1907	0.21	----
1936	0.53	0.37	1906	0.37	----
1935	0.49	0.36	1905	0.33	----
1934	0.54	0.36	1904	0.20	----
1933	0.39	0.36	1903	0.27	----
1932	0.53	0.36	1902	0.33	----
1931	0.56	0.36	1901	0.43	----
1930	0.52	0.36	1900	0.48	----
1929	0.32	0.35	1899	0.43	----
1928	0.47	----	1898	0.32	----
1927	0.45	----	1897	0.45	----
1926	0.54	----	1896	0.46	----

1895 | 0.39 | ----
1894 | 0.31 | ----
1893 | 0.28 | ----
1892 | 0.33 | ----
1891 | 0.37 | ----
1890 | 0.30 | ----
1889 | 0.53 | ----
1888 | 0.44 | ----
1887 | 0.31 | ----
1886 | 0.39 | ----
1885 | 0.37 | ----
1884 | 0.36 | ----
1883 | 0.44 | ----
1882 | 0.46 | ----
1881 | 0.49 | ----
1880 | 0.41 | ----

DATASET 18

Dataset 18 (VostokTempCO2Synch412Kyrs). This dataset contains the following Vostok variables: (Age)—from 414,085 to 2,342 years before calendar year 1989; (CO2v)—irregular-scale Vostok atmospheric carbon dioxide concentrations in parts per million (ppm); (TDv)—irregular-scale Vostok atmospheric temperature deviations in degrees Celsius (°C) relative to the Vostok base Kyr1 = 0.00 °C.[376]

Age | CO2v | TDv
2,342 | 284.7 | -0.64
3,634 | 272.8 | 1.01
3,833 | 268.1 | 0.09
6,220 | 262.2 | 0.37
7,327 | 254.6 | 0.04
8,113 | 259.6 | 1.64
10,123 | 261.6 | -0.12
11,013 | 263.7 | 0.27
11,326 | 244.8 | 0.16
11,719 | 238.3 | -1.53
13,405 | 236.2 | -3.13
13,989 | 225.3 | -1.85
17,695 | 182.2 | -7.37
19,988 | 189.2 | -7.37
22,977 | 191.6 | -7.49
26,303 | 188.5 | -6.94
27,062 | 191.7 | -8.36
31,447 | 205.4 | -7.49
33,884 | 209.1 | -6.88
39,880 | 209.1 | -5.69
44,766 | 189.3 | -6.62
47,024 | 188.4 | -6.29
48,229 | 210.1 | -5.97
49,414 | 215.7 | -5.35
51,174 | 190.4 | -3.43
57,068 | 221.8 | -3.73
57,799 | 210.4 | -4.51
63,687 | 195.4 | -5.95
65,701 | 191.4 | -8.01
66,883 | 195.0 | -7.24
72,849 | 227.4 | -5.44
75,360 | 229.2 | -4.79
78,995 | 217.1 | -5.55
80,059 | 221.8 | -3.85
82,858 | 231.0 | -2.66
84,929 | 241.1 | -2.70
85,727 | 236.4 | -2.92
86,323 | 228.1 | -3.24
87,180 | 214.2 | -3.96
88,051 | 217.0 | -3.98
89,363 | 208.0 | -5.25
91,691 | 224.3 | -3.64
92,460 | 228.4 | -4.37
95,349 | 232.1 | -2.94
99,842 | 225.9 | -3.03
100,833 | 230.9 | -2.96
101,829 | 236.9 | -1.98
103,372 | 228.2 | -3.31
105,213 | 236.9 | -3.79
106,203 | 230.7 | -3.14
108,308 | 238.2 | -6.25
108,994 | 245.7 | -5.81
110,253 | 251.3 | -5.12
111,456 | 256.8 | -5.22
112,577 | 266.3 | -5.32
113,472 | 261.4 | -4.41
114,082 | 274.6 | -3.92
114,738 | 273.3 | -4.06
116,175 | 262.5 | -2.40

117,519	267.6	-1.00	136,359	202.5	-6.96
118,396	273.8	-1.00	136,659	195.9	-6.93
119,273	272.0	-0.94	137,383	194.4	-7.42
120,002	265.2	-0.47	137,694	193.4	-7.04
120,652	277.7	0.51	138,226	190.2	-8.69
121,961	272.2	0.96	139,445	192.3	-8.01
122,606	276.5	0.72	141,312	196.5	-7.86
123,815	268.7	1.18	142,357	190.4	-7.87
123,858	266.6	1.12	145,435	197.0	-7.83
124,306	266.3	0.99	150,303	191.9	-6.94
124,571	279.8	1.09	154,471	189.0	-7.26
124,876	277.2	0.74	155,299	185.5	-7.60
125,746	273.8	0.90	160,494	204.4	-6.33
126,023	267.1	0.88	162,996	191.6	-6.19
126,475	262.5	1.37	165,278	183.8	-7.83
126,809	262.6	2.40	169,870	197.9	-6.40
127,445	275.4	2.86	172,596	197.8	-7.09
128,300	274.1	3.19	175,440	190.3	-6.56
128,399	287.1	3.63	176,271	190.1	-7.05
128,652	286.8	2.56	178,550	207.7	-7.69
129,007	282.7	1.92	180,779	213.2	-5.18
129,411	264.1	1.61	181,617	217.7	-6.27
129,755	263.4	2.42	183,355	199.8	-7.75
130,167	259.0	1.88	185,063	203.5	-7.63
131,789	240.4	-1.09	187,199	210.7	-6.81
133,334	224.0	-3.37	189,335	231.4	-6.19
134,205	208.9	-5.04	191,057	231.5	-3.34
135,003	204.6	-6.42	192,632	218.0	-5.54
135,683	198.1	-6.63	195,625	220.1	-5.26
135,976	201.8	-6.73	199,025	242.6	-2.62

202,212	251.0	-1.42		225,509	233.1	-4.82
203,191	239.1	-1.49		225,888	224.5	-6.09
204,283	247.7	-2.40		226,710	232.4	-6.02
205,148	244.4	-3.15		227,384	233.9	-7.02
205,715	232.2	-3.80		227,840	241.7	-6.42
206,119	228.7	-4.51		230,703	245.2	-5.49
207,991	238.2	-3.61		231,382	252.2	-4.66
209,414	242.2	-2.66		231,990	241.4	-4.07
210,022	244.6	-2.77		232,570	247.4	-3.74
210,830	247.3	-1.78		233,102	243.1	-3.01
211,005	252.0	-1.75		233,646	239.2	-2.53
212,281	257.4	-1.63		234,126	245.7	-2.32
214,153	251.2	-1.03		234,470	245.9	-1.76
215,041	241.4	-1.20		234,781	247.4	-2.03
215,593	240.3	-0.72		235,213	252.9	-1.68
215,879	242.7	-1.11		236,236	259.8	-0.95
216,459	247.5	-1.67		237,831	279.0	2.37
217,009	251.7	-0.75		238,935	263.8	1.32
217,271	251.2	-0.44		239,250	252.4	0.91
217,676	245.4	-0.72		239,545	249.9	-0.30
218,342	240.5	-1.66		240,201	230.4	-2.04
219,680	212.2	-2.71		240,577	219.4	-2.99
220,182	216.2	-3.79		242,068	214.7	-5.55
220,760	207.2	-3.89		243,653	200.2	-5.32
221,054	208.9	-4.56		244,215	213.9	-6.41
221,612	205.7	-5.55		244,863	195.4	-6.57
222,958	203.4	-6.42		245,483	196.7	-6.17
223,446	215.7	-6.89		247,447	199.0	-6.22
224,630	236.9	-6.45		248,087	201.9	-6.66
225,299	234.5	-5.61		248,980	204.0	-6.62

250,461 | 203.9 | -5.51
251,521 | 209.7 | -6.03
252,959 | 208.9 | -6.02
253,880 | 214.7 | -6.19
255,233 | 228.2 | -5.40
256,053 | 199.9 | -5.90
256,501 | 211.7 | -6.46
257,247 | 188.7 | -6.97
258,477 | 194.2 | -7.22
259,228 | 198.9 | -7.63
259,958 | 184.7 | -7.78
260,754 | 190.4 | -8.01
261,595 | 193.9 | -7.74
262,411 | 194.2 | -7.84
263,207 | 198.4 | -7.67
264,046 | 193.2 | -7.98
264,834 | 202.2 | -7.40
266,492 | 211.0 | -7.21
267,434 | 215.4 | -7.19
268,679 | 223.7 | -6.15
270,680 | 231.4 | -5.29
273,012 | 226.4 | -3.95
274,445 | 230.4 | -3.35
275,218 | 231.0 | -3.50
277,925 | 220.4 | -3.58
278,602 | 217.2 | -3.98
279,543 | 207.7 | -4.66
282,301 | 212.7 | -5.33
283,492 | 213.2 | -5.09
286,217 | 224.4 | -5.82

287,846 | 236.2 | -4.65
290,571 | 240.2 | -5.13
291,769 | 240.7 | -4.53
292,474 | 250.2 | -3.10
293,676 | 244.9 | -4.84
294,615 | 225.9 | -5.14
295,849 | 227.9 | -4.34
297,131 | 233.2 | -4.35
298,051 | 237.9 | -4.07
299,020 | 239.0 | -3.57
299,877 | 241.9 | -2.69
300,646 | 251.7 | -2.61
301,496 | 256.8 | -2.76
302,456 | 257.2 | -2.95
303,334 | 246.9 | -1.32
303,953 | 272.7 | -2.50
304,590 | 251.7 | -3.40
305,306 | 244.7 | -4.09
307,131 | 255.9 | -4.48
308,101 | 249.2 | -4.05
310,039 | 256.3 | -2.90
310,930 | 260.4 | -2.17
311,774 | 260.3 | -1.78
313,493 | 266.3 | -1.07
315,143 | 266.2 | -0.51
315,940 | 270.2 | 0.00
316,681 | 271.9 | 0.27
317,445 | 275.2 | 0.06
318,980 | 265.0 | 0.47
319,754 | 271.8 | 0.65

320,378	272.7	0.29
321,386	273.2	1.82
322,111	282.4	3.21
322,582	289.2	3.55
322,827	288.4	3.64
323,485	298.7	3.60
324,189	278.2	3.26
324,991	285.8	2.25
325,527	278.7	1.14
326,239	270.5	-0.24
327,114	255.7	-1.94
328,097	241.9	-2.38
329,267	239.7	-3.68
330,208	234.2	-4.51
332,293	250.2	-6.36
333,627	200.7	-7.57
335,290	205.2	-7.30
336,972	204.9	-6.86
340,165	220.4	-6.78
342,998	221.2	-6.14
344,735	216.2	-6.06
347,610	209.2	-6.42
350,765	193.0	-7.18
352,412	186.2	-7.24
356,838	201.2	-6.69
359,688	206.4	-5.53
362,766	201.9	-5.59
366,221	214.7	-4.69
369,563	229.7	-4.38
373,014	227.0	-5.58
374,561	240.0	-4.86
378,194	246.9	-3.24
379,633	245.9	-4.32
384,909	264.7	-3.79
386,579	259.3	-5.17
390,589	255.2	-4.89
392,451	250.2	-4.74
394,628	266.3	-4.15
396,713	274.7	-2.07
400,390	278.0	-1.14
405,844	279.7	0.60
409,022	283.7	1.59
410,831	276.3	1.99
414,085	285.5	1.30

DATASET 22

Dataset 22 (Hawaii temperature and CO2). This dataset contains the following Hawaiian variables: (Year)—calendar years from 1959 to 2014; (TDh)—Hawaiian atmospheric temperature deviations in degrees Celsius (°C) relative to the Vostok base Kyr1 = 0.00 °C; and (CO2h)—Hawaiian atmospheric CO_2 concentrations in parts per million (ppm).

Year	TDh	CO2h
1959	1.41	316.0
1960	0.49	316.9
1961	0.53	317.6
1962	0.24	318.5
1963	0.14	319.0
1964	-0.02	319.6
1965	-0.15	320.0
1966	1.00	321.4
1967	0.13	322.2
1968	0.68	323.0
1969	0.74	324.6
1970	0.03	325.7
1971	-0.41	326.3
1972	0.09	327.5
1973	0.84	329.7
1974	-0.57	330.2
1975	-0.35	331.1
1976	-0.73	332.1
1977	0.76	333.8
1978	0.09	335.4
1979	0.32	336.8
1980	1.25	338.7
1981	1.50	340.1
1982	1.01	341.4
1983	1.88	343.0
1984	1.29	344.6
1985	1.19	346.0
1986	2.30	347.4
1987	1.47	349.2
1988	1.06	351.6
1989	0.36	353.1
1990	1.12	354.3
1991	1.52	355.6
1992	1.94	356.4
1993	1.45	357.1
1994	1.57	358.8
1995	2.20	360.8
1996	1.48	362.6
1997	1.25	363.7
1998	2.08	366.7
1999	1.08	368.3
2000	1.27	369.5
2001	1.40	371.1
2002	1.80	373.2
2003	2.00	375.8
2004	1.13	377.5
2005	1.47	379.8
2006	1.27	381.9
2007	1.35	383.8
2008	0.50	385.6
2009	0.77	387.4
2010	1.71	389.8
2011	0.43	391.6
2012	0.69	393.8
2013	0.67	396.5
2014	0.97	398.6

DATASET 27

Dataset 27 (ThermalGap). This dataset contains the following variables: (Year)—calendar years from 412,096 BCE to 2014; (TDg)—global annual atmospheric temperature deviations in degrees Celsius (°C) relative to the Vostok base Kyr1 = 0.00 °C, and (CO2h)—Hawaiian atmospheric carbon dioxide concentrations in parts per million (ppm), from 1959 to 2014; (CO2v)—irregular-scale Vostok atmospheric carbon dioxide concentrations in parts per million, and (TDv)—irregular-scale Vostok atmospheric temperature deviations in degrees Celsius (°C) relative to The Vostok base Kyr1 = 0.00 °C, from 412,096 to 9,024 BCE.

Year	TDg	CO2h	CO2v	TDv
2014	1.32	398.6	--	--
2013	1.24	396.5	--	--
2012	1.21	393.8	--	--
2011	1.19	391.6	--	--
2010	1.30	389.8	--	--
2009	1.23	387.4	--	--
2008	1.13	385.6	--	--
2007	1.27	383.8	--	--
2006	1.23	381.9	--	--
2005	1.30	379.8	--	--
2004	1.15	377.5	--	--
2003	1.24	375.8	--	--
2002	1.25	373.2	--	--
2001	1.17	371.1	--	--
2000	1.04	369.5	--	--
1999	1.04	368.3	--	--
1998	1.25	366.7	--	--
1997	1.10	363.7	--	--
1996	0.97	362.6	--	--
1995	1.07	360.8	--	--
1994	0.93	358.8	--	--
1993	0.85	357.1	--	--
1992	0.83	356.4	--	--
1991	1.02	355.6	--	--
1990	1.03	354.3	--	--
1989	0.88	353.1	--	--
1988	0.99	351.6	--	--
1987	0.92	349.2	--	--
1986	0.78	347.4	--	--
1985	0.72	346.0	--	--
1984	0.76	344.6	--	--
1983	0.91	343.0	--	--
1982	0.73	341.4	--	--
1981	0.92	340.1	--	--
1980	0.87	338.7	--	--
1979	0.76	336.8	--	--

1978	0.69	335.4	--	--
1977	0.79	333.8	--	--
1976	0.52	332.1	--	--
1975	0.63	331.1	--	--
1974	0.57	330.2	--	--
1973	0.80	329.7	--	--
1972	0.67	327.5	--	--
1971	0.58	326.3	--	--
1970	0.68	325.7	--	--
1969	0.70	324.6	--	--
1968	0.59	323.0	--	--
1967	0.63	322.2	--	--
1966	0.60	321.4	--	--
1965	0.54	320.0	--	--
1964	0.45	319.6	--	--
1963	0.72	319.0	--	--
1962	0.68	318.5	--	--
1961	0.70	317.6	--	--
1960	0.60	316.9	--	--
1959	0.67	316.0	--	--
-9,024	--	--	263.7	0.27
-9,337	--	--	244.8	0.16
-9,730	--	--	238.3	-1.53
-11,416	--	--	236.2	-3.13
-12,000	--	--	225.3	-1.85
-15,706	--	--	182.2	-7.37
-17,999	--	--	189.2	-7.37
-20,988	--	--	191.6	-7.49
-24,314	--	--	188.5	-6.94
-25,073	--	--	191.7	-8.36
-29,458	--	--	205.4	-7.49
-31,895	--	--	209.1	-6.88
-37,891	--	--	209.1	-5.69
-42,777	--	--	189.3	-6.62
-45,035	--	--	188.4	-6.29
-46,240	--	--	210.1	-5.97
-47,425	--	--	215.7	-5.35

-49,185	--	--	190.4	-3.43
-55,079	--	--	221.8	-3.73
-55,810	--	--	210.4	-4.51
-61,698	--	--	195.4	-5.95
-63,712	--	--	191.4	-8.01
-64,894	--	--	195.0	-7.24
-70,860	--	--	227.4	-5.44
-73,371	--	--	229.2	-4.79
-77,006	--	--	217.1	-5.55
-78,070	--	--	221.8	-3.85
-80,869	--	--	231.0	-2.66
-82,940	--	--	241.1	-2.70
-83,738	--	--	236.4	-2.92
-84,334	--	--	228.1	-3.24
-85,191	--	--	214.2	-3.96
-86,062	--	--	217.0	-3.98
-87,374	--	--	208.0	-5.25
-89,702	--	--	224.3	-3.64
-90,471	--	--	228.4	-4.37
-93,360	--	--	232.1	-2.94
-97,853	--	--	225.9	-3.03
-98,844	--	--	230.9	-2.96
-99,840	--	--	236.9	-1.98
-101,383	--	--	228.2	-3.31
-103,224	--	--	236.9	-3.79
-104,214	--	--	230.7	-3.14
-106,319	--	--	238.2	-6.25
-107,005	--	--	245.7	-5.81
-108,264	--	--	251.3	-5.12
-109,467	--	--	256.8	-5.22
-110,588	--	--	266.3	-5.32
-111,483	--	--	261.4	-4.41
-112,093	--	--	274.6	-3.92
-112,749	--	--	273.3	-4.06
-114,186	--	--	262.5	-2.40
-115,530	--	--	267.6	-1.00
-116,407	--	--	273.8	-1.00

-117,284	--	--	272.0	-0.94
-118,013	--	--	265.2	-0.47
-118,663	--	--	277.7	0.51
-119,972	--	--	272.2	0.96
-120,617	--	--	276.5	0.72
-121,826	--	--	268.7	1.18
-121,869	--	--	266.6	1.12
-122,317	--	--	266.3	0.99
-122,582	--	--	279.8	1.09
-122,887	--	--	277.2	0.74
-123,757	--	--	273.8	0.90
-124,034	--	--	267.1	0.88
-124,486	--	--	262.5	1.37
-124,820	--	--	262.6	2.40
-125,456	--	--	275.4	2.86
-126,311	--	--	274.1	3.19
-126,410	--	--	287.1	3.63
-126,663	--	--	286.8	2.56
-127,018	--	--	282.7	1.92
-127,422	--	--	264.1	1.61
-127,766	--	--	263.4	2.42
-128,178	--	--	259.0	1.88
-129,800	--	--	240.4	-1.09
-131,345	--	--	224.0	-3.37
-132,216	--	--	208.9	-5.04
-133,014	--	--	204.6	-6.42
-133,694	--	--	198.1	-6.63
-133,987	--	--	201.8	-6.73
-134,370	--	--	202.5	-6.96
-134,670	--	--	195.9	-6.93
-135,394	--	--	194.4	-7.42
-135,705	--	--	193.4	-7.04
-136,237	--	--	190.2	-8.69
-137,456	--	--	192.3	-8.01
-139,323	--	--	196.5	-7.86
-140,368	--	--	190.4	-7.87
-143,446	--	--	197.0	-7.83

-148,314	--	--	191.9	-6.94
-152,482	--	--	189.0	-7.26
-153,310	--	--	185.5	-7.60
-158,505	--	--	204.4	-6.33
-161,007	--	--	191.6	-6.19
-163,289	--	--	183.8	-7.83
-167,881	--	--	197.9	-6.40
-170,607	--	--	197.8	-7.09
-173,451	--	--	190.3	-6.56
-174,282	--	--	190.1	-7.05
-176,561	--	--	207.7	-7.69
-178,790	--	--	213.2	-5.18
-179,628	--	--	217.7	-6.27
-181,366	--	--	199.8	-7.75
-183,074	--	--	203.5	-7.63
-185,210	--	--	210.7	-6.81
-187,346	--	--	231.4	-6.19
-189,068	--	--	231.5	-3.34
-190,643	--	--	218.0	-5.54
-193,636	--	--	220.1	-5.26
-197,036	--	--	242.6	-2.62
-200,223	--	--	251.0	-1.42
-201,202	--	--	239.1	-1.49
-202,294	--	--	247.7	-2.40
-203,159	--	--	244.4	-3.15
-203,726	--	--	232.2	-3.80
-204,130	--	--	228.7	-4.51
-206,002	--	--	238.2	-3.61
-207,425	--	--	242.2	-2.66
-208,033	--	--	244.6	-2.77
-208,841	--	--	247.3	-1.78
-209,016	--	--	252.0	-1.75
-210,292	--	--	257.4	-1.63
-212,164	--	--	251.2	-1.03
-213,052	--	--	241.4	-1.20
-213,604	--	--	240.3	-0.72
-213,890	--	--	242.7	-1.11

-214,470	--	--	247.5	-1.67
-215,020	--	--	251.7	-0.75
-215,282	--	--	251.2	-0.44
-215,687	--	--	245.4	-0.72
-216,353	--	--	240.5	-1.66
-217,691	--	--	212.2	-2.71
-218,193	--	--	216.2	-3.79
-218,771	--	--	207.2	-3.89
-219,065	--	--	208.9	-4.56
-219,623	--	--	205.7	-5.55
-220,969	--	--	203.4	-6.42
-221,457	--	--	215.7	-6.89
-222,641	--	--	236.9	-6.45
-223,310	--	--	234.5	-5.61
-223,520	--	--	233.1	-4.82
-223,899	--	--	224.5	-6.09
-224,721	--	--	232.4	-6.02
-225,395	--	--	233.9	-7.02
-225,851	--	--	241.7	-6.42
-228,714	--	--	245.2	-5.49
-229,393	--	--	252.2	-4.66
-230,001	--	--	241.4	-4.07
-230,581	--	--	247.4	-3.74
-231,113	--	--	243.1	-3.01
-231,657	--	--	239.2	-2.53
-232,137	--	--	245.7	-2.32
-232,481	--	--	245.9	-1.76
-232,792	--	--	247.4	-2.03
-233,224	--	--	252.9	-1.68
-234,247	--	--	259.8	-0.95
-235,842	--	--	279.0	2.37
-236,946	--	--	263.8	1.32
-237,261	--	--	252.4	0.91
-237,556	--	--	249.9	-0.30
-238,212	--	--	230.4	-2.04
-238,588	--	--	219.4	-2.99
-240,079	--	--	214.7	-5.55

-241,664	--	--	200.2	-5.32
-242,226	--	--	213.9	-6.41
-242,874	--	--	195.4	-6.57
-243,494	--	--	196.7	-6.17
-245,458	--	--	199.0	-6.22
-246,098	--	--	201.9	-6.66
-246,991	--	--	204.0	-6.62
-248,472	--	--	203.9	-5.51
-249,532	--	--	209.7	-6.03
-250,970	--	--	208.9	-6.02
-251,891	--	--	214.7	-6.19
-253,244	--	--	228.2	-5.40
-254,064	--	--	199.9	-5.90
-254,512	--	--	211.7	-6.46
-255,258	--	--	188.7	-6.97
-256,488	--	--	194.2	-7.22
-257,239	--	--	198.9	-7.63
-257,969	--	--	184.7	-7.78
-258,765	--	--	190.4	-8.01
-259,606	--	--	193.9	-7.74
-260,422	--	--	194.2	-7.84
-261,218	--	--	198.4	-7.67
-262,057	--	--	193.2	-7.98
-262,845	--	--	202.2	-7.40
-264,503	--	--	211.0	-7.21
-265,445	--	--	215.4	-7.19
-266,690	--	--	223.7	-6.15
-268,691	--	--	231.4	-5.29
-271,023	--	--	226.4	-3.95
-272,456	--	--	230.4	-3.35
-273,229	--	--	231.0	-3.50
-275,936	--	--	220.4	-3.58
-276,613	--	--	217.2	-3.98
-277,554	--	--	207.7	-4.66
-280,312	--	--	212.7	-5.33
-281,503	--	--	213.2	-5.09
-284,228	--	--	224.4	-5.82

-285,857	--	--	236.2	-4.65
-288,582	--	--	240.2	-5.13
-289,780	--	--	240.7	-4.53
-290,485	--	--	250.2	-3.10
-291,687	--	--	244.9	-4.84
-292,626	--	--	225.9	-5.14
-293,860	--	--	227.9	-4.34
-295,142	--	--	233.2	-4.35
-296,062	--	--	237.9	-4.07
-297,031	--	--	239.0	-3.57
-297,888	--	--	241.9	-2.69
-298,657	--	--	251.7	-2.61
-299,507	--	--	256.8	-2.76
-300,467	--	--	257.2	-2.95
-301,345	--	--	246.9	-1.32
-301,964	--	--	272.7	-2.50
-302,601	--	--	251.7	-3.40
-303,317	--	--	244.7	-4.09
-305,142	--	--	255.9	-4.48
-306,112	--	--	249.2	-4.05
-308,050	--	--	256.3	-2.90
-308,941	--	--	260.4	-2.17
-309,785	--	--	260.3	-1.78
-311,504	--	--	266.3	-1.07
-313,154	--	--	266.2	-0.51
-313,951	--	--	270.2	0.00
-314,692	--	--	271.9	0.27
-315,456	--	--	275.2	0.06
-316,991	--	--	265.0	0.47
-317,765	--	--	271.8	0.65
-318,389	--	--	272.7	0.29
-319,397	--	--	273.2	1.82
-320,122	--	--	282.4	3.21
-320,593	--	--	289.2	3.55
-320,838	--	--	288.4	3.64
-321,496	--	--	298.7	3.60
-322,200	--	--	278.2	3.26

-323,002	--	--	285.8	2.25
-323,538	--	--	278.7	1.14
-324,250	--	--	270.5	-0.24
-325,125	--	--	255.7	-1.94
-326,108	--	--	241.9	-2.38
-327,278	--	--	239.7	-3.68
-328,219	--	--	234.2	-4.51
-330,304	--	--	250.2	-6.36
-331,638	--	--	200.7	-7.57
-333,301	--	--	205.2	-7.30
-334,983	--	--	204.9	-6.86
-338,176	--	--	220.4	-6.78
-341,009	--	--	221.2	-6.14
-342,746	--	--	216.2	-6.06
-345,621	--	--	209.2	-6.42
-348,776	--	--	193.0	-7.18
-350,423	--	--	186.2	-7.24
-354,849	--	--	201.2	-6.69
-357,699	--	--	206.4	-5.53
-360,777	--	--	201.9	-5.59
-364,232	--	--	214.7	-4.69
-367,574	--	--	229.7	-4.38
-371,025	--	--	227.0	-5.58
-372,572	--	--	240.0	-4.86
-376,205	--	--	246.9	-3.24
-377,644	--	--	245.9	-4.32
-382,920	--	--	264.7	-3.79
-384,590	--	--	259.3	-5.17
-388,600	--	--	255.2	-4.89
-390,462	--	--	250.2	-4.74
-392,639	--	--	266.3	-4.15
-394,724	--	--	274.7	-2.07
-398,401	--	--	278.0	-1.14
-403,855	--	--	279.7	0.60
-407,033	--	--	283.7	1.59
-408,842	--	--	276.3	1.99
-412,096	--	--	285.5	1.30

ABSTRACTS OTHER BOOKS

THINK SMARTER...

Think Smarter with Nemonik Thinking. (Schade, 2016). This is the operating manual for your mind that you should have received at birth. Nemonik thinking is a smarter way of thinking that aims to maximize your success by evaluating seventeen nemoniks, which are memorized keywords describing all the perceived aspects of your mind, reality, and their interaction. Success is obtaining what you seek and escaping what you suffer. Therefore, it is goal oriented. To maximize your success, nemonik thinking mobilizes your hidden genius, accelerates your thinking, improves your memory, reveals opportunities and threats, creates questions and ideas, and reduces your stress levels. It is like playing a musical keyboard with seventeen keys producing an infinite repertoire of smart strategies. Nemonik thinking is unique because it is the first exhaustive and transferable way of thinking. Comparisons with Sir Richard Branson's way of thinking show that it is extremely productive. Unfortunately, the educational system conditions students still with pass-fail grades to win. Winning is defeating opponents in competition. Therefore, it is conflict oriented. The compulsion to win inhibits the truth and, therefore, fosters the corrupted way of conventional thinking. Conventional thinking creates the malignant cognitive virus CS7. In turn, that virus consolidates conventional thinking with cognitive dissonance and groupthink. Conventional thinking is time consuming. Hence, the less time you have, the greater the necessity to study nemonik thinking. You might be the best thinker in the world, but only nemonik thinking could make you the smartest thinker you can be.

Download a free eBook version
@ nemonik-thinking.org

GLOSSARY...

Glossary of Nemonik Thinking (Schade 2016). Nemonik thinking is a competitive advantage because it mobilizes your hidden genius, accelerates your thinking, improves your memory, prevents blind-spots, and reveals opportunities, while its constant preparedness reduces stress levels. Definitions associated with the mind and reality are inherently hypothetical, fuzzy, and intertwined. Nevertheless, to improve our understanding of the way we think, we have to identify, differentiate, and define those components. Therefore, this glossary provides descriptions for the concepts associated with nemonik thinking. To become skilled in nemonik thinking, it is recommended to study— *Think Smarter with Nemonik Thinking (Schade, 2016).*

Download a free eBook version
@ nemonik-thinking.org

DICTIONARY...

Dictionary Nemonik Thinking (Schade 2016). Nemonik
thinking mobilizes your hidden genius, accelerates your
thinking, improves your memory, reveals opportunities and
threats, creates questions and ideas, and reduces your stress
levels. Nemonik thinking divides the mind into 17 nemonik
regions. Those regions defragment information, which
facilitates the storage, maintenance, recall, and processing of
associated information from memory. However, the
boundaries of those nemonik regions are fuzzy. Therefore,
the aim of this dictionary is to differentiate them by providing
keywords for each nemonik concept. The first part of this
dictionary translates nemonik concepts into common
keywords (e.g. *advance* into attack, bypass, etc.). In contrast,
the second part translates common keywords into nemonik
concepts (e.g. attack, bypass, etc. into *advance*). This
dictionary shows that the complexity of conventional
thinking comprises thousands of keywords that can be
simplified to 17 nemoniks. This reduction will increase the
speed of your thinking. To become skilled in nemonik
thinking, it is recommended to study—*Think Smarter with
Nemonik Thinking (Schade, 2016).*

Download a free eBook version
@ nemonik-thinking.org

LAO ZI'S DAO DE JING

Lao Zi's Dao De Jing—The Way (Schade, planned 2017). In one curt sentence, Lao Zi explains the core of his book—Use it to obtain what you seek and to escape what you suffer. His inspirational guideline introduces the sophisticated yet simple principle of Dao. This principle explains the Universe, the meaning of life, and our place in nature. For more than two and a half thousand years, *Dao De Jing* has been shrouded in mystery. Many scholars have studied that intriguing manuscript by peeling away layer after layer of meaning to unravel its cryptic secrets. Nevertheless, this interpretation shows that *Dao De Jing* preserved its ancient secrets within a prosaic collection of aphorisms. These mysteries are revealed for the first time ever in a clearly understandable way, imparting forgotten knowledge about the Universe and the art of living. To become skilled in nemonik thinking, it is recommended to study—*Think Smarter with Nemonik Thinking (Schade, 2016).*

Download a free eBook version
@ nemonik-thinking.org

SUN ZI'S THE ART OF WAR

Sun Zi's Bin Fa—The Art of War (Schade, planned 2017). Sun Zi (~6th cent. BC) was a Chinese warrior-philosopher who wrote the military classic *Bing Fa* or *The Art of War.* Although his book is about war, his strategies apply to every facet of daily life. Sun Zi deals with the art of positioning yourself in space, matter, and time. He addresses the questions raised by nemonik thinking of where, what, and when to advance, stay, retreat, accumulate, preserve, dispose, act, wait, prepare, accept, reject, reveal, and conceal. Think smarter and incorporate Sun Zi's strategies in your thinking. To become skilled in nemonik thinking, it is recommended to study— *Think Smarter with Nemonik Thinking (Schade, 2016).*

Download a free eBook version
@ nemonik-thinking.org

DECLARATION OF INDEPENDENCE

I, Dr Auke Schade, declare that this study and the development of nemonik thinking were funded by private resources. No part of this study, or the development of nemonik thinking, was supported, financially or otherwise, by any third party including individuals, stakeholders, charities, commercial, academic, political, ideological, military, religious, and secret organizations. Consequently, I am an independent researcher and do not have to please anyone.

The main global problems are symptoms of humanity's dramatically failing way of thinking. Although a huge and immediate threat, climate change is only one of the many symptoms. Seen the lethargic response of leaders to global warming, it would be unwise to rely on the global establishment for adequate action. Turning the tide in time will require huge sacrifices and resources. Therefore, support from any individual or organisation will be welcomed, as long as it will not comprise my academic integrity.

Now, after the completion of this study and the development of nemonik thinking, I feel even free to approach the oil and coal industries for funding. Confirmation of the bilateral climate-change hypothesis would transform them from villains into heroes. Their industrial CO_2 might save us from living on a frozen planet.

ENDNOTES

[1] For more information, see the appendix Nemonik Thinking.

[2] The term solar heat would be too limiting. Although relatively small, heat is also generated by domestic and industrial human activities.

[3] The nemonik accelerator is an active tool to accelerate our thinking. It is based on Hegel's dialectic that describes the development of knowledge. His dialectic holds that each thesis evokes an antithesis, which is followed by a synthesis, which becomes the new thesis. However, Hegel's' dialectic is a passive description, ratchet than an active tool. See the Glossary.

[4] A thesis is a description of reality. An antithesis is a description of reality that contradicts a thesis. A synthesis is a description of reality that merges a thesis and an antithesis into a new thesis.

[5] The nemonik accelerator is based on Hegel's dialectic concerning the development of knowledge. Hegel's dialectic focuses on observation, while the nemonik accelerator focuses on the next action to be taken.

[6] File: Vostok423Kyrs-TemperatureKyrs.xlsm / Dataset 1.0 (VostokRawTemp)|1| / cells A4-5 /

[7] File: Global135Yrs-TemperatureYrs.xlsm / Dataset 2.1 (AnnualMeanTemp) / cell C136 /

[8] File: Global135Yrs-TemperatureYrs.xlsm / Integrate (synchronized) / cell E2 /

[9] See the section: Global versus Vostok °C window-scale.

[10] 'Accept' (support) and 'reject' are nemoniks associated with the perception of information.

[11] Unless indicated otherwise, all carbon dioxide concentrations (CO_2) are atmospheric CO_2 concentrations expressed in parts per million (ppm).

[12] File: CO2-Temperature1959-2014 / Annual CO2 Data / cell E57 /

[13] File: Vostok423Kyrs-TemperatureCO2.xlsm / ThicknessCO2Layer / cell B2 /

[14] A millibar is one-thousands of a bar, which is a measure of atmospheric pressure equal to 100 Pascal's.

[15] File: Vostok423Kyrs-TemperatureCO2.xlsm / ThicknessCO2Layer / cell A1 /

[16] File: CO2-Temperature1959-2014.xlsm / HawaiiCO2-GlobalTemp/ cells D3 /

[17] File: Vostok423Kyrs-TemperatureCO2.xlsm / VostokCO2 (Kyr423-Kyr1) / cells C286 // & // Vostok423Kyrs-TemperatureCO2.xlsm / VostokCO2 (Kyr10-Kyr1) / cell C10 /

[18] File: CO2-Temperature1959-2014 / HawaiiCO2-GlobalTemp / cells A70-F72 /

[19] File: Vostok423Kyrs-TemperatureCO2 / VostokCO2 (Kyr423-Kyr1) / cells C2-8 /

[20] File: Vostok423Kyrs-TemperatureCO2 / VostokCO2 (Kyr10-Kyr1) / cell D10 /

[21] File: Vostok423Kyrs-TemperatureCO2(Ranked-rpTestCorrelationTDv-CO2v) / p-value / cells B10003-10014 // & // File: HawaiiGlobalTemperatureCO2(Ranked-rpTestCorrelationTDh-TDg).xlsm / p-value / B10003-10014 /

[22] File: CO2-Temperature1959-2014.xlsm / CO2LeadTemp (lag1) / cells M65-N68 / The predictions are retrospective, because the predictions are based on the dataset that provide the information for the predictions. It is far from perfect, but it is the best what can be done with the limited amount of data.

[23] File: ThermalGap.xlsm / VostokGlobalTempCO2Scatter / This graph is simplified for readability. For more detail, see the complete graph in the section concerning statistical analyses.

[24] File: ThermalGap.xlsm / VostokGlobal-TempCO2(data) /

[25] File: ThermalGap / VostokCO2(PredictVostokTemp)|25 / cell E288 // & // HawaiiCO2(PredictGlobalTemp)|26 / cell C3 /

[26] File: ThermalGap / HawaiiCO2(PredictGlobalTemp) / cell E69 // This is the estimated temperature. The observed temperature in 2014 equals 1.32 °C.

[27] File: ThermalGap.xlsm / ThermalGap / cell B2-E2 /

[28] File: _Vostok423Kyrs-TemperatureCO2 / VostokTempCO2 (graph) / cell B3 /

[29] File: / Vostok423Kyrs-TemperatureCO2 / VostokCO2 (Kyr10-Kyr1) / cell A2 /

[30] The slope of the linear function will decrease over time, because the glacial pressure to reduce the temperature will increase, while the CO2 increases with global industrialization.

[31] File: _Vostok423Kyrs-TemperatureKyrs / ThermalStability (graph) /

[32] File: Vostok423Kyrs-TemperatureKyrs.xlsm / IG-duration-Tables 1.4.# / cells B16-D30 // Critical duration is the duration to reach statistical significance ($\alpha = 0.05$, $z = 1.645$).

[33] The duration of the current interglacial is about 10,000 years. $(10,000 - 8,500 = 1,500$ years) & $(10,000 - 6,500 = 3,500$ years)

[34] File: ThermalGapInterglacial.xlsm / TempCO2 (5kyrs) /

[35] File: ThermalGap.xlsm / ThermalEffectArtificialCO2 /

[36] File: 150804ObamaGreenhouse gases.doc

[37] File: ThermalGap.xlsm / ThermalGap / cell B2-E2 /

[38] File: ThermalGap(swTest-LinearFunctionsTDv-CO2v) / Probability/ cells D279-285, E289 /

[39] File: 150713IceageIn15Years.doc

[40] File: ThermalGap.xlsm / ObamaDecrease / cell C7 /

[41] File: Vostok423Kyrs-TemperatureKyrs.xlsm / IG-duration-Tables 1.4.# / cells B1-D14 /

[42] See the appendix "Think Smarter with Nemonik thinking."

[43] File: Vostok423Kyrs-TemperatureKyrs.xlsm / Dataset 1.0 (VostokRawTemp) /

[44] File: Vostok423Kyrs-TemperatureKyrs.xlsm / Dataset 1.1 (base 1850-1989) /

[45] Zero degrees Celsius is the temperature of melting water on ice.

[46] File: Vostok423Kyrs-TemperatureKyrs.xlsm / Figure 1.1.1 (base 1850-1989) / cells D3314-3320 /

[47] File: Vostok423Kyrs-TemperatureKyrs.xlsm / Dataset 1.1 (base 1850-1989) / cell F3314-3320 /. The first data intervals belong to the one base period (1850-1989 = 0.00°C). Therefore, n = 3,303, rather than 3,311.

[48] File: Vostok423Kyrs-TemperatureKyrs.xlsm / Dataset 1.1 (base 1850-1989) / cell H10 & I10 /

[49] File: Vostok423Kyrs-TemperatureKyrs.xlsm / Dataset 1.1 (base 1850-1989) / cell H3321 & I3321 /. The data interval at 393,636 years BP (663 years) is longer than the last data interval at 422,135 years (630 years).

[50] File: Vostok423Kyrs-TemperatureKyrs.xlsm / Figure 1.1.1 (base 1850-1989) / cells G3314-3320 /

[51] File: Vostok423Kyrs-TemperatureKyrs.xlsm / Figure 1.1.1 (base 1850-1989) / cells H3314-3319 /

[52] File: Vostok423Kyrs-TemperatureKyrs.xlsm / Figure 1.1.1 (base 1850-1989) /

[53] File: Vostok423Kyrs-TemperatureKyrs.xlsm / Figure 1.1.1 (base 1850-1989) / cell D3317 /

[54] File: Vostok423Kyrs-TemperatureKyrs.xlsm / TD (Weighted) / column L /

[55] File: Vostok423Kyrs-TemperatureKyrs.xlsm / TD (Weighted) / column M /

[56] File: Vostok423Kyrs-TemperatureKyrs / Dataset 1.2 (base 1850-1989) /

[57] File: Vostok423Kyrs-TemperatureKyrs.xlsm / Figure 1.2.1 (base 1850-1989) / cells C426-432 /

[58] File: Vostok423Kyrs-TemperatureKyrs.xlsm / Figure 1.2.1 (base 1850-1989) /

[59] File: Vostok423Kyrs-TemperatureKyrs.xlsm / Figure 1.2.1 (base 1850-1989) / cell C429 /

[60] File: Vostok423Kyrs-TemperatureKyrs.xlsm / Dataset 1.1 (base 1850-1989) / cell F3322 /

[61] File: Vostok423Kyrs-TemperatureKyrs.xlsm / Dataset 1.3 (VostokMill) /

[62] File: Vostok423Kyrs-TemperatureKyrs.xlsm / Fig 1.3.1 (base Kyr1 / cells C426-432 /

[63] File: Vostok423Kyrs-TemperatureKyrs.xlsm / Fig 1.3.1 (base Kyr#1) /

[64] File: Vostok423Kyrs-TemperatureKyrs.xlsm / Fig 1.3.1 (base Kyr1 / cell C429 /

[65] File: Vostok423Kyrs-TemperatureKyrs.xlsm / Fig 1.3.2 (Trend) / cells C426-432 /

[66] File: Vostok423Kyrs-TemperatureKyrs.xlsm / Fig 1.3.2 (Trend) /

[67] File: Vostok423Kyrs-TemperatureKyrs.xlsm / Fig 1.3.2 (Trend) / cells C429 and R2 /. It is emphasized that the linear trend in the graph starts at year zero. Therefore, the

positive sign of the slope indicates a decreasing trend across time, rather than an increasing one.

[68] File: Vostok423Kyrs-TemperatureKyrs.xlsm / Fig 1.3.2 (Trend) / cell R2 /

[69] File: Vostok423Kyrs-TemperatureKyrs.xlsm / Dataset 1.4 (VostokDetrend) /

[70] File: Vostok423Kyrs-TemperatureKyrs.xlsm / Dataset 1.4 (VostokDetrend) / cells C426-432 /

[71] File: Vostok423Kyrs-TemperatureKyrs.xlsm / Fig 1.3.2 (Trend) / cell C432 /

[72] File: Vostok423Kyrs-TemperatureKyrs.xlsm / Fig 1.4.1 (Detrend) /

[73] File: Vostok423Kyrs-TemperatureKyrs.xlsm / Fig 1.4.1 (Detrend) / cell C429 /

[74] File: Vostok423Kyrs-TemperatureKyrs.xlsm / Histogram423Kyrs /

[75] File: Vostok423Kyrs-TemperatureKyrs.xlsm / Dataset 1.4 (Table) / cells B2-F10 /

[76] STD = Standard deviation of the sample.

[77] n = number of cases in the sample.

[78] File: Vostok423Kyrs-TemperatureKyrs.xlsm / IG-duration-Tables 1.4.# / cells B1-D14 /

[79] File: Vostok423Kyrs-TemperatureKyrs.xlsm / IG-duration-Tables 1.4.# / cell B26 / / For critical z-score see Moore, G. P., & McCabe, D. S. (2003). Introduction to the Practice of Statistics. New York: W. H. Freeman and Company. (T2, Table A).

[80] STD = Standard deviation of the sample.

[81] File: Vostok423Kyrs-TemperatureKyrs.xlsm / IG-duration-Tables 1.4.# / cells B16-D30 /

[82] File: Vostok423Kyrs-TemperatureKyrs.xlsm / IG-duration-Tables 1.4.# / cells B32-D40 / For p-value see

Moore, G. P., & McCabe, D. S. (2003). *Introduction to the Practice of Statistics*. New York: W. H. Freeman and Company. Table A.

[83] File: Vostok423Kyrs-TemperatureKyrs.xlsm / IG-duration-Tables 1.4.# / cell E36-37 /

[84] File: Vostok423Kyrs-TemperatureKyrs.xlsm / IG-duration-Tables 1.4.# / cells F27-28 /

[85] File: Vostok423Kyrs-TemperatureKyrs.xlsm / Dataset 1.4 (VostokDetrend) / cells C426-432 /

[86] File: Vostok423Kyrs-TemperatureKyrs.xlsm / Declines-Computation / cells F426-432 /

[87] File: Vostok423Kyrs-TemperatureKyrs.xlsm / Declines-Fig 1.4.3 /

[88] File: Vostok423Kyrs-TemperatureKyrs.xlsm / DeclineWindow (Table) / cells B1-E9 /

[89] File: Vostok423Kyrs-TemperatureKyrs.xlsm / DeclineWindow /

[90] File: Vostok423Kyrs-TemperatureKyrs.xlsm / IG-Duration / cell L12 /

[91] File: Vostok423Kyrs-TemperatureKyrs.xlsm / IG-duration-Tables 1.4.# / cells D22 /

[92] File: Vostok423Kyrs-TemperatureKyrs.xlsm / IG-duration-Tables 1.4.# / cells D23 /

[93] File: Vostok423Kyrs-TemperatureKyrs.xlsm / IG-duration-Tables 1.4.# / cells D27 /

[94] File: Vostok423Kyrs-TemperatureKyrs.xlsm / Declines-Computation / cell F429 /

[95] File: Vostok423Kyrs-TemperatureKyrs.xlsm / Declines-Computation / cell F426 /

[96] File: Vostok423Kyrs-TemperatureKyrs.xlsm / Declines-Computation / cells F430 /

[97] File: Vostok423Kyrs-TemperatureKyrs.xlsm /
DeclineWindow / cells J2 and L2 /

[98] File: Vostok423Kyrs-TemperatureKyrs.xlsm /
ThermalStability /

[99] The *range* and *slope* could be affected significantly by a single
outlier. Therefore, those statistics are less reliable
measures of thermal variation and stability.

[100] File: Vostok423Kyrs-TemperatureKyrs.xlsm /
ThermalStability (Table) / cells B1-D14 /

[101] File: Vostok423Kyrs-TemperatureKyrs.xlsm /
ThermalStability / (B16 & F16) to (B419 & F419) /

[102] File: Vostok423Kyrs-TemperatureKyrs.xlsm /
ThermalStability / cell D433-436 /. For p-value see
Moore, G. P., & McCabe, D. S. (2003). *Introduction to the
Practice of Statistics*. New York: W. H. Freeman and
Company. Table A.

[103] File: Vostok423Kyrs-TemperatureKyrs.xlsm /
ThermalStability / cell G433-436 /

[104] File: Vostok423Kyrs-TemperatureKyrs.xlsm /
ThermalStability (graph) /

[105] File: Vostok10Kyrs-TemperatureYrs.xlsm /
ThermalStability (graph) / cell D16 /

[106] File: Vostok10Kyrs-TemperatureYrs.xlsm /
ThermalStability / cell D433 /

[107] File: Vostok423Kyrs-TemperatureKyrs.xlsm /
CorrelationTDxTSI (graph) /

[108] File: Vostok423Kyrs-
TemperatureKyrsRandomTestCorr.xlsm / Graph / cells
C426-432 /

[109] File: Vostok423Kyrs-
TemperatureKyrsRandomTestCorr.xlsm / Graph / cells
D426-432 /

[110] File: Vostok423Kyrs-TemperatureKyrs(Ranked-rpTestCorrelationTDxSW-TSISW).xlsm / p-value / cells B10003-10014 /

[111] File: Vostok423Kyrs-TemperatureKyrs(Ranked-rpTestCorrelationTDxSW-TSISW).xlsm / p-value / cells B10003-10014 /

[112] File: Vostok423Kyrs-TemperatureKyrs.xlsm / ThermalStability / cell D431 /

[113] The semi-annual temperature deviations are not true annual temperature deviations. In the original Vostok dataset, they are the means of multi-annual data intervals. Therefore, regression towards the mean is likely to have affected the semi-annual temperature deviations. Nevertheless, it is the best information available.

[114] File: Vostok10Kyrs-TemperatureYrs.xlsm / Dataset 1.10 (annual) /

[115] File: Vostok10Kyrs-TemperatureYrs.xlsm / Dataset 1.10 (annual) / cells C10003-10009 /

[116] File: Vostok10Kyrs-TemperatureYrs.xlsm / Dataset 1.10 (annual) /

[117] File: Vostok10Kyrs-TemperatureYrs.xlsm / Dataset 1.10 (annual) / cell C10006 /

[118] File: Vostok423Kyrs-TemperatureKyrs.xlsm / Dataset 1.2 (base 1850-1989) / cell C2 /

[119] File: Vostok423Kyrs-TemperatureKyrs.xlsm / Fig 1.3.2 (Trend) / cell R2 /

[120] File: Vostok10Kyrs-TemperatureYrs.xlsm / Fig 1.10.3 (10 kyrs-detrend) /

[121] File: Vostok10Kyrs-TemperatureYrs.xlsm / 10 Kyrs (Table) / cells B1-F12 /

[122] File: Vostok423Kyrs-TemperatureKyrs.xlsm / Dataset 1.1 (base 1850-1989) / cell D202 & F202 /

[123] File: Vostok10Kyrs-TemperatureYrs.xlsm / Fig 1.10.3 (10 kyrs-detrend) /

[124] File: Vostok10Kyrs-TemperatureYrs.xlsm / Fig 1.10.3 (10 kyrs-detrend) / cell G10006 /

[125] File: Vostok10Kyrs-TemperatureYrs.xlsm / Fig 1.10.3 (10 kyrs-detrend) / cell G10007 /

[126] File: Vostok10Kyrs-TemperatureYrs.xlsm / Histogram10Kyrs /

[127] File: Vostok10Kyrs-TemperatureYrs.xlsm / Fig 1.10.3 (10 kyrs-detrend) / cell G10003 /

[128] This comparison is made to show how easy data can be misinterpreted. The Vostok and Global statistics cannot be compared because they have different base periods. Although the conclusion is correct, the premises are incorrect. This will be addressed later.

[129] File: Vostok10Kyrs-TemperatureYrs.xlsm / Fig 1.10.3 (10 kyrs-detrend) / cell E4 /

[130] File: Vostok423Kyrs-TemperatureKyrs.xlsm / Dataset 1.1 (base 1850-1989) / cell F202 /

[131] File: Vostok423Kyrs-TemperatureKyrs.xlsm / Dataset 1.1 (base 1850-1989) / cells K3314-3320 /

[132] File: Vostok10KyrsGlobalDatasets.xlsm / Vostok10KyrSW /

[133] File: Vostok10Kyrs-TemperatureYrs.xlsm / 10 Kyrs (Table) / cells B2-F11 /

[134] File: Vostok10KyrsGlobalDatasets.xlsm / Vostok10KyrSW / cells D10003-10010 /

[135] File: Vostok10KyrsGlobalDatasets.xlsm / Vostok10KyrSW /

[136] File: Vostok10KyrsGlobalDatasets.xlsm / Vostok10KyrSW(histogram) /

137 File: Global135Yrs-TemperatureYrs.xlsm / Dataset 2.0 (GlobalRawTemp) /

138 File: Global135Yrs-TemperatureYrs.xlsm / Dataset 2.2 (GlobalAnnTemp)|10| /

139 File: Global135Yrs-TemperatureYrs.xlsm / Dataset 2.2 (GlobalAnnTemp)|10| / cells C138-144 /

140 File: Global135Yrs-TemperatureYrs.xlsm / Dataset 2.2 (GlobalAnnTemp)|10| /

141 File: Global135Yrs-TemperatureYrs.xlsm / GlobalHistogram /

142 File: Vostok10Kyrs-TemperatureYrs.xlsm / Dataset (annual-Kyr1) /

143 File: Global135Yrs-TemperatureYrs.xlsm / Dataset 2.2 (GlobalAnnTemp)|10| / cells C138-144 /

144 File: Vostok10Kyrs-TemperatureYrs.xlsm / Dataset (annual-Kyr1)/ cells C1003-1009 / The mean temperature deviation = -0.01 °C, which is slightly different form the millennial mean = 0.00 °C. This is due to rounding.

145 File: Global135Yrs-TemperatureYrs.xlsm / Integrate/

146 File: Global135Yrs-TemperatureYrs.xlsm / Dataset (annual-Kyr1) / cell C2 /

147 File: Global135Yrs-TemperatureYrs.xlsm / Global-FreezeSort / cell E138 /

148 File: Vostok423Kyrs-TemperatureKyrs.xlsm / Dataset 1.1 (base 1850-1989) / cells D2-9 /

149 File: Global135Yrs-TemperatureYrs.xlsm / GlobalExtension /

150 File: Global135Yrs-TemperatureYrs.xlsm / GlobalExtension / cells C168-174 /

151 File: Global135Yrs-TemperatureYrs.xlsm / GlobalExtension / cells G168-174 /

[152] File: Global135Yrs-TemperatureYrs.xlsm / GlobalExtension /

[153] File: Global135Yrs-TemperatureYrs.xlsm / GlobalExtension (table) / cells H1-L11 /

[154] File: Global135Yrs-TemperatureYrs.xlsm / GlobalExtension / See graphs / It is emphasized that Microsoft's notation (R^2) for the explained variance is incorrect. The correct notation for the explained variance is the correlation coefficient squared (r^2) or (R).

[155] File: Global135Yrs-TemperatureYrs.xlsm / GlobalExtension (table) / cell M5 /

[156] File: Global135Yrs-TemperatureYrs.xlsm / Integrate (extended) /

[157] File: Vostok10Kyrs-TemperatureYrs.xlsm / Integrate (extended) / cells C1029-1035 / Due to rounding, the mean temperature deviation of -0.01 °C is slightly different form the millennial mean = 0.00 °C.

[158] File: Global135Yrs-TemperatureYrs.xlsm / Integrate (extended) / cells E1029-1035 /

[159] File: Global135Yrs-TemperatureYrs.xlsm / Integrate (extended) /

[160] File: Global135Yrs-TemperatureYrs.xlsm / Integrate (extended) / cells D1030 /

[161] File: Global135Yrs-TemperatureYrs.xlsm / Integrate (extended) / cells F1030 /

[162] File: Global135Yrs-TemperatureYrs.xlsm / Integrate (extended) / cell D1037 /

[163] File: Global135Yrs-TemperatureYrs.xlsm / Integrate (extended) / cell D1029 /

[164] File: Global135Yrs-TemperatureYrs.xlsm / GlobalExtension (dataset) / cell D168 /

[165] File: Global135Yrs-TemperatureYrs.xlsm / Integrate (extended) / D1036 /

[166] File: Global135Yrs-TemperatureYrs.xlsm / Integrate (synchronized) /

[167] File: Global135Yrs-TemperatureYrs.xlsm / Integrate (synchronized) / cells C1029-1035 / The mean temperature deviation = -0.01 °C, which is slightly different form the millennial mean = 0.00 °C. This is due to rounding.

[168] File: Global135Yrs-TemperatureYrs.xlsm / Integrate (synchronized) / cells E1029-1035 /

[169] File: Global135Yrs-TemperatureYrs.xlsm / Integrate (synchronized) /

[170] File: Global135Yrs-TemperatureYrs.xlsm / Integrate (extended) / cells D1030 /

[171] File: Global135Yrs-TemperatureYrs.xlsm / Integrate (synchronized) / cell E2 /

[172] File: Vostok423Kyrs-TemperatureKyrs.xlsm / Dataset 1.1 (base 1850-1989) / cells K3314-3320 /

[173] File: Vostok423Kyrs-TemperatureKyrs.xlsm / Dataset 1.1 (base 1850-1989) / cells K3317-3318 /

[174] File: Global135Yrs-TemperatureYrs.xlsm / Dataset 2.1 (AnnualMeanTemp) / cell C141 /

[175] File: Vostok10KyrsGlobalDatasets.xlsm / GlobalSW /

[176] File: Vostok10KyrsGlobalDatasets.xlsm / GlobalSW / cells C138-144 /

[177] File: Vostok10KyrsGlobalDatasets.xlsm / GlobalSW / cells D138-145 /

[178] File: Vostok10KyrsGlobalDatasets.xlsm / GlobalSW /

[179] File: Vostok10KyrsGlobalDatasets.xlsm / GlobalSW(histogram) /

[180] File: Vostok10KyrsGlobalDatasets.xlsm /
GlobalVostok10KyrsSW /

[181] File: Vostok10KyrsGlobalDatasets.xlsm /
(Vostok10KyrSW / cells D10009-10010) & (GlobalSW /
cells D144-145)

[182] File: Vostok10KyrsGlobalDatasets.xlsm /
GlobalVostok10KyrsSW / cells C10030-10036 /

[183] File: Vostok10KyrsGlobalDatasets.xlsm /
GlobalVostok10KyrsSW / cells F10030-10036 /

[184] File: Vostok10KyrsGlobalDatasets.xlsm /
GlobalVostok10KyrsSW (graph) /

[185] File: Vostok10KyrsGlobalDatasets.xlsm /
GlobalVostok10KyrsSW / cell C10033 /

[186] File: Vostok10KyrsGlobalDatasets.xlsm /
GlobalVostok10KyrsSW / cell C10030 /

[187] File: Vostok10KyrsGlobalDatasets.xlsm /
GlobalVostok10KyrsSW / cell F10033 /

[188] File: Vostok10KyrsGlobalDatasets.xlsm /
GlobalVostok10KyrsSW / cell I10036 /

[189] File: Vostok10KyrsGlobalDatasets.xlsm /
GlobalVostok10KyrsSW/ cell H10036; C10036; F10033
/

[190] File: Vostok10KyrsGlobalDatasets.xlsm /
GlobalVostok10KyrsSW / cell I10037 /

[191] Window-scale 50 years.

[192] Window-scale 50 years. File:
Vostok10KyrsGlobalDatasets.xlsm / GlobalSW|15| /
cell D145 /

[193] File: Vostok10KyrsGlobalDatasets.xlsm / VostokGlobal
(table) / cells B1-F14 /

[194] File: Vostok10KyrsGlobalDatasets.xlsm / VostokGlobal
(table) / cell B12 /

[195] File: Vostok10KyrsGlobalDatasets.xlsm / GlobalSW / column B /

[196] File: Vostok10KyrsGlobalDatasets.xlsm / VostokSW (trend 86yrs) / cells K1-M11 // and // GlobalSW|15| / cell D145 / Window-scale 50 years, global temperature /

[197] File: Vostok10KyrsGlobalDatasets.xlsm / GlobalSW (trend) / cells I145-J145 /

[198] File: Vostok10KyrsGlobalDatasets.xlsm / VostokSW (trend 86yrs) / cells B9893; E10035; H10035; H10036 /

[199] File: Vostok10KyrsGlobalDatasets.xlsm / VostokSW (trend 86yrs) / cell H10036 /

[200] File: Vostok10KyrsGlobalDatasets.xlsm / VostokGlobal (table) / cells G1-I12 /

[201] File: Vostok10KyrsGlobalDatasets.xlsm / VostokGlobalSW(TrendDurat) / cell I61-J61 /

[202] File: Vostok10KyrsGlobalDatasets.xlsm / VostokGlobal (table) / cell B62 /

[203] File: Vostok10KyrsGlobalDatasets.xlsm / VostokGlobal (table) / cells I61; L61-62 /

[204] See the section: Global versus Vostok °C window-scale.

[205] File: Vostok423Kyrs-TemperatureCO2/ VostokCO2 (Kyr423-Kyr1) /

[206] File: Vostok423Kyrs-TemperatureCO2.xlsm / VostokCO2 (Kyr423-Kyr1) / cells C286-292 /

[207] File: Vostok423Kyrs-TemperatureCO2.xlsm / VostokCO2 (Kyr423-Kyr1) / cells D286-292 /

[208] File: Vostok423Kyrs-TemperatureCO2.xlsm / VostokCO2 (Kyr423-Kyr1) /

[209] File: Vostok423Kyrs-TemperatureCO2.xlsm / VostokTemp (base Kyr1) / column D / This same correction was used for the millennial-scale temperatures.

See file: Vostok423Kyrs-TemperatureKyrs.xlsm / Figure 1.2.1 (base 1850-1989) / cell C2 / Kyr1 = -0.46 °C /

[210] File: Vostok423Kyrs-TemperatureCO2.xlsm / VostokTempCO2 (compute temp) / column E /

[211] File: Vostok423Kyrs-TemperatureCO2.xlsm / VostokTempCO2 (Kyr423-Kyr1) /

[212] File: Vostok423Kyrs-TemperatureCO2.xlsm / VostokCO2 (Kyr423-Kyr1) / cells C286-292 /

[213] File: Vostok423Kyrs-TemperatureCO2.xlsm / VostokCO2 (Kyr423-Kyr1) / cells D286-292 /

[214] File: Vostok423Kyrs-TemperatureCO2.xlsm / VostokTempCO2 (Kyr423-Kyr1) /

[215] File: Vostok423Kyrs-TemperatureCO2.xlsm / VostokTempCO2 (Kyr423-Kyr1) / cells A4; A2-D3 /

[216] File: Vostok423Kyrs-TemperatureCO2.xlsm/ VostokTempCO2 (histogram) /

[217] File: Vostok423Kyrs-TemperatureCO2.xlsm / VostokTempCO2 (compute temp) / cells B3604-C3604 /

[218] File: Vostok423Kyrs-TemperatureCO2.xlsm / VostokCO2 (Kyr423-Kyr1) / cells C286-292 /

[219] File: Vostok423Kyrs-TemperatureCO2.xlsm / VostokCO2 (Kyr423-Kyr1) / cells D286-292 /

[220] File: Vostok423Kyrs-TemperatureCO2.xlsm / VostokTempCO2 (scatterplot) /

[221] File: Vostok423Kyrs-TemperatureCO2(Ranked-rpTestCorrelationTDv-CO2v) /

[222] File: Vostok423Kyrs-TemperatureCO2(Ranked-rpTestCorrelationTDv-CO2v) / p-value / cells B10003-10014 /

[223] File: HawaiiTemp(1955-2015).xlsm / HawaiiTemp(Celsius) /

[224] File: HawaiiTemp(1955-2015).xlsm /
HawaiiTemp(Celsius) | 19 | /

[225] File: HawaiiTemp(1955-2015).xlsm / HawaiiTemp(base
Vostok Kyr1) /

[226] File: HawaiiTemp(1955-2015).xlsm / HawaiiGlobalTemp
/ cells E63-69 /

[227] File: HawaiiTemp(1955-2015).xlsm / HawaiiGlobalTemp
/ cells C63-69 /

[228] File : HawaiiTemp(1955-2015).xlsm / HawaiiGlobalTemp
/

[229] File: HawaiiTemp(1955-2015).xlsm / HawaiiTemp(Vostok
base Kyr1) / cells O63-69 /

[230] File: HawaiiTemp(1955-2015).xlsm / HawaiiGlobalTemp
(VostokKyr1) /

[231] File: HawaiiGlobalTemperatureCO2(Ranked-
rpTestCorrelationTDh-TDg).xlsm /

[232] File: HawaiiGlobalTemperatureCO2(Ranked-
rpTestCorrelationTDh-TDg).xlsm / p-value / B10003-
10014 /

[233] File: HawaiiTemp1955-2015 / HawaiiTemp(graph) /

[234] File: HawaiiTemp(1955-2015).xlsm /
HawaiiTemp(histogram) /

[235] File: HawaiiCO2(1959-2014).xlsm / HawaiiCO2 /

[236] File: HawaiiCO2(1959-2014).xlsm / HawaiiCO2 (graph) /

[237] File: HawaiiCO2(1959-2014).xlsm / HawaiiCO2 (graph)/
cells C59-65 /

[238] File: HawaiiCO2(1959-2014).xlsm / HawaiiCO2 (graph)/

[239] File: HawaiiCO2(1959-2014).xlsm /
HawaiiCO2(histogram) /

[240] File: HawaiiTempCO2(1955-2014).xlsm /
HawaiiTempCO2 /

[241] File: HawaiiTempCO2(1955-2014).xlsm /
HawaiiTempCO2 / cells C59-65 /

[242] File: HawaiiCO2(1959-2014).xlsm / HawaiiCO2 (graph)/
cells C59-65 /

[243] File: HawaiiTempCO2(1955-2014).xlsm /
HawaiiTempCO2 /

[244] File: HawaiiTemperatureCO2(Ranked-
rpTestCorrelationTDh-CO2h) /

[245] File: HawaiiTemperatureCO2(Ranked-
rpTestCorrelationTDh-CO2h) / p-value / B10003-10014
/

[246] File: HawaiiTempCO2(1955-2014).xlsm /
HawaiiTempCO2 (scatter) /

[247] File: CO2-Temperature1959-2014 / Annual CO2 Data /

[248] File: CO2-Temperature1959-2014.xlsm / HawaiiCO2-
GlobalTemp /

[249] File: CO2-Temperature1959-2014.xlsm / HawaiiCO2-
GlobalTemp/ cells C60-66 /

[250] File: CO2-Temperature1959-2014.xlsm / HawaiiCO2-
GlobalTemp/ cells D60-66 /

[251] File: CO2-Temperature1959-2014 / HawaiiCO2-
GlobalTemp / D69 /

[252] File: CO2-Temperature1959-2014/ HawaiiCO2-
GlobalTemp /

[253] File: CO2-Temperature1959-2014/ HawaiiCO2-
GlobalTempScatter|24| /

[254] File: CO2-Temperature1959-2014(Ranked-
rpTestCorrelationTDg-CO2h) /

[255] File: CO2-Temperature1959-2014(Ranked-
rpTestCorrelationTDg-CO2h) / p-value / cells B10003-
10014 /

[256] File: CO2-Temperature1959-2014.xlsm / HawaiiCO2-GlobalTemp / cells D69 /

[257] File: Vostok423Kyrs-TemperatureCO2,xlsm / VostokCO2 (Kyr423-Kyr1) / cell A2 /

[258] File: Vostok423Kyrs-TemperatureCO2.xlsm / VostokCO2 (Kyr423-Kyr1) / cells D14 /

[259] File: Vostok423Kyrs-TemperatureCO2.xlsm / VostokCO2 (Kyr423-Kyr1) / cells C14 /

[260] File: CO2-Temperature1959-2014.xlsm / HawaiiCO2-GlobalTemp/ cells C60-66 /

[261] File: CO2-Temperature1959-2014.xlsm / HawaiiCO2-GlobalTemp/ cells D60-66 /

[262] File: CO2-Temperature1959-2014.xlsm / CO2LeadTemp (Table) / cells P1-T10 /

[263] File: CO2-Temperature1959-2014.xlsm / CO2LeadTemp (lag1) /

[264] File: CO2-Temperature1959-2014.xlsm / CO2LeadTemp / cells I65-N68 /

[265] File: CO2-Temperature1959-2014-CO2LeadsTemp(rpTestCorrelationCO2pTDc).xlsm /

[266] File: CO2-Temperature1959-2014-CO2LeadsTemp(rpTestCorrelationCO2pTDc).xlsm / p-value / cells B10003-10014 /

[267] File: CO2-Temperature1959-2014.xlsm / CO2LeadTemp(graph) /

[268] File: CO2-Temperature1959-2014-CO2LeadsTemp-rpTest-PercentCorrectLag1.xlsm /

[269] File: CO2-Temperature1959-2014-CO2LeadsTemp-rpTest-PercentCorrectLag1.xlsm / p-value / cells B10003-10013 /

[270] File: CO2-Temperature1959-2014-CO2LeadsTemp-rpTest-PercentCorrectLag10.xlsm /

[271] File: CO2-Temperature1959-2014-CO2LeadsTemp-rpTest-PercentCorrectLag10.xlsm / p-value / cells B10003-10013 /

[272] File: ThermalGap.xlsm / VostokTempCO2 (Kyr423-Kyr11) / cells B282-283 /

[273] File: ThermalGap.xlsm / VostokTempCO2 (Kyr423-Kyr11) /

[274] File: ThermalGap.xlsm.xlsm / VostokTempCO2 (Kyr423-Kyr11) / cells C280-286 /

[275] File: ThermalGap.xlsm.xlsm / VostokTempCO2 (Kyr423-Kyr11) / cells D280-286 /

[276] File: ThermalGap.xlsm / VostokTempCO2 (Kyr423-Kyr11) /

[277] File: ThermalGap.xlsm / / VostokTempCO2 (Kyr423-Kyr11)|25 / cells A3, A278 /

[278] File: ThermalGap.xlsm / VostokTempCO2 (Kyr423-Kyr11) / function from graph / cell B290 /

[279] File: ThermalGap.xlsm / VostokTempCO2 (Kyr423-Kyr11)|25 / cells D291-293 /

[280] File: ThermalGap.xlsm / ThermalDecrease|28| / cell B25 /

[281] File: ThermalGap.xlsm / VostokTempCO2 (Kyr423-Kyr11) / cells F280-286 /

[282] File: ThermalGap.xlsm / VostokCO2 (graph) /

[283] File: ThermalGapVostok(Ranked-rpTestCorrelationTDv-TDvCO2v).xlsm /

[284] File: ThermalGapVostok(Ranked-rpTestCorrelationTDv-TDvCO2v).xlsm / p-value / cells B10003-B10014 /

[285] File: ThermalGap.xlsm / VostokTempCO2 (Kyr423-Kyr11)|25 / cells D291-293 /

[286] File: ThermalGap.xlsm / HawaiiCO2-GlobalTemp|24| /cell D2 /

[287] File: ThermalGap.xlsm / HawaiiCO2(PredictGlobalTemp) /

[288] File: ThermalGap.xlsm.xlsm / HawaiiCO2(PredictGlobalTemp) / cells C60-66 /

[289] File: ThermalGap.xlsm.xlsm / HawaiiCO2(PredictGlobalTemp) / cells D60-66 /

[290] File: ThermalGap.xlsm / HawaiiCO2-GlobalTempScatter|24| /

[291] File: ThermalGap.xlsm / HawaiiCO2(PredictGlobalTemp)|26 / Linear function from graph /

[292] File: ThermalGap.xlsm / HawaiiCO2(graph) /

[293] File: ThermalGapGlobal(Ranked-rpTestCorrelationTDg-CO2h).xlsm /

[294] File: ThermalGapGlobal(Ranked-rpTestCorrelationTDg-CO2h).xlsm / p-value / cells B10003-B10014 /

[295] File: ThermalGap.xlsm / VostokGlobal-TempCO2(data) /

[296] File: ThermalGap.xlsm / VostokGlobalTempCO2Scatter|27| /

[297] File: ThermalGap(swTest-LinearFunctionsTDv-CO2v) / Vostok / cells D286-287 /

[298] File: ThermalGap(swTest-LinearFunctionsTDv-CO2v).xlsm / Probability / cells D286-287 /

[299] File: ThermalGap(swTest-LinearFunctionsTDv-CO2v) / Probability/ cells D279-285, E289 /

[300] File: ThermalGap.xlsm / HawaiiCO2-GlobalTempScatter|24| / cell C2 /

[301] File: ThermalGap.xlsm / HawaiiCO2(PredictGlobalTemp) / cell E68 /

[302] File: ThermalGap.xlsm / VostokCO2(PredictVostokTemp) / cell E288 /

[303] File: ThermalGap.xlsm / ThermalGap / cell B1 /

[304] File: ThermalGapVostok(Ranked-rpTestCorrelationTDv-CO2v).xlsm /

[305] File: ThermalGapVostok(Ranked-rpTestCorrelationTDv-CO2v).xlsm / p-value / cells B10003-B10014 /

[306] File: ThermalGapGlobal(Ranked-rpTestCorrelationTDg-CO2h) /

[307] File: ThermalGapGlobal(Ranked-rpTestCorrelationTDg-CO2h) / p-value / cells B10003-10014 /

[308] File: CO2-Temperature1959-2014.xlsm / HawaiiCO2-GlobalTempScatter|24| / cell D2 // & // File: CO2-Temperature1959-2014.xlsm / HawaiiCO2(1959-2014) / cells C62 /

[309] File: ThermalGap.xlsm / VostokTempCO2 (Kyr423-Kyr11) / cells E279-285 /

[310] File: ThermalGap.xlsm / VostokTempCO2 (Kyr423-Kyr11) / cells F279-285 /

[311] File: ThermalGap.xlsm / VostokTempCO2 (Kyr423-Kyr11) / cells F279-285 /

[312] File: 150804ObamaGreenhouse gases.doc // & // File: ThermalGap / ObamaDecrease / cell B3 /

[313] File: ThermalGap / ObamaDecrease / cell D5 /

[314] File: ThermalGap / ObamaDecrease / cell C5 /

[315] The term solar heat would be too limiting. Although relatively small, heat is also generated by domestic and industrial human activities.

[316] File: ThermalGapInterglacial.xlsm / TempCO2(10Kyrs)|18| / cells D11-17 /

[317] File: ThermalGapInterglacial.xlsm / TempCO2(10Kyrs)|18| / cells E11-17 /

[318] File: ThermalGapInterglacial.xlsm / TempCO2(10Kyrs)|18| /

319 File: ThermalGapInterglacial.xlsm / TempTemp(10kyrs)|18| / cells G11-17 /

320 File: ThermalGapInterglacial.xlsm / TempTemp(10kyrs)|18| /

321 File: ThermalGapInterglacial.xlsm / TempTemp(10kyrs)|18| /

322 File: ThermalGapInterglacial.xlsm / TempTemp(10kyrs)|18| / cell K2 /

323 File: ThermalGap.xlsm / ThermalEffectArtificialCO2 /

324 File: ThermalGap(swTest-LinearFunctionsTDv-CO2v) / Vostok / cells D286-287 /

325 File: ThermalGap.xlsm / HawaiiCO2-GlobalTempScatter|24| / cell D2 /

326 File: ThermalGap.xlsm / HawaiiCO2-GlobalTempScatter|24| / cell C2 /

327 File: ThermalGap.xlsm / VostokCO2(PredictVostokTemp) / cell E288 /

328 File: ThermalGap.xlsm / ThermalGap / cell B1 /

329 File: ThermalGap.xlsm / VostokCO2(PredictVostokTemp) / cells E288-F288 /

330 File: ThermalGap.xlsm / VostokCO2(PredictVostokTemp) / E288 /

331 File: ThermalGap.xlsm / VostokTempCO2 (Kyr423-Kyr11) / cell B284 /

332 File: ThermalGap.xlsm / VostokCO2(PredictVostokTemp) / cells E288-F288 /

333 File: ThermalGap.xlsm / VostokTempCO2 (Kyr423-Kyr11) / column E /

334 File: Vostok423Kyrs-TemperatureCO2.xlsm / VostokCO2 (Kyr423-Kyr1) / cell C282 /

335 File: ThermalGap.xlsm / HawaiiCO2-GlobalTempScatter|24| / cell C2 /

[336] File: CO2-Temperature1959-2014.xlsm / CO2LeadTemp / cells I65-N68 /

[337] File: ThermalGap.xlsm / HawaiiCO2-GlobalTemp|24| / C2 & D2 /

[338] File: 150804ObamaGreenhouse gases.doc // and // File: ThermalGap.xlsm / ObamaDecrease / cells B2-4 /

[339] File: ThermalGap.xlsm / ObamaDecrease / cell C9 /

[340] File: ThermalGap.xlsm / ObamaDecrease / C5 /

[341] File: ThermalGap.xlsm / ObamaDecrease / D5 /

[342] File: ThermalGap.xlsm / ObamaDecrease / D3 & C7 /

[343] File: ThermalGap.xlsm / ObamaDecrease / C6 & D6 /

[344] File: ThermalGap.xlsm / ThermalDecrease|28| /

[345] File: ThermalGap.xlsm / ThermalDecrease(graph 2)|28| /

[346] File: ThermalGap.xlsm / ThermalDecrease|28| / cell C2 /

[347] File: ThermalGap.xlsm / ThermalDecrease|28| / cell D4 /

[348] File: ThermalGap.xlsm / ThermalDecrease|28| / cell I2 /

[349] File: ThermalGap.xlsm / ThermalDecrease(graph 3)|28| /

[350] File: ThermalGap.xlsm / ThermalDecrease|28| / cell F8 /

[351] File: ThermalGap.xlsm / ThermalDecrease(graph 4)|28| /

[352] File: ThermalGap.xlsm / ThermalDecrease|28| / cell B8 /

[353] File: ThermalGap.xlsm / ThermalDecrease(graph 5)|28| /

[354] File: ThermalGap.xlsm / ThermalDecrease|28| / Function from graph & cell B19 /

[355] File: ThermalGap.xlsm / ThermalDecrease|28| / Function from graph & cell B23 /

[356] File: ThermalGap.xlsm / ZeroArtificialCO2(graph)|28| /

[357] File: ThermalGap.xlsm / ThermalDecrease|28| /

[358] File: ThermalGap.xlsm / ThermalDecrease|28| / cell B24 /

[359] File: ThermalGap.xlsm / ThermalDecrease|28| / cell C24 /

[360] File: ThermalGap.xlsm / ObamaDecrease(graph)|28| /

[361] File: ThermalGap.xlsm / ThermalDecrease|28| / cells C21, D4, C20 /

[362] File: ThermalGap.xlsm / ZeroTemperature(graph)|28| /

[363] File: ThermalGap.xlsm / ThermalDecrease(graph 5)|28| / cells J22 & D4 /

[364] File: ThermalGap.xlsm / ThermalDecrease|28| / cell B8 /

[365] File: ThermalGap.xlsm / ZeroTemperature(graph)|28| / cells B16, B2, B6, C2, E6 /

[366] File: 150804ObamaGreenhouse gases.doc

[367] File: ThermalGap.xlsm / ThermalDecrease|28| / cells C21, D4, C20 /

[368] See the section: Global versus Vostok °C window-scale.

[369] File: Statistics / DescriptiveStatistics / cells B11, B17 /

[370] File: CO2-Temperature1959-2014.xlsm / HawaiiCO2-GlobalTemp (Scatter) / cell D2 // & // File: CO2-Temperature1959-2014.xlsm / HawaiiCO2(1959-2014) / cells C62 /

[371] File: Statistics / DescriptiveStatistics / cells B20, D20 /

[372] It is acknowledged that this definition is simplified. Other standard distributions are used for parametric statistics.

[373] File: Vostok423Kyrs-TemperatureKyrs.xlsm / Dataset 1.1 (base 1850-1989) /

[374] File: Vostok423Kyrs-TemperatureKyrs.xlsm / Dataset 1.4 (VostokDetrend) /

[375] File: Vostok10KyrsGlobalDatasets.xlsm / GlobalSW /

[376] File: Vostok423Kyrs-TemperatureCO2.xlsm / VostokTempCO2 (Kyr423-Kyr1) /